STATISTICAL METHODS IN ECONOMETRICS

Ramu Ramanathan

Department of Economics
University of California, San Diego
La Jolla, California

Academic Press
San Diego New York Boston
London Sydney Tokyo Toronto

Find Us on the Web!   http: //www.apnet.com

To my family—

Vimala, Sadhana, Pradeep, and Sridhar

Academic Press Rapid Manuscript Reproduction

This book is printed on acid-free paper. ∞

Copyright © 1993 by ACADEMIC PRESS

All Rights Reserved.

No part of this publication may be reproduced or transmitted in any form or by any means, electronic or mechanical, including photocopy, recording, or any information storage and retrieval system, without permission in writing from the publisher.

Academic Press
A Division of Harcourt Brace & Company
525 B Street, Suite 1900, San Diego, California 92101-4495

United Kingdom Edition published by
Academic Press Limited
24–28 Oval Road, London NW1 7DX

Ramanathan, Ramu, date
    Statistical methods in econometrics / Ramu Ramanathan
        p.   cm.
    Includes bibliographical references and indexes.
    ISBN: 0-12-576830-3
    1. Economics–Statistical methods.   2. Econometrics.   I. Title.
HB137R36   1993
330'.01'5195–dc20   92-43101
                                                                              CIP

PRINTED IN THE UNITED STATES OF AMERICA
97 98 IBT 9 8 7 6 5 4 3 2

CONTENTS

12. Nonspherical Disturbances 330

APPENDICES

PREFACE

This book is designed to fill the gap between two types of texts that are currently being used in econometrics courses. Books that cater to a wide audience are skimpy on multivariate and multiparameter analyses because they involve a knowledge of linear algebra. Books that deal with multivariate analysis, on the other hand, assume a considerable knowledge of basic probability and statistics. Econometrics makes extensive use of matrix algebra, but a typical graduate student has not been taught statistics with the linear algebra prerequisite.

This book is appropriate for a beginning course on mathematical statistics and econometrics in which the foundations of probability and statistical theory are developed with a view to applying them to econometric methodology. Because econometrics generally deals with the study of several unknown parameters, this book places greater emphasis than is common on the estimation and hypothesis testing involving several parameters. Accordingly, the multivariate normal and distribution of quadratic forms are emphasized more. The Lagrange multiplier tests, which have gained popularity in recent years, are discussed here in considerable detail along with the traditional likelihood ratio and Wald tests. Characteristic functions and their properties are exploited more fully than in most books. Also, asymptotic distribution theory, which is typically given only cursory treatment in other books, is discussed in detail.

The book assumes a working knowledge of advanced calculus (including integral calculus), basic probability and statistics, and linear algebra. Important properties from matrix algebra are, however, summarized in Appendix A, but no proofs are presented.

The book is organized in three parts. Part I consists of an introductory chapter followed by several chapters that provide the foundations of probability theory. Part II deals with the theory of sampling, the specific implications of having a large sample, the principles behind the methods of estimating unknown parameters and studying their properties, and testing hypotheses on parameters. Part III covers basic econometrics in which topics in probability and statistics are brought together to address issues that are special to economics. At the beginning of each part, there is a brief description of the material in the chapters for that part. In addition to the usual exercises at the end of each chapter, there are numerous "Practice Problems" scattered throughout the book. These are usually short and pertain to the section in which the problems are given. Solving all the problems and exercises in each chapter will be extremely useful in understanding the material.

As the emphasis on this book is more on theory and methodology (especially on probability and statistics), the number of empirical applications and associated data are somewhat limited. Interested readers and instructors are referred to my econometrics book *Introductory Econometrics with Applications*, Second Edition (1992), published by Harcourt Brace Jovanovich, for numerous applications and data tables. A diskette containing all the data sets in that book, along with an easy to use econometrics program, can be purchased (MS-DOS version only) by sending a check payable to Ramu Ramanathan on a U.S. bank in the amount of $35 for U.S. and Canadian destinations and $50 for other countries.

I would like to acknowledge several people who have helped improve the book. My colleagues Richard Carson, Robert Engle, Clive Granger, Glenn Sueyoshi, and Hal White have read parts of the manuscript (Glenn read all of the manuscript) and made many useful suggestions. Present and former graduate students Bruno Broseta, Francis Lim, Chien-Fu Lin, Joao Issler, Sheila Najberg, Pu Shen, Farshid Vahid-Araghi, and Jeff Wooldridge also made comments on earlier drafts. Professors Tae Jun Seo (Southern Methodist University), David Brownstone (University of California, Irvine), and Glenn Sueyoshi (University of California, San Diego) served as detailed reviewers for the manuscript. Their input has been immensely valuable in improving the book. Finally, I am indebted to my colleague John Conlisk for his "theorem list" on linear algebra which was of great help in writing the appendix on matrices.

I would also like to thank my UNIX guru, Michael O'Hagan, for all the help in setting up the chapter files so that a "camera ready" version of the manuscript could be printed. The department of economics word processing specialists, Paula Lindsay and Meredee O'Brien, prepared the bibliography and the Appendix tables with expert diligence. Paula and Meredee have been invaluable in incorporating all the manuscript editor's corrections and helping me prepare the final manuscript. I am also indebted to the editor, Rick Roehrich, for his constant support and encouragement and to Bill LaDue who was in charge of the production. Finally, I acknowledge with thanks help from my daughter Sadhana who painstakingly proofread the manuscript and also compiled the indices. As for errors that remain, I alone take responsibility for them. I would, however, very much appreciate readers' comments on the book as well as their pointing out typographical and other errors.

Part I

PROBABILITY THEORY

This part consists of five chapters. The introductory chapter provides an overview of probability theory, mathematical statistics, and their applications to econometrics. Chapter 2 presents the definitions of probability and related concepts and develops certain basic results. Random variables and probability distributions associated with them are discussed in Chapter 3. Several measures that characterize statistical distributions of single random variables are also discussed here. Chapter 4 presents a number of special probability distributions used in statistics. The concepts developed in the first four chapters are extended in considerable detail in Chapter 5 to bivariate and multivariate random variables and distributions.

1

INTRODUCTION

The discipline of econometrics uses statistical techniques to develop and apply tools for (1) estimating economic relationships, (2) testing hypotheses involving economic behavior, and (3) forecasting the behavior of economic variables. Econometric theorists focus their attention on developing the analytical tools, examining assumptions behind certain methods, and studying the consequences of applying them to inappropriate situations. Applied econometrics deals mainly with the application of the methods in a specific context. This distinction, however, is arbitrary. A specific applied problem might demand that new techniques be developed to address the issues that arise.

In this chapter, we give a brief overview of what econometric methodology is all about and we introduce a number of basic concepts, all of which are defined and analyzed more formally in later chapters.

An econometrician typically formulates a **theoretical model**, that is, a framework for analyzing economic behavior using some underlying logical structure. The model might arise out of formal economic theory, other studies, past experience, intuition about actual behavior that the model is expected to portray, and so on. A well-specified model would be a reasonable approximation to the actual process that generates the observed data. This process (known as the **data generating process** or **DGP**) would involve the interactions of behavior among numerous economic agents.

The econometric model might be formulated as a **single equation model**, such as a cost or production function of a firm, or a **simultaneous equation model** comprising a system of equations that characterizes the interdependence among variables (for example, a complete macro economic model). A single equation model generally has the form

$$Y = F(X_1, X_2, \ldots, X_k, u)$$

where Y is a variable of primary interest, referred to as the

3

dependent variable (also as **regressand** or **endogenous variable**), X_i's are variables that have **causal effects** on the dependent variable (and are referred to as the **independent variables** or **regressors** or **exogenous variables**), and u is an unobserved variable (referred to as a **random variable** or **stochastic variable**) that captures uncertainties in the formulation. The X's may also be random variables but are generally taken to be observable along with Y. As an example, Y could be the earnings of an employee in a firm, X_1 might be the employee's age, X_2 the number of years of education, X_3 the number of years of work experience, and so on. If we draw a sample of workers and measure the attributes described above, not all of them will have exactly the same relationship between Y and the X's. An estimated relation will instead be a "statistical average." To allow for this fact, an **econometric model** will be formulated with an additional variable (denoted by u in the above equation) that captures the uncertainty in the relationship. A particularly simple form of an econometric model is given by the following equation in which all the variables appear linearly.

$$Y = \beta_1 X_1 + \beta_2 X_2 + \cdots + \beta_k X_k + u$$

where the β's are unknown **parameters** to be estimated from the data. The independent variables denoted by X_i in the above equation might also be past values of the dependent and independent variables. To illustrate, suppose Y_t is the consumption expenditure of a family, evaluated at time t. Families typically maintain their past standard of living, but adjust it if their financial position changes. Let X_t be the family's income at time t. If income fell from time period $t-1$ to time period t, we would expect consumption expenditure to be adjusted downward to accommodate the reduction in income. The following econometric model captures the underlying behavior specified here.

$$Y_t = \beta_1 + \beta_2 Y_{t-1} + \beta_3 (X_t - X_{t-1}) + u_t$$

We would expect β_2 to be positive because of the assumption that consumers try to maintain their standard of living. β_3 is also likely to be positive because an increase in income would induce additional consumption. The random error term u_t is included to capture the uncertainty in the postulated behavioral equation.

A simultaneous equation model is typically of the form

$$F_i(Y_1, Y_2, \ldots, Y_G, X_1, X_2, \ldots, X_K, u_1, u_2, \ldots, u_G) = 0$$

where the Y's are the endogenous variables, X's are the exogenous variables, and the u's are unobserved stochastic variables. The subscript i represents the index of an equation ($i = 1, 2, \ldots, G$). To illustrate, consider the following macro econometric model.

$$C_t = \alpha_0 + \alpha_1 C_{t-1} + \alpha_2 Y_t^d + u_{1t}$$

$$I_t = \beta_0 + \beta_1 r_t + \beta_2 r_{t-1} + \beta_3 Y_t^d + u_{2t}$$

$$r_t = \gamma_0 + \gamma_1 Y_t + \gamma_2 M_t + \gamma_3 M_{t-1} + u_{3t}$$

$$Y_t^d = Y_t - T_t$$

$$Y_t = C_t + I_t + G_t$$

where Y is net national product, C is consumption, I is net investment, G is government expenditure, T is taxes, M is money supply, and Y^d is disposable income, all measured at the time period t. All of these variables are measured in real terms. The other variable in the system is the corporate bond rate (r). The endogenous variables in the system are C, I, r, Y^d, and Y, which are jointly determined within the system. T, G, and M are given exogenously. The variables M_{t-1}, I_{t-1}, and Y_{t-1}^d are known as **predetermined variables** because at time t their values are known. The variables u_1, u_2, and u_3 represent unobservable error terms that capture the uncertainty in the relationships. Based on data on all the observable variables, the econometrician would be interested in estimating the unknown parameters (α's, β's, γ's, and the parameters that represent the statistical properties of the random variables u_1, u_2, and u_3).

The fourth equation is an accounting identity and the last equation is an equilibrium condition equating aggregate demand to the net national product. These two equations do not contain any unknown parameters.

In order to estimate the econometric model specified by an investigator, data on the observable variables (that is, the exogenous and endogenous variables) must be obtained. Such data might be **time series**, if the analyst is interested in modeling the behavior of economic variables (such as the unemployment and inflation rates) over time, or **cross section**, if the focus is on economic behavior of a number of units (individuals, firms, states, countries, and so on) at a given point in time. Some studies might require **panel data** which are time series data for a cross section of economic units.

An investigator might find that the type of data available does not match the theoretical specification exactly. For example, a great deal of economic theory deals with the interest rate, but there is no such thing as a single interest rate. When one studies investment behavior, the appropriate rate might be the prime rate, the corporate bond rate, or another rate that applies to borrowers. If, on the other hand, the focus is on the demand for housing, the home mortgage rate would be appropriate. Other problems that might arise in obtaining the data are changes in definitions, new products requiring the measurement of new variables, changes in prices and other market situations, and so forth. Thus a considerable amount of care should be exercised in obtaining data and in being aware of their limitations.

Once a model has been formulated and the data gathered, the next step in an empirical study is to estimate the unknown parameters of the model and to subject the model to a variety of diagnostic tests to make sure that one obtains robust conclusions, that is, conclusions that are not sensitive to model specification. To achieve this goal, an investigator may have to reformulate the models and perhaps use alternative techniques to estimate them. Methods of hypothesis testing would be useful not only at this diagnostic testing stage but also to test the validity of a body of theory.

Unlike the natural sciences where a researcher can usually conduct a controlled experiment in a laboratory, economics most frequently deals with nonexperimental data generated by a complicated process involving the interactions of the behavior of numerous economic and political agents. This imposes a great deal of uncertainty in the models and methods used by econometricians. In particular, estimated relations are not precise, hypothesis testing can lead to the error of rejecting a true hypothesis or that of accepting a false hypothesis, and forecasts of variables often turn out to be far from their actual values. This uncertainty makes statistical methodology very important in econometrics.

The mathematical or statistical model that an econometrician formulates is meant to be a description of the DGP. The actual data, however, are treated as one of several possible realizations of events. The **theory of probability** is used to construct a framework that portrays the likelihood of one type of realization or another. Not surprisingly, the probability framework invariably depends on a number of unknown **parameters**. In order to estimate the parameters, an analyst typically obtains a sample of observations and uses them in conjunction with a **probability model**. **Statistical inference** deals with the methods of obtaining these estimates, measuring their precision, and testing hypotheses on the parameters in question.

2

BASIC PROBABILITY

The theory of probability is over three centuries old and has its origin in games of chance such as throwing dice, playing cards, and the roulette wheel. Although one cannot predict the outcome of any particular trial exactly, it has been found that it is possible to predict, with a fair degree of accuracy, the frequency of occurrence of one or more outcomes of a trial, in the long run or over a large number of replications. As an example, suppose we roll a single die over and over again and note the frequency with which the number five appears. We are likely to find that, on average, it occurs one-sixth of the time. Probability theory characterizes observed phenomena like the one just described with a formal framework of analysis. Although the original development of probability theory had a great deal to do with the games of chance, it is now a valuable tool in a wide variety of disciplines. In this chapter we discuss the basics of probability theory and related concepts.

2.1 SAMPLE SPACE, SAMPLE POINTS, AND EVENTS

The starting point of an investigation is typically an **experiment** that might be as simple as rolling a pair of dice or as complicated as conducting a large-scale survey of households or firms. An experiment is a **random experiment** if it satisfies the following conditions: (i) all possible distinct outcomes are known ahead of time; (ii) the outcome of a particular trial is not known a priori; and (iii) the experiment can be duplicated, in principle, under ideal conditions. The totality of all possible outcomes of the experiment is referred to as the **sample space** (denoted by S) and its distinct individual elements are called the **sample points** or **elementary events**. Thus, when a coin is tossed twice, the sample space (denoting the occurrence of a head by H and that of a tail by T) consists of the four sample points HH, HT, TH, and TT, that is, $S = \{ HH, HT, TH, TT \}$.

An **event** is a subset of a sample space and is a set of sample points that represents several possible outcomes of an experiment. For instance, when a pair of dice is rolled, "total score of 9" is an event represented by the sample points (3,6), (4,5), (5,4), and (6,3), where the first number is the outcome of the first die and the second number is the outcome of the second die. The **impossible event** or **null event** is denoted by ∅. A sample space with a finite or **countably infinite** sample points (with a one to one correspondence to positive integers) is called a **discrete space**. An example of such a space is rolling a die twice which has the 36 sample points (1,1), (1,2), . . . , (6,6). A **continuous space** is one with an **uncountably infinite** number of sample points (that is, it has as many elements as there are real numbers). When measured accurately, possible values of the height of a person, the temperature in a room, and so on are examples of continuous spaces.

In Figure 2.1, tossing a coin twice is represented by the four points (0,0), (0,1), (1,0), and (1,1), where 1 refers to a head and 0 to a tail. The set consisting of the points (1,0) and (0,1) constitutes the event "a head and a tail." Such a pictorial representation of sample spaces and outcomes as sets is known as a **Venn diagram** and is a convenient tool of analysis.

Practice Problems

2.1   If a pair of dice is rolled together, the outcomes are of the form (1,1), (1,2), and so on. Use a Venn diagram to represent the sample space and the events "total score is five," and "total score is ten."

$$\Sigma_1 = 5 \qquad \Sigma_2 = 10$$

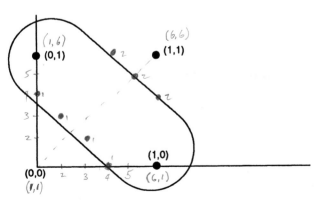

Figure 2.1   Sample space for tossing a coin twice

2.2   A coin is tossed three times. Represent the sample space and
the event "at least two heads."

[handwritten: $SAMPLE SPACE = 8\ EVENTS\ (DISTINCT)$]
[handwritten: $@LEAST\ 2H = 4/8 EVENTS = \frac{1}{2}$]

[handwritten table:]
H H H   H H T   H T T   T T T
H T H   T H T
T H T   T T H

[handwritten: $(3H)\ (2H)\ (1H)\ (0H)$]

2.2   SOME RESULTS FROM SET THEORY

Because events are generally represented by sets, to calculate the
probabilities of various compound events it is useful to understand
certain concepts on sets and the algebra of sets (known as **Boolean
algebra**). A number of concepts are first defined and several postulates of Boolean algebra are then stated. This is followed by a discussion of sets made up of subsets of S called σ–**fields**.

Definition 2.1

The **sample space** *is denoted by S. A = S implies that the events
in A must always occur. The* **empty set** *is a set with no elements
and is denoted by ∅. A = ∅ implies that the events in A do not
occur.*

The set of all elements not in A is called the **complement** *of A
and is denoted by A^c. Thus, A^c occurs if and only if A does not
occur. [Sometimes A^c is also denoted by $S - A$.]*

The set of all points in either a set A or a set B or both is called the
union *of the two sets and is denoted by ∪. A ∪ B means that
either the event A or the event B or both occur. Note: $A \cup A^c = S$.*

The set of all elements in both A and B is called the **intersection**
*of the two sets and is represented by ∩. A ∩ B means that both the
events A and B occur simultaneously.*

*$A \cap B = \varnothing$ implies that A and B cannot occur together. A and B
are then said to be* **disjoint** *or* **mutually exclusive**. *Note:
$A \cap A^c = \varnothing$.*

$A \subset B$ means that A is contained in B or that A is a **subset** *of B,
that is, every element of A is an element of B. In other words, if an
event A has occurred, then B must have occurred also.*

Figure 2.2 is a Venn diagram representation of some of the
above sets.

Example 2.1

Toss a coin twice. The sample space is $S = \{ HH,\ HT,\ TH,\ TT \}$.

A: Exactly one head: (*HT, TH*).

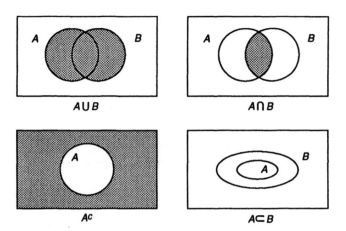

Figure 2.2   Venn diagram representation of union, intersection, etc.

LET — A: EXACTLY ONE HEAD: (HT, TH)
  — *B*: At least one tail: (*TH, HT, TT*).
UNION $A \cup B$ = (*HT, TH, TT*). Note: $A \subset B$.
INTERSECTION $A \cap B$ = (*HT, TH*) = *A*.

Example 2.2

  A: Set of all wealthy people.

  B: Set of all intelligent people.

  $A \cup B$: either wealthy or intelligent or both.

  $A \cap B$: both intelligent and wealthy.

  $A \cap B^c$: wealthy but not intelligent.

  $A^c \cup B^c$: either not wealthy or not intelligent or both.

  Sometimes it is useful to divide the elements of a set *A* into several subsets that are disjoint. Such a division is known as a **partition**. If A_1 and A_2 are such partitions, then $A_1 \cap A_2 = \emptyset$ and $A_1 \cup A_2 = A$. This can be generalized to *n* partitions; $A = \cup_1^n A_i$ with $A_i \cap A_j = \emptyset$ for $i \neq j$.

LET A = FINITE UNION OF
  A_n DISJOINT EVENTS

Boolean Algebra

  The set operations of union, intersection, and complementation satisfy a number of postulates that are enumerated below.

- **Identity:** There exist unique sets $\emptyset$ and S such that, for every set A, $A \cap S = A$ and $A \cup \emptyset = A$. $\Rightarrow A = S$?

- **Complementation:** For each A we can define a unique set A^c such that $A \cap A^c = \emptyset$ and $A \cup A^c = S$.

  ? IF $A = S$ $A \cup A^c = ?$?

- **Closure:** For every pair of sets A and B, we can define unique sets $A \cup B$ and $A \cap B$.

- **Commutative:** $A \cup B = B \cup A;$ $A \cap B = B \cap A.$

- **Associative:** $(A \cup B) \cup C = A \cup (B \cup C).$
  Also, $(A \cap B) \cap C = A \cap (B \cap C).$

- **Distributive:** $A \cap (B \cup C) = (A \cap B) \cup (A \cap C).$
  Also, $A \cup (B \cap C) = (A \cup B) \cap (A \cup C).$

These postulates can be verified by Venn diagrams. Figure 2.3 illustrates the result that $A \cup (B \cap C) = (A \cup B) \cap (A \cup C)$. The diagram on the left is $A \cup (B \cap C)$. In the diagram on the right, the horizontal lines represent $A \cup B$ and the vertical lines represent $A \cup C$. Their intersection has small squares and is the same as the left-hand side.

$(A \cup B)^c =$ $= (A \cup B)$

Practice Problems

2.3   Use Venn diagrams and verify the following (known as **De Morgan's Laws**):

  (a) $(A \cup B)^c = A^c \cap B^c$, that is, the complement of a union is the intersection of the complements. This can be extended to

  $$\left[\bigcup_{i=1}^{i=\infty} A_i \right]^c = \bigcap_{i=1}^{i=\infty} A_i^c$$

  $= A^c$ $= A^c \cap B^c$
  $= B^c$ $= (A \cup B)^c$

  (b) $(A \cap B)^c = A^c \cup B^c$, that is, the complement of an intersection is the union of the complements. This can be extended to

  $$\left[\bigcap_{i=1}^{i=\infty} A_i \right]^c = \bigcup_{i=1}^{i=\infty} A_i^c$$

  $= A \cap B$
  $= (A \cap B)^c$

2.4   Verify using Venn diagrams that

  $$A \cap (B \cup C) = (A \cap B) \cup (A \cap C)$$

$= A^c$

$= B^c$

$= A^c \cup B^c$

$= (A \cap B)^c$

$= A \cap (B \cup C)$
$= (A \cap B) \cup (A \cap C)$

$= (B \cup C)$

$= (A \cap B)$

$= (A \cap C)$

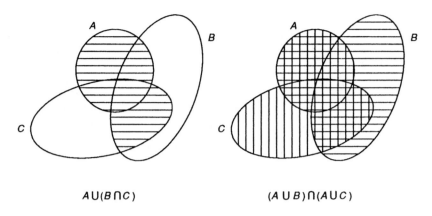

$$A \cup (B \cap C) \qquad\qquad (A \cup B) \cap (A \cup C)$$

Figure 2.3   Distributive property of sets

Borel Fields and σ-fields

It is clear from the operations on sets that by combining sets (or events) we obtain other sets (or events). In order to make sure that in whatever manner we combine events, the result is always an event, it is essential to impose some mathematical structure on the set of all events (called **field** and denoted by $\mathcal{F}$). Otherwise, attribution of probabilities to events may sometimes not make sense. Sets which have the required mathematical structure are known as σ-**fields** (or as σ-**algebras**) associated with S.

Definition 2.2

Let $\mathcal{F}$ be a nonempty set of subsets of S that is also nonempty. $\mathcal{F}$ is said to be a σ–**field** if the following two conditions hold:

(1)   if $A \in \mathcal{F}$, then $A^c \in \mathcal{F}$. *CLOSED WRT COMPLEMENTATION*

(2)   if $A_i \in \mathcal{F}$ for $i = 1, 2, \ldots$, then $[\cup_{i=1}^{i=\infty} A_i] \in \mathcal{F}$.

The first condition implies **closure** under complementation (that is, A^c is also in the field) and the second condition implies closure under countable unions.

Certain simple properties are easily derived from this definition.

? Closure under countable unions ??

$$Let\ A = \left[\bigcup_{i=1}^{\infty} A_i \right] \Rightarrow PARTITIONING\ OF\ SET\ A$$
$$THEN (A \in F)\ A\ IS\ FULLY\ CONTAINED$$
$$W/IN\ THE\ FIELD \Rightarrow \sigma\text{-}F$$

Theorem 2.1

Definition 2.2 implies the following: (1) $S \in \mathcal{F}$, (2) $\emptyset \in \mathcal{F}$, and (3) if $A_i \in \mathcal{F}$ for $i = 1, 2, \ldots$, then $[\cap_{i=1}^{i=\infty} A_i] \in \mathcal{F}$.

Proof: The first property follows from the fact that A and A^c being in $\mathcal{F}$ implies that $S = A \cup A^c$ is also in $\mathcal{F}$. Also, $\emptyset = S^c \in \mathcal{F}$. Finally, by De Morgan's law,

$$\cap_{i=1}^{i=\infty} A_i = [\cup_{i=1}^{i=\infty} A_i^c]^c$$

which, by Definition 2.2, is a member of $\mathcal{F}$, thus establishing part (3) of the theorem.

It follows from the above that a σ-field is a set of subsets of S that is closed under complementation, countable unions, and countable intersections. Also note that the smallest σ-field is $(S, \emptyset)$.

Example 2.3

In the coin-tossing experiment in Example 2.1, consider the set $A = \{ S, \emptyset, HH, (HT, TH, TT) \}$. It is easily verified that A is a σ–field. However, the set $B = \{ (HH, TT) \}$ is not a σ–field because it does not contain S, $\emptyset$, or $[(HH, TT)]^c$.

Practice Problem

2.5   In Example 2.1, construct the **power set** associated with S, that is, the set of all subsets of S, and verify that it is a σ–field.

Example 2.4

In some situations, we may have more than one σ-field associated with a sample space. To illustrate, suppose x refers to the annual income of households in a certain population and $x_1, x_2, \ldots, x_N$ are the actual values of the N households in the population. We can form different subsets of this population by alternatively grouping them in intervals of \$5,000 or \$10,000. Thus, the income distribution for 0–5000, 5001–10000, 10001–15000, and so on, will yield a different σ-field as compared to the one formed by the distribution for 0–10000, 10001–20000, and so on. Another way of formulating the income distribution is by having the same percentage of the population in each income group. For example, the lowest 10 percent of the population, the next 10 percent of the population, and so on. All these subdivisions will result in different σ-fields. In Section 2.3 we will encounter a convenient representation of the distribution of income.

Borel Fields

In probability theory, the usefulness of a σ-field will be apparent when we construct it for the real line $\mathcal{R} = \{x : -\infty < x < \infty\}$. Consider the set $A_x = \{z : z \le x\} = \{(-\infty, x]\}$. The complementary set is $A_x^c = \{z : z \in \mathcal{R} \text{ and } z > x\}$. For different values of x, A_x and A_x^c constitute a family of sets (and belong to a wider class known as **Borel sets**). Starting from A_x, if we take countable unions and intersections of A_x and A_x^c, we can obtain a σ-field on $\mathcal{R}$. Such a σ-field is called a **Borel field** (denoted by $\mathcal{B}$).

Practice Problems

2.6   Verify that for $x < z$ the interval $(x, z]$ is a member of $\mathcal{B}$ by expressing it in terms of A_x defined above.

2.7   Let $A_n(x) = (x, x + (1/n)]$. What is the set $B = \cap_{n=1}^{n=\infty} A_n(x)$? Is $B \in \mathcal{B}$?

Measurable Spaces

In characterizing the attributes of a set or a space, we often want to have numerically quantifiable measures as otherwise summary characteristics are impossible to obtain. For example, corresponding to the set A suppose we define a set function $\mu(A)$ that is simply the number of elements in A if the number is finite and $+\infty$ otherwise. This is a **counting measure** and is a special case of measures in general. Formally, a **measure** is a nonnegative countably additive set function μ defined on $\mathcal{F}$ that has the following properties:

(1) $\mu(A) \ge \mu(\varnothing) = 0$ for all $A \in \mathcal{F}$.

(2)   if $A_i \in \mathcal{F}$ are disjoint sets (that is, $A_i \cap A_j = \varnothing$ for all $i \ne j$), then $\mu(\cup_1^n A_i) = \Sigma_i \mu(A_i)$.

Thus, $\mu : \mathcal{F} \to \mathcal{R}$. A special case of such a measure that has the property $\mu(S) = 1$ is called a **probability measure** (more on this in the next section). Another example of a measure is the length of a real interval. It is called a **Lebesgue measure** (λ) and is defined on the Borel field $\mathcal{B}$ as

$$\lambda\{(a, b]\} = b - a \quad \text{for all} \quad a < b$$

The pair $(S, \mathcal{F})$ is known as a **measurable space,** that is, a space on which a measure can be assigned. For an advanced treatment of measure theory, see Halmos (1950). For a detailed treatment of Borel and σ-fields in the context of probability theory see Spanos (1989).

2.3   PROBABILITY: DEFINITIONS AND CONCEPTS

The probability of an event is defined in a number of ways, all of which are useful in calculating probabilities.

Definition 2.3 (Axiomatic definition)

The **probability** *of an event $A \in \mathcal{F}$ is a real number such that*

   (1) $P(A) \geq 0$ *for every $A \in \mathcal{F}$,*

   (2)   *the probability of the entire sample space S is 1, that is, $P(S) = 1$, and*

   (3)   *if $A_1, A_2, \ldots, A_n$ are mutually exclusive events (that is, $A_i \cap A_j = \emptyset$ for all $i \neq j$), then $P(A_1 \cup A_2 \cup \cdots A_n) = \Sigma_i P(A_i)$, and this holds for $n = \infty$ also.*

The triple $(S, \mathcal{F}, P)$ is referred to as the **probability space** and P is a **probability measure.** [For an advanced treatment of this topic, see Loève (1960)]. It is readily noted that $P(\cdot)$ is simply a set function that maps elements in $\mathcal{F}$ to the unit interval $[0, 1]$.

Although the axiomatic definition of probability is rigorous, it does not directly tell us how to assign probabilities to elementary events. That is accomplished by two other definitions given below. All three definitions are then used to calculate probabilities of various events.

Definition 2.4 (Classical definition)

If an experiment has n ($n < \infty$) mutually exclusive and equally likely outcomes, and if n_A of these outcomes have an attribute A (that is, the event A occurs in n_A possible ways), then the probability of A is n_A/n, denoted as $P(A) = n_A/n$.

As an example, consider the experiment of throwing a pair of dice. The sample space consists of the 36 sample points: (1,1), (1,2) , . . . , (6.6). Each is equally likely and hence the probability of each of these outcomes is 1/36. Next consider the event A, "total score is 5." This can occur in the four mutually exclusive ways, (1,4), (2,3), (3,2), and (4,1). Hence $P(A) = 4/36$.

Definition 2.5 (Frequency definition)

Let n_A be the number of times the event A occurs in n trials of an experiment. If there exists a real number p such that $p = \lim_{n \to \infty} (n_A / n)$, then p is called the probability of A and is denoted as P(A).

Thus, the probability of an event is its limiting frequency when an experiment is repeated indefinitely. In practice, however, it is not feasible to conduct any experiment an infinite number of times. The usefulness of this definition is, therefore, when the number of observations is large. As an illustration, Table 2.1 has the distribution of household income in the United States for 1987 (known as the **frequency distribution**). Suppose we form the income intervals 0–4,999, 5,000–9,999, and so on, and calculate the fraction of the households that falls in each income group (see Table 2.1). The percentage of households that falls in the income group 40,000–49,999 (9.9 in Table 2.1) can be taken to be the probability that a household drawn at random will have income falling in that interval. This concept is graphically illustrated in Figure 2.4. The horizontal axis represents household income (in thousands). The proportion is plotted against the midpoint of the intervals in the form of a bar diagram (known as the **histogram**) in which the areas of the bars equal the corresponding percentages. If the size of the population is sufficiently large and the intervals small enough, we can approximate the percentages with a smooth curve as shown in the diagram.

Example 2.5

In the experiment of tossing a coin twice, the sample space is $S = \{ TT, TH, HT, HH \}$. We can construct a σ-field by choosing the outcomes one or two at a time, obtaining their complements, and including the sets $\emptyset$ and S. One such σ-field is (verify that fact) $\mathcal{F} = \{ \emptyset, S, (TT), (TH, HT, HH), (HH), (TT, TH, HT), (TH, HT), (TT, HH) \}$. It is easy to verify that the corresponding probabilities of the elements of $\mathcal{F}$ are (0, 1, 1/4, 3/4, 1/4, 3/4, 1/2, 1/2). Figure 2.5 is a graphical representation of the probability space $(S, \mathcal{F}, P)$ for this example.

Subjective Probability

In several instances, individuals use personal judgments to assess the relative likelihoods of various outcomes. For instance, one often hears the expression "the odds are 2 to 1 that candidate X will be elected." This is an example of **subjective (or personal)**

Table 2.1 Distribution of Household Income in the United States[†]

Income	Number	Percentage
under 5,000	6,129	6.8
5,000 - 9,999	10,689	12.0
10,000 - 14,999	10,166	11.3
15,000 - 19,999	9,303	10.4
20,000 - 24,999	8,617	9.6
25,000 - 29,999	7,733	8.6
30,000 - 34,999	7,078	7.9
35,000 - 39,999	6,089	6.8
40,000 - 49,999	8,867	9.9
50,000 - 74,999	10,085	11.2
75,000 - 99,999	2,938	3.3
100,000 and over	1,984	2.2
	89,678	100.0

[†]   Income is annual for households for the year 1987, and the number of households is in thousands. Source: *Statistical Abstracts of the United States*, 1991, pp. 451–452.

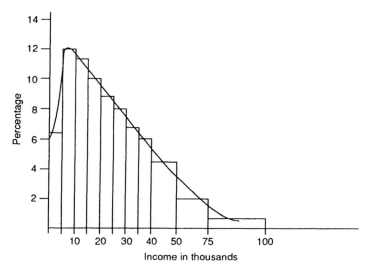

Figure 2.4   Histogram for frequency distribution of incomes

probability which is based on "educated guesses" or intuition. In statistical inference, the practicality of this approach stems from using prior beliefs or new information in updating previous model specifications. In this book we provide an introduction to this **Bayesian updating** later in this and other chapters. For more details on this, refer to Cyert and DeGroot (1987), Barnett (1973), and Leamer (1978).

The axiomatic definition of probability enables us to derive a number of properties of probabilities, and these are discussed next.

Theorem 2.2: $P(A^c) = 1 - P(A)$.

Proof: $A \cup A^c = S$ and $A \cap A^c = \varnothing$. By the second and third axioms, $P(A) + P(A^c) = 1$. Therefore, $P(A^c) = 1 - P(A)$.

Theorem 2.3: $P(A) \leq 1$.

Proof: $P(A^c) \geq 0$ by the first axiom. From this and Theorem 2.2, $P(A) \leq 1$.

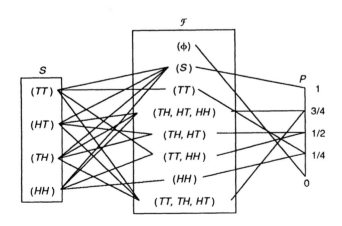

Figure 2.5   Graphical illustration of a probability space

Theorem 2.4: $P(\varnothing) = 0$.

    Proof: $S^c = \varnothing$. $P(S \cup \varnothing) = 1 = P(S) + P(\varnothing)$, which implies that $P(\varnothing) = 0$.

Theorem 2.5: If $A \subset B$, then $P(A) \le P(B)$.

    Proof: If $A \subset B$, then B can be expressed as $B = A \cup (A^c \cap B)$, which are disjoint. Hence, $P(B) = P(A) + P(A^c \cap B) \ge P(A)$ because $P(A^c \cap B) \ge 0$.

Theorem 2.6: $P(A \cup B) = P(A) + P(B) - P(A \cap B)$.

    Proof: The set B can be partitioned as $B = (A \cap B) \cup (A^c \cap B)$ and therefore $P(B) = P(A \cap B) + P(A^c \cap B)$. Hence $P(A^c \cap B) = P(B) - P(A \cap B)$. The set $A \cup B$ can be partitioned as $A \cup B = A \cup (A^c \cap B)$. Therefore $P(A \cup B) = P(A) + P(A^c \cap B)$. It follows that $P(A \cup B) = P(A) + P(B) - P(A \cap B)$.

Practice Problem

2.8   Verify using Venn diagrams that $P(A \cup B \cup C) = P(A) + P(B) + P(C) - P(A \cap B) - P(A \cap C) - P(B \cap C) + P(A \cap B \cap C)$.

Conditional Probability

    Very often, we wish to calculate the probabilities of events when it is known that another event has occurred. It would be useful to know how the probability of an event alters as a result of the occurrence of another event. The following definition presents the rule for calculating such conditional probabilities.

Definition 2.6

    Let A and B be two events in a probability space (S, $\mathcal{F}$, P) such that P(B) > 0. The **conditional probability** *of A given that B has occurred, denoted by P(A|B), is given by P(A $\cap$ B)/P(B).*

    Thus, we are looking at the subspace in which the event B has already occurred. By dividing by $P(B)$, we are normalizing the probability values so that they add up to 1 in the subspace. It should be noted that the original probability space $(S, \mathcal{F}, P)$ remains unchanged even though we focus our attention on the subspace in question which is $[S, \mathcal{F}, P(\cdot \mid B)]$.

Example 2.6 [Ramanathan (1992), Example 2.1]

A high school has 1000 student drivers of whom 600 attended a driver's training course. A survey of all these drivers was conducted over a one-year period to find out how many of them were involved in at least one accident in which they were at fault. The results are presented in Table 2.2. What is the probability that a student who took the course was not involved in any accident?

Table 2.2   Hypothetical Data on Accidents and Driver Training

	Attended	Did not attend	Total
Had accidents	30	70	100
No accidents	570	330	900
Total	600	400	1000

Denote by A the event that a student was involved in at least one accident and by T the event that a student took the training course. Then, by the classical definition, $P(A \cap T) = 30/1000 = 0.03$. $P(A \cap T^c) = 70/1000 = 0.07$. $P(A^c \cap T) = 570/1000 = 0.57$. $P(A^c \cap T^c) = 330/1000 = 0.33$. Note that the four probabilities sum to 1. The event T is the union of the disjoint events $(T \cap A)$ and $(T \cap A^c)$. Therefore, by the axiomatic approach, $P(T) = P(T \cap A) + P(T \cap A^c) = 0.60$. The conditional probability that a student was not involved in any accident given that he or she had driver training is therefore given by $P(A^c \mid T) = 0.57/0.60 = 0.95$.

The above result could have been obtained directly from the table by noting that out of the 600 students who took the course 570 had no accidents. Therefore the desired probability is 570/600 = 0.95. However, the set theoretic approach is used here to illustrate its usefulness in other situations where it may not be possible to construct a table.

Practice Problem

2.9   Calculate $P(A)$, $P(T^c)$, and $P(A \mid T^c)$.

Theorem 2.7 (Bayes theorem)

If A and B are two events with positive probabilities, then

$$P(A \mid B) = \frac{P(A) P(B \mid A)}{P(B)}$$

Proof: The proof is quite straightforward and follows from the definition of conditional probability.

$$P(A \mid B) = \frac{P(A \cap B)}{P(B)} \qquad P(A) = \frac{P(A \cap B)}{P(B/A)}$$

$$P(A \cap B) = P(B \mid A) P(A)$$

Combining the two equations, we have the desired result.

Although the theorem appears to be trivial, its interpretation and use in practical situations has spawned a whole new approach to statistical inference. To motivate the idea behind the approach, suppose that the event A is a possible strategy of the management of a company and the event B is a possible strategy by the employee union. Knowing the probability of occurrence of B given that A has occurred, we might want to infer the probability of occurrence of the preceding event, namely, that the company would adopt strategy A. Thus, we might be interested in a **Bayesian updating** of A from the conditional probability of B given A. A more common use of the approach is when there are several possible states of nature (call them $A_1, A_2, \ldots A_n$) associated (usually in a subjective manner) with different probabilities. The rule for updating the probabilities of the states of nature is given in the following theorem.

Theorem 2.8 (extended Bayes theorem)

If $A_1, A_2, \ldots, A_n$ constitute a partition of the sample space, so that $A_i \cap A_j = \varnothing$ for $i \neq j$ and $\cup_i A_i = S$, and $P(A_i) \neq 0$ for any i, then for a given event B with $P(B) > 0$,

$$P(A_i \mid B) = \frac{P(A_i) P(B \mid A_i)}{\Sigma_i P(A_i) P(B \mid A_i)}$$

Proof: By the previous theorem,

$$P(A_i \mid B) = \frac{P(A_i) P(B \mid A_i)}{P(B)}$$

Because A_i's constitute a partition of S, B can be written as $B = \cup_i (B \cap A_i)$ where each is disjoint from the others. Hence $P(B) = \Sigma_i P(B \cap A_i) = \Sigma_i P(A_i) P(B \mid A_i)$. Substitution of this in the above expression establishes the theorem.

Example 2.7

Suppose that a person with lung cancer is given a chest X-ray and the probability is 0.99 that the cancer will be detected, and that if a person without lung cancer is given a chest X-ray the probability that he or she will be incorrectly diagnosed as having lung cancer is 0.001. Suppose further that 1 percent of the resident patients of a hospital have lung cancer. If one of these persons (chosen at random with equal probability) is diagnosed as having lung cancer on the basis of a chest X-ray, what is the probability that he or she actually has lung cancer?

Although the answer to this is quite simple if we prepare a table like the one in Example 2.6, let us use set theory and the axiomatic approach. Let D = diagnosed as having lung cancer; A = actually having lung cancer.

Given: $P(D \mid A) = 0.99$, $P(D \mid A^c) = 0.001$, $P(A) = 0.01$.

Question: $P(A \mid D) = ?$

$$P(A \mid D) = \frac{P(D \mid A)P(A)}{P(D)} = \frac{P(D \mid A)P(A)}{P(D \cap A) + P(D \cap A^c)}$$

$$= \frac{P(D \mid A)P(A)}{P(D \mid A)P(A) + P(D \mid A^c)P(A^c)}$$

$$= \frac{0.99 \times 0.01}{(0.99 \times 0.01) + (0.001 \times 0.99)} = 0.9091$$

Statistical Independence

Suppose that the conditional probability of A given B is the same as the unconditional probability of A, that is, $P(A \mid B) = P(A)$. This means that in assessing the probability of A there is no informational content in the knowledge that B has occurred. As an example, let A be the outcome of the roll of a die and B be the outcome of the roll of a second die. The probability that the roll of the second die results in a 4 is 1/6 *regardless of the outcome of the first die*, and this holds for any other outcome of the rolls. Thus, the events A and B occur independently. This notion is formalized in the following definition.

Definition 2.7

Two events A and B with positive probabilities are said to be

statistically independent *if and only if* $P(A \mid B) = P(A)$. *Equivalently,* $P(B \mid A) = P(B)$ *and* $P(A \cap B) = P(A) \cdot P(B)$.

We note from the definition that in the case of <u>statistical independence</u> the joint probability of two events is equal to the product of the individual probabilities and conversely, if the joint probability is the product of individual probabilities then the events are independent. In Example 2.6, $P(A \cap T) = 0.03$. $P(A)P(T) = 0.06 \neq P(A \cap T)$. Hence driver training and accidents are <u>not independent</u>.

The distinction between two events being *statistically independent* and being *mutually exclusive* must be clearly understood. If A and B are mutually exclusive, then $A \cap B = \varnothing$ and hence the corresponding probability is zero, but independence requires that $P(A \cap B)$ be the product of the individual (non-zero) probabilities.

Sequences and Limiting Sets of Events

When conducting a random experiment, we often encounter sequences of sets of events. For instance, consider the set $A_n = [x - n, x + n]$ for a fixed x. For different values of n this defines a <u>sequence of sets</u>. The <u>outcomes of repetitions of trials also lead to a</u> sequence of events. In such cases it is useful to know what happens as the number of trials becomes indefinitely large. Sequences also arise when an experiment involves drawing observations over time. In this case we would be interested in knowing what happens "in the long run." The notion of sequences of sets and their limits is formalized in the rest of this section. Applications of the concepts are presented in the next chapter.

i. e.
$\longrightarrow$ DISTRIBUTIONAL FORMS, ARE AN INTERPRETATION
OF SEQUENCES OF SETS, FROM THE LIMIT OF
EXPERIMENTAL EVENTS,

Definition 2.8

A sequence of sets $A_1, A_2, A_3, \ldots$ *is called* **monotone increasing** *if* $A_1 \subset A_2 \subset A_3 \subset \ldots$ *and* **monotone decreasing** *if* $A_1 \supset A_2 \supset A_3 \supset \ldots$. *The limit set is defined as follows:*

Monotone increasing: $\lim\limits_{n \to \infty} A_n = \bigcup_1^{\infty} A_n$.

Monotone decreasing: $\lim\limits_{n \to \infty} A_n = \bigcap_1^{\infty} A_n$.

Theorem 2.9

If $A_1, A_2, \ldots A_n, \ldots$ *is a monotone sequence, then*

$$P(\lim_{n \to \infty} A_n) = \lim_{n \to \infty} P(A_n)$$

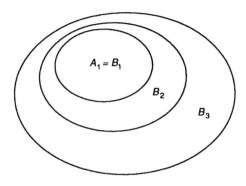

Figure 2.6   Monotone increasing sets

Proof: We will prove this only for the monotone increasing case $A_1 \subset A_2 \subset A_3 \ldots$. The other case is similar. Define disjoint sets $B_1 = A_1$, $B_2 = A_2 \cap A_1^c$ (that is, all points in A_2 that are outside A_1), $B_3 = A_3 \cap A_2^c$, and so on (see Figure 2.6). Thus, $A_n = A_{n-1} \cup B_n = \cup_1^n B_i$ by induction. $P(A_n) = P(\cup_1^n B_i) = \Sigma_1^n P(B_i)$. Hence, as $n \rightarrow \infty$, $\lim [P(A_n)] = \Sigma_1^\infty P(B_i)$. By monotonicity, $\lim (A_n) = \cup_1^\infty A_i$, which is equal to $\cup_1^\infty B_i$ because $A_n = \cup_1^n B_i$. Hence $P[\lim (A_n)] = P(\cup_1^\infty B_i) = \Sigma_1^\infty P(B_i)$. We showed earlier that $\lim [P(A_n)] = \Sigma_1^\infty P(B_i)$. Hence, $\lim P(A_n) = P[\lim (A_n)]$.

EXERCISES

2.1   Let

   A = all persons who voted for a particular candidate
   B = all wealthy people
   C = all landlords
   D = all Republicans
   E = all private (nonpublic) employees

Write the following statement in set theoretic notation. "A person who voted for the candidate must be a wealthy landlord or a Republican nonpublic employee."

2.2   A physician checks the weights of a number of patients and classifies them as "very overweight," "slightly overweight," "near normal weight," "slightly underweight," and "very underweight." Draw a Venn diagram to represent the outcomes when two patients are weighed and identify the subsets "neither of them is underweight," and "at least one of them is very overweight."

2.3   Let B be an event and $A_1, A_2, \ldots, A_n$ be n mutually exclusive events. Define $A = \cup_1^n A_i$. Also assume that $P(A) > 0$ and $P(B \mid A_i) = p$ for all i. Show that $P(B \mid A)$ is also equal to p. [A Venn diagram might help.]

2.4   A husband and wife are each 70 years old. The probability that the husband will die in the next year is 0.1, and the probability that the wife will die in the next year is 0.05. The probability that the husband will die given that the wife dies is 0.4. What is the probability that at least one of them will die in the next year? What is the probability that the wife will die given that the husband has died?

2.5   The probability of a male, age 60, dying within one year is 0.025 and the probability of a female, age 55, dying within one year is 0.01 and the two events are independent. If a man and his wife are ages 60 and 55 respectively, what is the probability that (1) neither will die within a year, (2) at least one will die within a year, (3) at least one will live through one year?

2.6   A rental car company operates between two cities A and B. In city A it has a Fords and b Chevrolets. In city B it has c Fords and d Chevrolets. A customer picks a car at random from city A, drives it to city B, and leaves it there. A second customer then chooses a car at random from city B. What is the probability that it is a Ford?

2.7   The probability that a person will watch a movie on TV is 0.60. If a person is watching, the probability that the show is taped is one-third. If a person is not watching, the probability that the show will be taped is 0.9. What is the probability that the show will be taped? What is the probability that a show is being watched given that it is being taped?

2.8   With probability 0.7 I set my wrist watch alarm to alert me to get ready for class. If the alarm is set, I am on time to class with probability 0.99. If it is not set, I am on time only 60 percent of the time. What is the probability that I will be on time?

Given that I was on time, what is the probability that the alarm was set?

2.9 The probability that a person chosen at random will have high blood pressure is 0.05. Given that a person has high blood pressure, the probability that he drinks alcohol is 0.250, whereas 75 percent of those without high blood pressure drink alcohol. Given that a person is an alcohol drinker, what is the probability that he has high blood pressure?

2.10 Let x be any point in the interval $(0, a)$ and $P(x \le x_0) = x_0/a$ for any x_0 in that interval. Define the set of all points in the interval $(x_0 - 1/n, x_0 + 1/n)$ as A_n. Show that A_n is a monotone sequence. Is it increasing or decreasing? What is the limiting set? Compute $P(A_n)$. Derive the implications of Theorem 2.9 applied to the sequence of sets.

3

RANDOM VARIABLES
AND DISTRIBUTIONS

The probability model represented by $(S, \mathcal{F}, P)$ is too general to be of practical use. For a complete description, we need to enumerate every element of $\mathcal{F}$ and the associated probability (see Figure 2.5). This makes the mathematical manipulation of probabilities cumbersome. A more useful approach would be to measure attributes of events quantitatively and use them in calculating probabilities of events. In the previous chapter we saw several examples in which events and their probabilities are associated with variables whose values are measured by an experimenter. The total score when a pair of dice is rolled, the number of heads when a coin is tossed several times, annual household income, and so on are examples of such variables. Variables of this type are known as **random variables** or **stochastic variables** and are fundamental to the theory of probability and statistics. Not all variables can be called a random variable, however. To illustrate, consider the experiment of tossing a coin twice and the associated probability space $(S, \mathcal{F}, P)$ presented in Example 2.5. Let X denote the number of heads in the two trials. It can take only the values 0, 1, and 2. $X(\cdot)$ can thus be thought of as a set function that maps the sample space S into $\mathcal{R}_x = \{0, 1, 2\} \in \mathcal{R}$. Thus we have,

$$\{TT\} \rightarrow 0 \qquad \{HT, TH\} \rightarrow 1 \qquad \{HH\} \rightarrow 2$$

Denoting the inverse mapping by $X^{-1}(\cdot)$, we have $X^{-1}(1) = \{HT, TH\} \in \mathcal{F}$, and so on for the others. Thus there is a correspondence between $\mathcal{R}_x$ and S such that the event structure is preserved. This is not true, however, for the set function Y defined as $Y(TT) = Y(HT) = 1$, $Y(HH) = Y(TH) = 0$, because the inverse mapping is $Y^{-1}(1) = \{TT, HT\}$ and $Y^{-1}(0) = \{HH, TH\}$, which are not elements of the σ-field $\mathcal{F}$ defined in Example 2.5. Thus, with respect to that σ-field, $Y(\cdot)$ does not preserve the event structure, whereas $X(\cdot)$ does (we can, however, construct a different σ-field with respect to which $Y(\cdot)$ will be event-preserving). For a variable to be called a

[handwritten margin note: ? R.V. w/ CONVERGENT SEQUENCE OF SETS?]

27

[margin: R.V. + PRESERVATION OF EVENT STRUCTURE!!]

random variable the preservation of the event structure is important as otherwise inconsistencies will arise. The formal definition of a random variable is given below.

Definition 3.1

In simple terms, a **random variable** *(also referred to as a* **stochastic variable***) is a real-valued* set function *whose value is a real number determined by the outcome of an experiment. The* **range** *of a random variable is the set of all the values it can assume. More formally, in measure theoretic terms, a random variable X is a real-valued function that maps S into $\mathcal{R}$ and satisfies the condition that for every Borel set $B \in \mathcal{B}$, the inverse image $X^{-1}(B) \in \mathcal{F}$, where*   *[margin: ↳ BOREL FIELD]*

[margin: OF THE SET B]
$$X^{-1}(B) = \{ s : s \in S \text{ and } X(s) \in B \}$$
[margin: S IS IN EVENT SPACE S]

[margin: X(s) – THE "REAL VALUED SET-FUNCTION" – IS IN THE BOREL SET]

A random variable is therefore a real-valued function (and hence is not really a variable) that maps S into the real line $\mathcal{R}$ and assigns a real number to each $s \in S$. Furthermore, the term "random" is really inappropriate because the function $X(\cdot)$ does not return a random value. What distinguishes a random variable from other types of variables is the fact that, for any given set $B \in \mathcal{B}$ the corresponding events must be in $\mathcal{F}$. [see Spanos (1989), Section 4.1, for a more detailed discussion of these points.] *i.e. EVENT PRESERVING*

Note that in the triple $(S, \mathcal{F}, P)$, the sample space S now corresponds to the real line $\mathcal{R}$ and the σ-field $\mathcal{F}$ now corresponds to the Borel Field $\mathcal{B}$. Corresponding to the probability measure $P(\cdot)$ it is possible to define a set function, call it $P_x(\cdot)$, that maps the Borel field $\mathcal{F}$ into the closed unit interval [0, 1]. For example, in the coin tossing case, the appropriate set function is $P_x(X=0) = 1/4$, $P_x(X=1) = 1/2$, $P_x(X=2) = 1/4$, $P_x(X=1 \text{ or } X=2) = 3/4$, $P_x(X=1 \text{ and } X=2) = 0$, and so on. The random variable X would now enable us to work with the new probability space $(\mathcal{R}, \mathcal{B}, P_x)$ which is more amenable for mathematical manipulation.

3.1   DISTRIBUTION FUNCTION

If the sample space is countably or uncountably infinite, the probability set function P_x is still not workable. It would therefore be useful to construct a point function that can be defined over continuous intervals also and has the same information content as the probability set function. Such a function is defined below.

Definition 3.2

The real-valued function $F(x)$ such that $F(x) = P_x\{(-\infty, x]\} = P(X \le x)$ for each $x \in \mathcal{R}$ is called the **distribution function,** *also known as the* **cumulative distribution** *(or* **cumulative density) function,** *or* **CDF.**

$\underbrace{P(X \le \overset{\curvearrowright}{x})}$ —"SPECIFIED VALUE"
$\!\downarrow$ R.V.

$F(x)$ summarizes the probability defined on the Borel set $A_x = (-\infty, x]$ introduced in Section 2.2. It gives the probability that a random variable assumes values less than or equal to a specified value. Note that the random variable X in conjunction with the CDF transforms the triple $(S, \mathcal{F}, P)$ into $(\mathcal{R}, \mathcal{B}, F)$.

As an example of a CDF, consider the experiment of rolling a die once, and let X be the score. The possible values for X are $1, 2, \ldots 6$, and each is equally likely with a probability of 1/6. We readily see that $F(x)$ is as given below (see Figure 3.1 for a graph of it).

SPECIFIED VALUE $= P(X \le x) = P(R.V. \le x)$

x	$F(x)$
< 1	0
$i \le x < i+1$	$i / 6$ for $i = 1, 2, \ldots 5$
$x \ge 6$	1

for any specified
value, greater than
or equal to 6 $\Rightarrow$ $P(X \le 6) = 1$

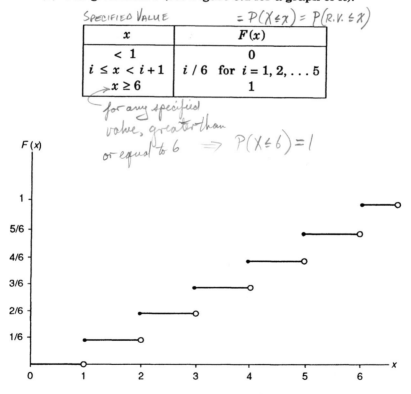

Figure 3.1   Graph of the CDF for rolling a single die

Theorem 3.1: $P(a < X \le b) = F(b) - F(a)$.

Proof: Let I_1 be $(-\infty, a]$ and I_2 be $(a, b]$. Then I_1 and I_2 are disjoint and hence $P(I_1) + P(I_2) = P(I_1 \cup I_2)$. But $P(I_1 \cup I_2) = F(b)$ and $P(I_1) = F(a)$. Hence $P(a < X \le b) = P(I_2) = F(b) - F(a)$.

Given the CDF $F(x)$, this theorem enables us to assign probabilities to any open interval $(a, b]$. Other properties of the CDF are listed in the following practice problems and theorems.

Practice Problems

3.1   Show that $F(-\infty) = 0$ and $F(+\infty) = 1$, that is, that $F(x)$ is **normed**.

3.2   Show that $F(x)$ is monotone nondecreasing, that is, if $b > a$, then $F(b) \ge F(a)$.

Theorem 3.2:   *For each $x \in \mathcal{R}$, $F(x)$ is continuous to the right of x.*

Proof: Consider the interval $B_n = (x, x + (1/n)]$ for $n > 0$, which is open at the left and closed at the right. We have, $P(B_n) = F[x + (1/n)] - F(x)$. Also, $B_{n+1} \subset B_n$ and hence B_n is monotonic decreasing. Note that $\lim_{n \to \infty} B_n = \varnothing$, that is, the limit set of B_n is the empty set (because the interval is open at x). Therefore, $P(\lim B_n) = 0$. By Theorem 2.9, $P(\lim B_n) = \lim P(B_n)$. Hence,

$$0 = P\left[\lim_{n \to \infty} B_n\right] = \lim_{n \to \infty} \left[F\left(x + \frac{1}{n}\right) - F(x)\right] = F(x+) - F(x)$$

where $F(x+)$ is the right-hand limit of $F(x)$ at x. This establishes the theorem that $F(x)$ is right continuous at x.

Theorem 3.3:   *If $F(x)$ is continuous at $x \in \mathcal{R}$, then $P(X = x) = 0$.*

Proof: First define $B_n = (x - 1/n, x + 1/n]$. We note that $B_{n+1} \subset B_n$. Hence, by monotonicity (Theorem 2.9), as $n \to \infty$, $P(\lim B_n) = \lim P(B_n)$. But

$$\lim P(B_n) = \lim \left[F\left(x + \frac{1}{n}\right) - F\left(x - \frac{1}{n}\right)\right] = 0$$

because $F(x)$ is continuous at x. By monotonicity, $\lim B_n = x$, and hence $P(\lim B_n) = P(X = x)$. Hence the result that $P(X = x) = 0$ when $F(x)$ is continuous at x.

It will be noted from this theorem that a random variable for which $F(x)$ is continuous at all points of its range assigns a zero probability for any x.

3.2   DISCRETE DISTRIBUTIONS

A random variable X is said to have a **discrete distribution** if it can take only a finite number of different values $x_1, x_2, \ldots, x_n$, or a countably infinite number of distinct points. For the discrete case the probability measure P_x can be expressed in a simple form.

Definition 3.3

*For a discrete random variable X, let $f(x) = P_x(X = x)$. The function $f(x)$ is called the **probability function** (or as PF).*

The relationship between a CDF and a PF is straightforward. Because $F(x) = P(X \le x)$, we have $F(x) = \Sigma_{X \le x} f(X)$. We have already seen two examples of discrete random variables.

Example 3.1

Roll a die once. If each of the six possibilities is equally likely, then $P(X = 1) = P(X = 2) = \cdots = P(X = 6) = 1/6$. We thus have the following PF and CDF.

x	1	2	3	4	5	6
$f(x)$	1/6	1/6	1/6	1/6	1/6	1/6
$F(x)$	1/6	2/6	3/6	4/6	5/6	1

The above is a special case of the **uniform distribution on integers** for which X takes only the values 1, 2, ..., n, each with equal probability p. Because the probabilities must add up to 1, we have $p = 1/n$. Thus, for the uniform distribution on integers, $f(x) = 1/n$ for $x = 1, 2, \ldots, n$, and 0 elsewhere.

Example 3.2

Tossing a coin three times has the 8 outcomes *HHH, HHT,* and so on. If each of them is equally likely, we get the following distribution for the number of heads.

x	0	1	2	3
$f(x)$	1/8	3/8	3/8	1/8
$F(x)$	1/8	1/2	7/8	1

We present below other examples of discrete distributions. Before doing that, however, it should be pointed out that every probability function involves one or more parameters that will be denoted by θ (which may contain more than one element). The **parameter space**, that is, the set of values θ can take, is denoted by Θ. This point is emphasized by writing the probability function as $f(x;\theta)$, $\theta \in \Theta$ and noting that what we have is a **family of distributions**.

The Bernoulli Distribution

Consider an experiment which has only two outcomes, one named a *success* and the other named a *failure*. The sample space for this is $S = \{$ success, failure $\}$. Let $P(\text{success}) = p$, which implies that $P(\text{failure}) = 1-p$. Define the random variable X for which $X(\text{success}) = 1$ and $X(\text{failure}) = 0$. We then have $P(X=1) = p$ and $P(X=0) = 1-p$. This can be conveniently represented in the form $f(x;p) = p^x(1-p)^{1-x}$ for $x = 0, 1$ and $0 \leq p \leq 1$. This distribution is known as the **Bernoulli distribution** and the trials of the experiment are referred to as the **Bernoulli trials**. *SAMPLING*

The Binomial Distribution *WITH REPLACEMENT*

Example 3.2 is a special case of the **binomial distribution** which arises out of a number of Bernoulli trials. Let p be the probability of a success in a given trial and $q = 1-p$ be the probability of failure. Assume also that (1) the probability of a success is the same for each trial and (2) the trials are independent. Let X be the number of successes in n such trials. Then what is $f(x;p) = P(X = x)$?

First consider the special case of 4 trials and denote success by S and failure by F. A single success can occur in one of four ways (*SFFF, FSFF, FFSF, FFFS*). Each of these has the probability $p(1-p)^3$. Therefore, $f(1) = 4p(1-p)^3$. Two successes can occur in six distinct ways (*SSFF, SFSF, SFFS, FSSF, FSFS, FFSS*). The probability for this is $f(2) = 6p^2(1-p)^2$. Using the *factorial* notation $n! = n(n-1)\cdots 1$ this can also be written as

$$f(2) = \frac{4!}{2!\,2!}\, p^2(1-p)^2$$

By a similar argument,

$$f(3) = \frac{4!}{3!\,1!}\, p^3(1-p)$$

The general form of the density function for the binomial distribution is given by (see Freund, 1992, pp. 185–86)

$$f(x;\theta) = B(x;n,p) = \binom{n}{x}p^x q^{n-x} = \frac{n!}{x!\,(n-x)!}\,p^x q^{n-x}$$

$$x = 0, 1, \ldots, n \qquad 0 \le p \le 1 \qquad q = 1 - p$$

It can be shown that $\sum_{x=0}^{x=n}\binom{n}{x}p^x q^{n-x}$ is the expansion of $(p+q)^n$. Thus the binomial density is a typical term in the binomial expansion of $(p+q)^n$.

Figure 3.2 is a representation of the binomial distribution for $n = 10$ and $p = 0.25$. It is a **bar diagram** in which the x-axis represents the number of successes and the y-axis represents the probabilities. To give the appearance of continuity, the probability for x is attributed equally to all points in the interval $x - \frac{1}{2}$ to $x + \frac{1}{2}$.

Example 3.3

Suppose a manufacturer of TV tubes draws a random sample of 10 tubes. The production process is such that the probability that a single TV tube, selected at random, is defective is 10 percent. Calculate the probability of finding (a) exactly 3 defective tubes, (b) no more than 2 defectives.

(a) $P(X = 3) = \begin{bmatrix} 10 \\ 3 \end{bmatrix} (0.1)^3 (0.9)^7$

(b) $P(X \le 2) = \begin{bmatrix} 10 \\ 0 \end{bmatrix} (0.9)^{10} + \begin{bmatrix} 10 \\ 1 \end{bmatrix} (0.1)(0.9)^9 + \begin{bmatrix} 10 \\ 2 \end{bmatrix} (0.1)^2 (0.9)^8$

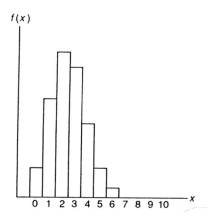

Figure 3.2 An example of the binomial distribution

Example 3.4

A company that markets Brand A cola drink claims that 65 percent of all residents of a certain area prefer its brand to Brand B. The company that makes Brand B employs an independent market research consultant to test the claim. The consultant takes a random sample of 25 persons and decides in advance to reject the claim if fewer than 12 people prefer Brand A. What is the probability that the market researcher will make the error of rejecting the claim even though it is correct?

In this binomial example, $n = 25$ and $p = 0.65$. Let $E_i =$ exactly i people prefer Brand A, and $E =$ fewer than 12 of them prefer Brand A. $E = \cup_0^{11} E_i$ and the E_i's are mutually exclusive. We want $P(E)$ when $p = 0.65$.

$$P(E) = \sum_0^{11} P(E_i) = \sum_0^{11} \begin{bmatrix} n \\ i \end{bmatrix} p^i q^{n-i} = \sum_0^{11} \begin{bmatrix} 25 \\ i \end{bmatrix} (0.65)^i (0.35)^{25-i}$$

Calculating this is obviously cumbersome. However, standard statistical tables provide tabulations of the values of the cumulative binomial distribution for different values of n, x, and p. Appendix Table B.1 has

$$1 - F(k-1) = \sum_{x=x'}^{x=n} \begin{bmatrix} n \\ x \end{bmatrix} p^x (1-p)^{n-x}$$

for different values of n and x' and selected values of $p \leq 0.5$. But this is not exactly what we want. The trick is to resort to the symmetry of the binomial distribution to get what we want. For example, let $x = 25 - i$. Then

$$\begin{bmatrix} 25 \\ i \end{bmatrix} (0.65)^i (0.35)^{25-i} = \begin{bmatrix} 25 \\ 25-x \end{bmatrix} (0.35)^x (0.65)^{25-x}$$

$$= \begin{bmatrix} 25 \\ x \end{bmatrix} (0.35)^x (0.65)^{25-x}$$

noting that $\begin{bmatrix} n \\ x \end{bmatrix} = \begin{bmatrix} n \\ n-x \end{bmatrix}$. Hence

$$P(E) = \sum_{x=14}^{x=25} \begin{bmatrix} 25 \\ x \end{bmatrix} (0.35)^x (0.65)^{25-x}$$

This has the form of the binomial with $p = 0.35$, which is tabulated in Table B.1. By referring to it we find that $P(E) = 0.0255$. Thus there is a 2.55% chance of erroneously concluding that Brand A is not preferred.

Practice Problems

3.3   The English teacher in a high school gives a test consisting of
20 multiple choice questions with four possible answers to each
of them, of which only one is correct. One of the students who
has not been studying decides to check off answers at random.
What is the probability that he will get half of the questions
right?

3.4   The owner of a mountain resort has 15 cabins available for
rent, and they are rented independently. The probability that
any one of them will be rented for a night is 0.8. Compute the
probability that at least 12 cabins will be rented in a night.

3.3   CONTINUOUS DISTRIBUTIONS

Unlike discrete random variables that take only specific values, a
continuous random variable can take any value in a real interval. In
this section we study a few cases of continuous distributions. The
next chapter has several other examples of discrete and continuous
distributions.

Definition 3.4

*For a random variable X if there exists a nonnegative function
f (x), defined on the real line, such that for any interval B,*

$$P(X \in B) \;=\; \int_B f(x)\,dx$$

then X is said to have a **continuous distribution** *and the function f (x) is called the* **probability density function** *or simply
the* **density function** *(or PDF).*

Practice Problem

3.5   For a continuous random variable verify the following.

$$F(x) \;=\; \int_{-\infty}^{x} f(u)\,du \qquad\qquad f(x) \;=\; F'(x)$$

$$\int_{-\infty}^{+\infty} f(u)\,du \;=\; 1 \qquad\qquad F(b) - F(a) \;=\; \int_{a}^{b} f(u)\,du$$

It follows from the above that the derivatives of a CDF are the
corresponding ordinates for the PDF.

Uniform Distribution on an Interval

A random variable X for which the density function $f(x; a, b)$ is a positive constant c in the interval $a \leq X \leq b$ is called the **uniform distribution on an interval**. For $f(x; a, b)$ to be a PDF,

$$\int_a^b f(x; a, b)\,dx = 1 = \int_a^b c\,dx = c(b-a)$$

Hence $f(x; a, b) = 1/(b-a)$ uniformly in $a \leq x \leq b$. Its distribution function is a straight line and is given by

$$F(x; a, b) = \int_a^x f(x; a, b)\,dx = \frac{x-a}{b-a} \qquad \text{for } a \leq x \leq b$$

Two special cases of the parameters are frequently used in actual applications; $a = 0$, $b = \theta$ and $a = -\theta$, $b = \theta$. Figure 3.3 graphs both the PDF and the CDF of the uniform distribution.

The Normal Distribution

The most widely used distribution in all applications of statistics is the **normal** (also known as the **Gaussian**) **distribution** which has the following density (exp is the exponential function):

$$f(x; \mu, \sigma) = \frac{1}{\sigma\sqrt{2\pi}} \exp\left[-\frac{(x-\mu)^2}{2\sigma^2}\right] \qquad -\infty < x < \infty$$

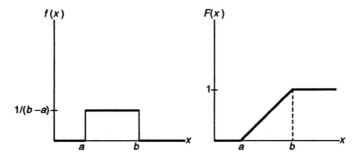

Figure 3.3   PDF and CDF of the continuous uniform distribution

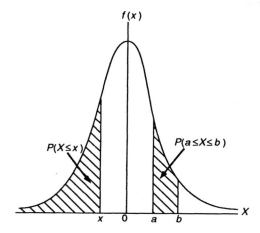

Figure 3.4 The standard normal distribution

The distribution is written as $N(\mu, \sigma^2)$, and we say $X \sim N(\mu, \sigma^2)$. The values of the two parameters μ and σ^2 are generally unknown. Their intuitive interpretation, estimation from data, tests of hypotheses, and applications will be discussed in later sections and chapters. It will be seen that a variety of distributions approximate the normal distribution. The special case of the normal distribution for $\mu = 0$ and $\sigma = 1$ is known as the **standard normal** and its density function is independent of the parameters.

$$f(x) = \frac{1}{\sqrt{2\pi}} e^{-x^2/2} \qquad -\infty < x < \infty$$

Figure 3.4 is a graph of the standard normal distribution which is symmetric around the origin and is bell-shaped. The CDF of the standard normal distribution is

$$F(x) = \int_{-\infty}^{x} \frac{1}{\sqrt{2\pi}} e^{-(y-\mu)^2/2} dy$$

This integral does not have a closed form solution but requires numerical integration. For selected values of z Table B.3 presents the area under the standard normal curve to the right of z [which is also $1 - F(z)$].

3.4   TRANSFORMATIONS OF RANDOM VARIABLES

In statistical inference, transformations of random variables are often carried out and hence we need the means by which we can derive the statistical distributions of the transformed variables from those of the original variables. Just as a requirement of a random variable was that it be event preserving, the transformation functions should also have the same property. In other words, transformations should be **measurable functions.**

Definition 3.5

A function $g(X)$ on R to $\mathcal{R}$ is called a **measurable function** *(or $\mathcal{F}$-measurable) if the set $\{ x : g(x) \le y \} \in \mathcal{F}$ for every real number $y \in \mathcal{R}$.*

Thus, a function $g(X)$ being measurable implies that we can express the probability of the event $\{g(X) \le y\}$ in terms of the probability of an event in $\mathcal{F}$ corresponding to X.

Theorem 3.4

Let $F_X(x)$ be the CDF of the random variable X and let $Y = g(X)$ be measurable, differentiable, and monotonic. Then the CDF of Y is given by [$h(Y)$ is the inverse of $g(X)$].

$$F_Y(y) = F_X[h(y)] \qquad \text{if } g(X) \text{ is monotonic increasing}$$

$$F_Y(y) = 1 - F_X[h(y)] \qquad \text{if } g(X) \text{ is monotonic decreasing}$$

Proof: We prove this only when $g(X)$ is monotonic increasing (verify the other case). $F_Y(y) = P(Y \le y) = P[g(X) \le y]$. Because the transformation is monotonic increasing, the event $g(X) \le y$ is identical to the event $X \le h(y)$ (the inverse function exists because of monotonicity). Hence

$$P[g(X) \le y] = P[X \le h(y)] = F_X[h(y)]$$

Theorem 3.5

Let the assumptions of Theorem 3.4 hold. In addition, let $f_X(x)$ be the PDF of X, and $dx/dy \ne 0$. Then the PDF of $Y = g(X)$ is given by [denoting the inverse function as $X = h(Y)$]

$$f_Y(y) = f_X[h(y)] \qquad \text{when } X \text{ is discrete}$$

$$f_Y(y) = f_X[h(y)] \mid dx/dy \mid \qquad \text{when } X \text{ is continuous}$$

Proof: The proof is trivial for the discrete case. We have, $P(Y = y) = P[X = h(y)] = f_X[h(y)]$. For a continuous random variable the PDF is the derivative of the CDF and hence,

$$f_Y(y) = \frac{d}{dy} F_Y(y)$$

But $F_Y(y) = F_X[h(y)]$ by Theorem 3.4. That is,

$$F_Y(y) = \int_{-\infty}^{h(y)} f_X(x)\,dx$$

Differentiating with respect to y and using the chain rule,

$$f_Y(y) = f_X[h(y)]\, h'(y)$$

But $h'(y) = dx/dy$. Because $f_Y(y)$ must be nonnegative, we have to use the absolute value of the derivative. Hence

$$f_Y(y) = f_X[h(y)]\, |dx/dy|$$

Example 3.5

Let X be a random variable distributed as $N(\mu,\sigma^2)$. Consider the transformation $Y = g(X) = (X-\mu)/\sigma$, which has the inverse $X = h(Y) = \mu + \sigma Y$. Also, $|dx/dy| = \sigma$. By Theorem 3.5, the density function of Y is given by

$$f_Y(y) = \frac{1}{\sigma\sqrt{2\pi}} e^{-y^2/2}\, \sigma = \frac{1}{\sqrt{2\pi}} e^{-y^2/2}$$

which is the standardized normal $N(0, 1)$. It is easy to verify that the general normal density function can be derived from the standard normal by the transformation $Y = \mu + \sigma X$.

Example 3.6

Let us apply Theorem 3.5 to derive the distribution of $Y = e^X$, when X is standard normal. Thus

$$f_X(x) = \frac{1}{\sqrt{2\pi}} e^{-x^2/2} \quad \text{and} \quad X = \ln Y$$

Therefore $f_Y(y)$ is given by

$$f_Y(y) = \frac{1}{y\sqrt{2\pi}} e^{-(\ln y)^2/2} \qquad y > 0$$

The above distribution is known as the **lognormal distribution** (more on this in the next chapter). It will be noted that the range of Y is the positive real line. This is because $Y = e^X$ which is nonnegative. Thus the support boundaries of $-\infty < x < +\infty$ are transformed into $y > 0$.

Practice Problem

3.6   Derive the transformations that would convert a uniform distribution on $[a, b]$ to $[0, \theta]$ and to $[-\theta, +\theta]$.

3.5   CHARACTERISTICS OF DISTRIBUTIONS

The probability density and the cumulative distribution functions determine the probabilities of random variables at various points or in different intervals. Very often we are interested in summary measures of where the distribution is located, how it is dispersed around some average measure, whether it is symmetric around some point, and so on. Numerical measures that characterize a distribution are useful in the calculation of probabilities. In this section we study a variety of measures of central location and dispersion.

The Stieltjes Integral

In the previous sections we used the integral (known as the **Riemann integral**) in the context of continuous random variables. In probability theory, a second type of integral (referred to as the **Stieltjes integral**) is widely used. Before developing that, let us review the definition of the Riemann integral. Consider the closed interval $[a, b]$ for any pair of real numbers such that $a < b$, and a single-valued function $g(x)$ bounded in $[a, b]$. Next subdivide $[a, b]$ into a number of intervals by inserting points, denoted by x_i, as follows.

$$a = x_0 < x_1 < x_2 < \cdots < x_n = b$$

The subdivision is referred to as a **partition** and the largest of the lengths of the intervals ($\Delta x_i = x_i - x_{i-1}$) as the **norm** of the partition, denoted by $\|\Delta x\|$. Let w_i be any point in $[x_{i-1}, x_i]$. Next construct the following sum (known as the **Riemann sum**)

$$\sum g(w_i)\Delta x_i = \sum g(w_i)(x_i - x_{i-1})$$

Note that each partition of $[a, b]$ yields a different Riemann sum. If the limit of the sums as the norm of the partition goes to zero exists, it is called the Riemann integral of $g(x)$. It is written as

$$\int_a^b g(x)dx = \lim_{\|\Delta x\| \to 0} \sum g(w_i)\Delta x_i$$

There is no reason why we should limit ourselves to multiplying $g(w_i)$ by just the length of the interval $[x_{i-1}, x_i]$. Suppose we replace Δx_i by $\Delta F(x) = F(x_i) - F(x_{i-1})$, where $F(x)$ is any single-valued function. Thus, if the limit exists, the analogous integral is

$$\int_a^b g(x)\,dF = \lim_{\|\Delta F(x)\| \to 0} \sum g(w_i)[F(x_i) - F(x_{i-1})]$$

The above integral is called the **Stieltjes integral**. In the context of the theory of probability, we would choose $F(x)$ to be the cumulative distribution function. The advantage of the Stieltjes integral with the CDF is that we do not have to distinguish between a continuous and a discrete random variable. The integral is well defined in both cases.

Mathematical Expectation

Suppose you roll a die and are paid the square of the outcome, that is, if you roll a three you get 9 dollars and so on. If you roll the die indefinitely, *on average*, how much can you expect to earn *per trial*? (instead of a square we could use any function of the score.) To answer questions of this type we introduce the concept of a **mathematical expectation**.

Definition 3.6

Let X be a random variable on $(S, \mathcal{F}, P)$ with $f(x)$ as the PF or PDF, and $g(x)$ be a single-valued function. If the Stieltjes integral $\int_{-\infty}^{+\infty} g(x)\,dF$ exists, it is called the **expected value** *(or* **mathematical expectation***) of $g(X)$ and is denoted by $E[g(X)]$. In the case of a discrete random variable this takes the form $E[g(X)] = \sum_i g(x_i)f(x_i)$ and in the continuous case, $E[g(X)] = \int_{-\infty}^{+\infty} g(x)f(x)dx$.*

Intuitively, we get a weighted average of $g(X)$, the weights being the corresponding probabilities. In the die throwing example given above, it is easily verified that $E(X^2) = 91/6$.

Mean of a Distribution

The special case of $g(X) = X$ is very interesting. The expected value of X is a measure of central location and is called the **mean of a distribution** (usually denoted by μ). Thus, $\mu = E(X)$.

Example 3.7 (*mean of a continuous uniform distribution*)

For the uniform distribution on an interval we have,

$$f(x) = \frac{1}{b-a} \qquad a < x < b$$

$$\mu = E(X) = \int_a^b \frac{x}{b-a} \, dx = \frac{1}{2} \left[\frac{x^2}{b-a} \right]_a^b = \frac{b+a}{2}$$

Example 3.8 (*mean of the normal distribution*)

Let X be a general normal variable with density

$$f(x) = \frac{1}{\sigma\sqrt{2\pi}} \exp\left[-\frac{(x-\mu)^2}{2\sigma^2} \right] \qquad 0 < x < \infty$$

$$E(X) = \int_{-\infty}^{\infty} x \, \frac{1}{\sigma\sqrt{2\pi}} \exp\left[-\frac{(x-\mu)^2}{2\sigma^2} \right] dx$$

Making the substitution $y = (x-\mu)/\sigma$ we have,

$$E(X) = \int_{-\infty}^{\infty} (\mu + \sigma y) \left[\frac{1}{\sigma\sqrt{2\pi}} e^{-y^2/2} \right] \sigma \, dy$$

$$= \mu \int_{-\infty}^{\infty} \frac{1}{\sqrt{2\pi}} e^{-y^2/2} \, dy + \sigma \int_{-\infty}^{\infty} y \, \frac{1}{\sqrt{2\pi}} e^{-y^2/2} \, dy$$

The first integrand is simply the density function of the standard normal and hence integrates to 1, making the first term μ. The second integrand is an **odd function** [that is, $g(-y) = -g(-y)$] and hence the second integral is zero. To see this more clearly, the second integral can be written as

$$\int_{-\infty}^{0} y \, \frac{1}{\sqrt{2\pi}} e^{-y^2/2} \, dy + \int_{0}^{\infty} y \, \frac{1}{\sqrt{2\pi}} e^{-y^2/2} \, dy$$

Setting $u = -y$ in the second integral, it becomes

$$\int_{0}^{-\infty} u \, \frac{1}{\sqrt{2\pi}} e^{-u^2/2} \, du = -\int_{-\infty}^{0} u \, \frac{1}{\sqrt{2\pi}} e^{-u^2/2} \, du$$

which cancels with the first integral making the net result zero. Therefore, $E(X) = \mu$ for $N(\mu, \sigma^2)$.

We now state a number of useful and easy to prove properties of the mathematical expectation.

Theorem 3.6

 1.   *If c is a constant, $E(c) = c$.*

 2.   *If c is a constant, $E[c\,g(X)] = c\,E[g(X)]$.*

 3. $E[u(X) + v(X)] = E[u(X)] + E[v(X)]$.

 4. $E(X - \mu) = 0$, *where* $\mu = E(X)$.

Practice Problem

3.7   Prove Theorem 3.6. Also, derive the value of b (a constant) for which $E[(X - b)^2]$ is minimum.

Moments of a Distribution

The mean of a distribution is the expected value of the random variable X. A generalization of this is to raise X to any positive integer power greater than 1, that is, set $m = 2, 3, \ldots$, and compute

$$E(X^m) = \int_{-\infty}^{\infty} x^m \, dF$$

If the integral exists, it is called the **mth moment around the origin** and is denoted by μ'_m. Moments can also be obtained around the mean and these are called **moments around the mean** or the **central moments** (denoted by μ_m). Thus,

$$\mu_m = E[(X - \mu)^m] = \int_{-\infty}^{\infty} (x - \mu)^m \, dF$$

Example 3.9

Consider the uniform distribution on the interval $[a, b]$. The moments are given by

$$E(X^m) = \int_a^b \frac{x^m}{b-a} dx = \left[\frac{x^{m+1}}{(m+1)(b-a)} \right]_a^b = \frac{b^{m+1} - a^{m+1}}{(m+1)(b-a)}$$

$$E[(X - \mu)^m] = \int_a^b \frac{(x-\mu)^m}{b-a} dx = \left[\frac{(x-\mu)^{m+1}}{(m+1)(b-a)} \right]_a^b$$

$$= \frac{(b-\mu)^{m+1} - (a-\mu)^{m+1}}{(m+1)(b-a)}$$

It should be reiterated that the mean and higher moments may not always exist. An example of a distribution for which the mean and higher moments do not exist is the **Cauchy distribution** which has the density function

$$f(x) = \frac{1}{\pi(1+x^2)} \qquad -\infty < x < \infty$$

That the mean does not exist is easily seen by noting that

$$\int_0^\infty \frac{x}{\pi(1+x^2)}\,dx = \left[\frac{1}{2\pi}\ln(1+x^2)\right]_0^\infty = \infty$$

Variance and Standard Deviation

The central moment of a distribution that corresponds to $m = 2$ is an extremely useful measure characterizing a distribution. This measure $E[(X-\mu)^2]$, where $E(X) = \mu$, is called the **variance of a distribution**, and is denoted by σ^2 or Var(X) provided μ and σ^2 both exist. [Note that Var(X) is nonnegative]. The positive square root of the variance is called the **standard deviation** (abbreviated as **s.d.**) and is denoted by σ. Thus we have,

$$\sigma^2 = E[(X-\mu)^2] = \text{Var}(X) = \int(x-\mu)^2\,dF$$

Intuitively, $X-\mu$ is the deviation of X from its mean. Hence $E[(X-\mu)^2]$ is an average of the squared deviation from the mean. Squaring magnifies large deviations and also treats positive and negative deviations symmetrically. Thus σ^2 is a measure of the dispersion of a distribution that can be written in another form also.

$$\sigma^2 = E[(X-\mu)^2] = E(X^2 - 2\mu X + \mu^2)$$
$$= E(X^2) - 2\mu E(X) + E(\mu^2) = E(X^2) - \mu^2 = \mu_2' - \mu^2$$

The usefulness of this alternative form is that we can obtain the variance easily once the first two moments around the origin have been computed (as is seen in the following example).

Example 3.10

For the uniform distribution on $[a, b]$ we have (using Example 3.9)

$$E(X^2) = \frac{b^3 - a^3}{3(b-a)} = \frac{1}{3}(a^2 + ab + b^2)$$

$$\sigma^2 = \frac{(a^2 + ab + b^2)}{3} - \frac{(a^2 + 2ab + b^2)}{4} = \frac{(b-a)^2}{12}$$

Example 3.11

The bakery section of a grocery store prepares a number of decorated cakes at the beginning of each day. For each cake sold the store makes a profit of 3 dollars and for each cake not sold it loses 9 dollars because of labor and material costs not recovered. Suppose the number of cakes demanded in a day has the uniform distribution $f(x) = 1/10$ for $x = 1, 2, \ldots, 10$. If the bakery stocks n cakes ($n \leq 10$), calculate the expected profit. To maximize expected profit what is the optimum n?

If $x < n$, the profit is $3x - 9(n-x) = 12x - 9n$. If $x \geq n$, then the profit is $3n$. Thus expected profit is

$$\Pi(n) = \sum_{x=1}^{n-1} (12x - 9n)\frac{1}{10} + 3n\left[1 - \frac{n-1}{10}\right]$$

$$= \frac{12}{10}\frac{n(n-1)}{2} - \frac{9n(n-1)}{10} + \frac{3n(11-n)}{10}$$

$$= \frac{1}{10}(36n - 6n^2)$$

To maximize profit, $\partial \Pi / \partial n = 0$ which gives $n = 3$. The store should thus stock 3 cakes a day. [Verify the second order condition.]

Example 3.12 [Freund (1962), Page 162, Exercise 2]

A contractor who bids on road construction jobs has found from experience that if his cost estimate for a job is C dollars, the lowest opposing bid may be looked upon as a random variable X having the uniform distribution with

$$f(x) = \frac{1}{C} \quad \text{for} \quad \frac{C}{2} \leq x \leq \frac{3C}{2}$$

and zero everywhere else. The contract is awarded to the lowest bidder. In order to maximize expected profits, what percent markup should the contractor *add* to his cost estimate in submitting his bids? More specifically, suppose his bid is Y dollars. What should the optimum Y be in terms of C?

Two possibilities arise; if $Y < X$ then the bid is successful and if $Y > X$ the bid is unsuccessful. If the bid is successful, the profit is $Y - C$ and the corresponding probability is

$$P(X > Y) = \int_{Y}^{3C/2} \frac{1}{C}\, dx = \frac{1}{C}\left[\frac{3C}{2} - Y\right]$$

If the bid is unsuccessful, the profit is zero and the corresponding probability is

$$P(X < Y) = \int_{C/2}^{Y} \frac{1}{C}\, dx = \frac{1}{C}\left[Y - \frac{C}{2}\right]$$

The expected profit is

$$E(\Pi) = (Y - C)\frac{1}{C}\left[\frac{3C}{2} - Y\right]$$

The first order condition for profit maximization is

$$\frac{\partial \Pi}{\partial Y} = \frac{1}{C}(Y - C)(-1) + \frac{1}{C}\left[\frac{3C}{2} - Y\right] = 0$$

The solution to this equation is $Y = 5C/4 = 1.25\,C$. This means that the required markup is 25 percent. [Verify the second order condition.]

Mean and Variance of a Normal Distribution

The calculation of the second moment of a normal distribution requires integrating x^2 multiplied by the density function, and that involves integration by parts. In Section 3.6 we present a much easier and general way of obtaining all the moments of a distribution that exist. It is shown there (see Example 3.15) that for a random variable distributed as $N(\mu, \sigma^2)$, the mean is μ and variance is σ^2. It readily follows that, for the standard normal, $E(X) = 0$ and $Var(X) = 1$. It is interesting to note from Example 3.5 that if X is distributed as $N(\mu, \sigma^2)$, then $Z = (X - \mu)/\sigma$ is $N(0, 1)$. This operation of subtracting the mean and dividing by the standard deviation is called **standardizing**. In the normal case, standardization results in another distribution in the same family. In general, this property does not hold for other distributions.

Example 3.13

A manufacturer of tires has found that the life of a certain kind of tire is a normal random variable with a mean of 30,000 miles and standard deviation 2,000 miles. The company wishes to guarantee it for M miles with a full refund if the tire does not last that long. Suppose it wants to make sure that the probability that a tire will be returned is no more than 0.10 (that is, no more than 10 percent of the tires sold will be returned). What value of M should be chosen?

Let X be the life of the tire. Then $X \sim N(30000, 2000^2)$. We want $P(X \le M) \le 0.10$, which is equivalent to $P(X \ge M) \ge 0.9$.

$$P(X \ge M) = P\left[\frac{X-\mu}{\sigma} \ge \frac{M-\mu}{\sigma}\right] \ge 0.90$$

But $Z = (X-\mu)/\sigma \sim N(0, 1)$. Hence, $P[Z \ge (M-\mu)/\sigma] \ge 0.90$. From Table B.3 of the Appendix, we note that the area to the right of 1.28 is 0.1003 and that to the right of 1.29 is 0.0985. Interpolating between these two numbers we find that the area to the right of 1.282 is 0.1. Since $N(0, 1)$ is symmetric around the origin, this implies that the area to the left of -1.282 is also 0.1. Hence the area to the right of -1.282 is 0.9. It follows that the required z is -1.282. This gives

$$M \le \mu - 1.282\,\sigma = 30000 - (1.282)\,2000$$

that is, $M \le 27,436$ miles. Note that the actual value varies with the s.d. σ, a higher σ being associated with a lower M.

Practice Problem

3.8   Suppose the annual income (X), in thousands, of a family is approximately normally distributed with mean 26 and variance 36. Compute the probability that a family drawn at random will have income between $22,000 and $29,000.

Mean and Variance of a Binomial Distribution

It is shown in Example 3.16 of Section 3.5 that the Binomial distribution $B(n, p)$ has mean np and variance $np(1-p)$ or npq.

Theorem 3.7

If $E(X) = \mu$ and $Var(X) = \sigma^2$, and a and b are constants, then $Var(a + bX) = b^2 \sigma^2$.

Proof: Let $Y = a + bX$. Then $E(Y) = a + bE(X) = a + b\mu$. Hence $Y - E(Y) = b(X - \mu)$.

$$\text{Var}(Y) = E[Y - E(Y)]^2 = E[b^2(X-\mu)^2] = b^2\sigma^2$$

Practice Problem

3.9   Let X be the binomial Bin(n, p) and $Y = X/n$ be the proportion of successes in n trials. Derive the mean, variance, and standard deviation of Y.

Chebyshev's Inequality

Theorem 3.8

Let b be a positive constant and $h(X)$ be a nonnegative measurable function of the random variable X. Then

$$P[h(X) \geq b] \leq \frac{1}{b} E[h(X)]$$

Corollary: (Chebyshev's inequality). For any constant $c > 0$ and $\sigma^2 = Var(X)$,

(1) $P[\,|X-\mu| \geq c] \leq \sigma^2/c^2$

(2) $P(|X-\mu| < c) \geq 1 - \dfrac{\sigma^2}{c^2}$

(3) $P(\,|X-\mu| \geq k\sigma) \leq 1/k^2$

Proof:

$$E[h(X)] = \int h(x)\,dF = \int_{h(x)\geq b} h(x)\,dF + \int_{h(x)<b} h(x)\,dF$$

$$\geq \int_{h(x)\geq b} h(x)\,dF$$

because $h(X)$ is a nonnegative function. When $h(X) \geq b$,

$$E[h(X)] \geq \int_{h(x)\geq b} b\,dF = bP[h(x)\geq b]$$

Therefore, $P[h(X) \geq b] \leq (1/b)E[h(X)]$.

Corollary: Let $h(X) = (X-\mu)^2$ and $b = c^2$. Then

$$P[(X-\mu)^2 \geq c^2] = P[\,|X-\mu| \geq c]$$

Hence the result. If $c = k\sigma$, the third alternative form follows.

Chebyshev's inequality states that for any distribution (with a finite μ) the probability that X falls outside a 2σ interval from the mean is $\leq 1/4$, outside a 3σ range is $\leq 1/9$, and so on. Thus we note that σ controls the dispersion of X.

Example 3.14

A machine manufactures nails of mean length 3 inches (that is, $\mu = 3$). However, as no machine is perfect, the actual length of a nail will deviate slightly from the mean. If at least 95 percent of the nails should be within 1/100 of an inch from the mean, what is the maximum value the standard deviation σ can take?

As the distribution of the length of a nail (X) is not given, we have to use Chebyshev's inequality. We want $P(\,|X-\mu\,|\,\leq 0.01)$ ≥ 0.95. By Chebyshev's inequality

$$P[\,|X-\mu\,|\,\leq 0.01] \;\geq\; 1 - \frac{\sigma^2}{(0.01)^2}$$

Thus the required condition is

$$1 - \frac{\sigma^2}{(0.01)^2} \;\geq\; 0.95 \quad \text{or} \quad \frac{\sigma^2}{(0.01)^2} \;\leq\; 0.05$$

that is, $\sigma^2 \leq (0.01)^2\, 0.05$. Thus σ must not exceed $(0.01)\sqrt{0.05} = 0.0022361$.

Approximate Mean and Variance of $g(X)$

In many applications, the mean and variance of a measurable function $g(X)$ of a random variable X may not be readily obtainable. As an example, consider $g(X) = \sqrt{X}$. If X had the standard normal density, the integral corresponding to $E[g(X)]$ does not have a tractable expression. In these cases, it is often useful to obtain approximate mean and variance for $g(X)$. Suppose X is a random variable defined on $(S, \mathcal{F}, P)$ with $E(X) = \mu$ and $\text{Var}(X) = \sigma^2$, and let $g(X)$ be a differentiable and measurable function of X. We first take a linear approximation of $g(X)$ in the neighborhood of μ. This is given by

$$g(X) \approx g(\mu) + g'(\mu)(X-\mu)$$

provided $g(\mu)$ and $g'(\mu)$ exist. Since the second term has zero expectation, $E[g(X)] \approx g(\mu)$. Using Theorem 3.7, we have

$$\text{Var}[g(X)] \approx \sigma^2[g'(\mu)]^2$$

In the case of $g(X) = \sqrt{X}$, we readily see that the approximate mean is $\sqrt{\mu}$ and the approximate variance is $\sigma^2/(4\mu)$. A word of caution should be added here. What we use here is a linear approximation which may be grossly inaccurate if the curvature of the function $g(X)$ is high. The reader must therefore be aware that the approximation might be dangerously misleading in some cases.

Mode of a Distribution

The point(s) for which $f(x)$ is maximum is (are) called the **mode**. It is the most frequently observed value of X. Thus, for example, in Figure 3.2, the mode is 2. In some practical situations, the mode can give misleading results. For instance, if an industrial plant has predominantly union workers, then the empirically observed modal wage might actually be the maximum wage (which is not a good central measure) because union wages tend to be higher than nonunion wages.

Median, Upper and Lower Quartiles, and Percentiles

A value of x such that $P(X < x) \leq \frac{1}{2}$, and $P(X \leq x) \geq \frac{1}{2}$ is called a **median of the distribution**. If the point is unique, then it is *the* median. Thus the median is the point on either side of which lies 50 percent of the distribution. A median is often preferred over the mean as an "average" measure because the arithmetic average can be misleading if extreme values are present. For example, suppose a coastal town has beach homes that sell for over five million dollars as well as homes away from the beach for under \$250,000. Averaging all homes will yield a high mean, whereas a median will better reflect the prevailing prices of homes. This is the reason that so many government statistics report median income, median price of a home, and so on.

Instead of $\frac{1}{2}$, we could use any probability between 0 and 1. The point(s) with an area $\frac{1}{4}$ to the left is (are) called the **lower quartile(s)**, and the point(s) corresponding to $\frac{3}{4}$ is (are) the **upper quartile(s)**. For any probability p, the values of x for which the area to the right is p are called the **upper pth percentiles** (also referred to as **quantiles**). Students who have taken the Scholastic Aptitude Test (SAT) or the Graduate Record Examination (GRE) should be familiar with the percentiles of an empirically observed distribution.

Practice Problem

3.10   Let X be a continuous random variable with density $f(x) = e^{-x}$ for $x > 0$ and zero everywhere else (this is called the **exponen-**

tial distribution). Derive the upper pth percentile of the distribution and deduce the median, upper, and lower quartiles.

Coefficient of Variation

The coefficient of variation is defined as the ratio σ/μ, where the numerator is the standard deviation and the denominator is the mean. It is a measure of the dispersion of a distribution *relative to its mean* and is useful in the estimation of relationships. The estimation of the relation between a dependent variable (Y) and one or more independent variables (X) is concerned with explaining how variations in X affect the variation in Y. If however, the coefficient of variation of X is small (say less than 5 percent), then X has not varied much and cannot be expected to explain why Y changes. The coefficient of variation is thus useful in identifying variables that have low dispersions.

Skewness and Kurtosis

If a continuous density $f(x)$ has the property that $f(\mu+a) = f(\mu-a)$ for all a (μ being the mean of the distribution), then $f(x)$ is said to be **symmetric around the mean**. For instance, $N(\mu,\sigma^2)$ is symmetric around the mean μ. If a distribution is not symmetric about the mean, then it is called **skewed**. Figure 3.5 gives two kinds of **skewness**, positive (that is, to the right with a long tail in that direction) and negative (that is, to the left). A commonly used measure of skewness is $E[(X-\mu)^3/\sigma^3]$. For a symmetric distribution such as the normal, this is zero.

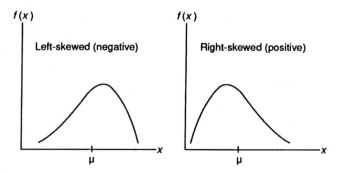

Figure 3.5 Skewness of a distribution

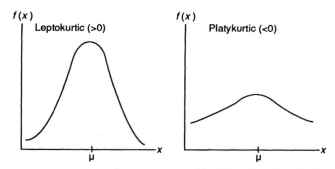

Figure 3.6   Excess kurtosis of a distribution

The peakedness of a distribution is called **kurtosis**. A narrow distribution is called **leptokurtic** and a flat distribution is called **platykurtic**. These are illustrated in Figure 3.6. One measure of kurtosis is $E[(X-\mu)^4/\sigma^4]$. For a normal distribution, kurtosis is 3. The value $E[(X-\mu)^4/\sigma^4] - 3$ is often referred to as **excess kurtosis**. Skewness and kurtosis are often used in econometrics to characterize forecast errors and nonnormality of the distributions of such errors. Note that the measures of skewness and kurtosis are defined over the standardized variables so that they would not be dependent on the scale of units of measurement or the location.

3.6   GENERATING FUNCTIONS

It was stated earlier that $\mu_m^{'} = E(X^m)$ is called the mth moment of a distribution. If $m = 1$, we have the mean, which is the first moment. $E(X^2)$ is the second moment around the origin which is $\sigma^2 + \mu^2$. Calculation of each of these moments is often cumbersome and therefore it would be desirable to derive the moments of a distribution from some general framework rather than compute $E(X^m)$ for each m. This is achieved through two functions discussed presently.

The Moment Generating Function

Definition 3.7

*The function $m(t) = E(e^{Xt}) = \int_{-\infty}^{+\infty} e^{xt} dF$ is called the **moment generating function** (also known as the **Laplace transform**) of X. The function is defined for those values of t for which the integral exists.*

The Characteristic Function

Definition 3.8

The function $\phi(t) = E(e^{iXt})$ *where* $i^2 = -1$, *that is, the complex i, is called the* **characteristic function (CF)** *of X.*

Using power series expansions of e^{ixt}, $\sin xt$, and $\cos xt$, it is easily verified that $e^{ixt} = \cos xt + i \sin xt$. Therefore

$$\phi(t) = \int_{-\infty}^{\infty} \cos xt \ dF \ + \ i \int_{-\infty}^{\infty} \sin xt \ dF$$

Each of the integrals is called a **Fourier transform** of $f(x)$. Because $|\cos xt|$ and $|\sin xt|$ do not exceed 1 and a probability density integrates (or sums) to 1, each of the Fourier integrals exists and hence $\phi(t)$ always exists. Thus the characteristic function always exists even though the moment generating function may not exist. The power series expansion of e^{iXt} is

$$e^{iXt} = \sum_{r=0}^{\infty} \frac{(iXt)^r}{r!}$$

Therefore, the characteristic function can also be written as

$$\phi(t) = \sum_{r=0}^{\infty} \frac{E[(iXt)^r]}{r!} = \sum_{r=0}^{\infty} \frac{i^r t^r}{r!} \mu_r'$$

where $\mu_r' = E(X^r)$ is the rth moment of X. We thus note that the rth moment, if it exists, is generated as the coefficient of $(i^r t^r)/r!$ in the infinite series expansion of $\phi(t)$. Thus if we can obtain the characteristic function of a distribution, we can get all the moments of a distribution in one sweep.

Theorem 3.9

A characteristic function is uniformly continuous on the real line.

Proof: The proof is given only for a continuous random variable.

$$\phi(t+h) - \phi(t) = \int e^{ixt}(e^{ixh} - 1)f(x)dx$$

Because $|e^{ixt}| \le 1$, $|\phi(t+h) - \phi(t)| \le \int |e^{ixh} - 1| f(x)dx$. For a given ε, choose A large such that $\int_{|x|>A} f(x)dx < \varepsilon/4$ and a corresponding h small so that for $|x| \le A$, $|e^{ixh} - 1| < \varepsilon/(4A)$. We have,

$$|\phi(t+h) - \phi(t)| \le \int_{-A}^{+A} |e^{ixh} - 1| f(x)dx + \int_{|x|>A} |e^{ixh} - 1| f(x)dx$$

The first integral is $< \varepsilon / 2$ because $f(x) \leq 1$, and when $|x| \leq A$, $|e^{ixh} - 1| < \varepsilon/(4A)$. When $|x| > A$, $|e^{ixh} - 1| \leq 2$, which is attained when $xh = \pi$. Therefore the second integral is less than $\varepsilon / 2$, making the total less than ε. Hence $\phi(t)$ is uniformly continuous (because A and h do not depend on t).

Assuming that $\phi(t)$ is differentiable under the integral sign we get

$$\phi'(t) = \int_{-\infty}^{\infty} e^{ixt} \, ix \, dF$$

It follows that $\phi'(0) = iE(X)$. By proceeding similarly, we note that

$$\phi^r(0) = \left[\frac{d^r\phi(t)}{dt} \right]_0 = i^r \mu'_r$$

provided that the rth derivative exists. Thus, moments exist if and only if the corresponding derivatives of $\phi(t)$ exist at $t = 0$, and when they do, the technique used here is extremely useful in deriving the moments of a distribution.

It is useful to mention that, because the characteristic function always exists, even a distribution such as the Cauchy for which the moments do not exist has a well-defined characteristic function. It can be shown that the characteristic function for the Cauchy distribution is $\phi(t) = e^{-|t|}$. But, because $\phi(t)$ is not differentiable at $t = 0$, its first and higher moments do not exist.

Example 3.15

Let us now derive the CF of a standard normal distribution (Z) and then deduce the same for $X \sim N(\mu, \sigma^2)$. We want $E(e^{iZt})$ which is

$$\phi_Z(t) = \frac{1}{\sqrt{2\pi}} \int_{-\infty}^{\infty} e^{izt} \, e^{-z^2/2} \, dz = \frac{1}{\sqrt{2\pi}} \int_{-\infty}^{\infty} e^{-[(z-it)^2 - t^2]/2} \, dz$$

$$= e^{-t^2/2} \left[\int_{-\infty}^{\infty} \frac{1}{\sqrt{2\pi}} e^{-(z-it)^2/2} \, dz \right] = e^{-t^2/2}$$

because the integrand is the normal density with mean it and variance 1 and hence integrates to 1. Thus, $\phi_Z(t) = e^{-t^2/2}$ for a standard normal.

To obtain the CF for a general normal distribution $X \sim N(\mu, \sigma^2)$ first note that $X = \mu + \sigma Z$.

$$\phi_X(t) \ = \ E\left[e^{i(\mu+\sigma Z)t}\right] \ = \ e^{i\mu t} \, E\left[e^{iZ\sigma t}\right] \ = \ e^{i\mu t} \, \phi_Z(\sigma t)$$

$$= \ e^{i\mu t} \, e^{-\sigma^2 t^2/2} \ = \ e^{i\mu t - (\sigma^2 t^2/2)}$$

The CF of $N(\mu, \sigma^2)$ is therefore $e^{i\mu t - (\sigma^2 t^2/2)}$. We can use the CF to derive the mean and variance of the general normal. Differentiating $\phi_X(t)$ with respect to t we get

$$\phi_X'(t) \ = \ [e^{i\mu t - (\sigma^2 t^2/2)}] \, (i\mu - \sigma^2 t)$$

Therefore, $i\, E(X) = \phi_X'(0) = i\mu$. It follows from this that $E(X) = \mu$. Differentiating once more,

$$\phi_X''(t) \ = \ [e^{i\mu t - (\sigma^2 t^2/2)}] \, (i\mu - \sigma^2 t)^2 \ + \ [e^{i\mu t - (\sigma^2 t^2/2)}] \, (-\sigma^2)$$

Hence,

$$i^2 E(X^2) \ = \ \phi_X''(0) \ = \ i^2\mu^2 + i^2\sigma^2$$

Therefore, $E(X^2) = \mu^2 + \sigma^2$ and $\text{Var}(X) = \sigma^2$. This establishes the result stated earlier, namely, that $N(\mu, \sigma^2)$ has mean μ and variance σ^2.

Example 3.16

For the binomial distribution, $B(n, p)$, we have

$$\phi(t) = E(e^{iXt}) = \sum_0^n e^{ixt} \begin{bmatrix} n \\ x \end{bmatrix} p^x q^{n-x} = \sum_0^n \begin{bmatrix} n \\ x \end{bmatrix} (pe^{it})^x q^{n-x} = (q + pe^{it})^n$$

which makes use of the binomial expansion. Differentiating $\phi(t)$ with respect to t we have, $\phi'(t) = n(q + pe^{it})^{n-1} pe^{it} i$. Hence (noting that $q + p = 1$), $iE(X) = \phi'(0) = inp$. The mean of a binomial is therefore np. Differentiating $\phi'(t)$ we have,

$$\phi''(t) \ = \ n(q + pe^{it})^{n-1} pe^{it} i^2 \ + \ n(n-1)(q + pe^{it})^{n-2} (pe^{it} i)^2$$

Hence,

$$i^2 E(X^2) \ = \ \phi''(0) \ = \ i^2 np + n(n-1)p^2 i^2$$

Therefore, $E(X^2) = n^2 p^2 + np(1-p)$ from which we have, $\text{Var}(X) = np(1-p) = npq$. The binomial distribution therefore has mean np and variance npq.

It should be noted that the approach used here for computing the mean and variance avoids having to perform cumbersome integrations or summations.

Practice Problems

3.11   Noting that $\phi^r(0) = i^r \mu'_r$, derive the skewness and kurtosis for $N(0, 1)$.

3.12   Compute $\phi(t)$ for the standard exponential distribution with the density function $f(x) = e^{-x}$ for $x > 0$ and zero elsewhere. Use that to derive the mean, variance, skewness, and kurtosis.

The following theorems, stated without proof, have important implications. Proofs may be found in Lukacs (1970, pp. 28–33).

Theorem 3.10 (uniqueness theorem)

Two distribution functions are identical if and only if their characteristic functions are also identical.

This theorem implies that corresponding to each characteristic function $\phi(t)$ there exists a unique distribution function $F(x)$ having that characteristic function, and vice versa.

Theorem 3.11 (inversion theorem)

Let $F(x)$ be the distribution function, assumed continuous, and $\phi(t)$ be the corresponding characteristic function. Then, for any $x, h \in \mathcal{R}$,

$$F(x+h) - F(x-h) = \lim_{T \to \infty} \frac{1}{2\pi} \int_{-T}^{T} \left[\frac{1 - e^{-ith}}{it} \right] e^{-itx} \phi(t)dt$$

provided that $x-h$ and $x+h$ $(h > 0)$ are continuity points of $F(x)$.

Theorem 3.12 (inversion formula)

If a characteristic function $\phi(t)$ is absolutely integrable over $(-\infty, +\infty)$, then the corresponding distribution function $F(x)$ is absolutely continuous and the corresponding density function (which is continuous) is

$$f(x) = F'(x) = \frac{1}{2\pi} \int_{-\infty}^{\infty} e^{-itx} \phi(t)dt$$

It is evident from the above theorems that corresponding to each characteristic function $\phi(t)$ there exists a unique distribution function $F(x)$ having that characteristic function and that it is often possible to obtain the corresponding density function $f(x)$ also. It will be seen in later chapters that this property is extremely useful in identifying the distribution of measurable functions of one or more random variables.

For an excellent treatment of characteristic functions and their properties, the reader is referred to Lukacs (1970).

EXERCISES

3.1   My wife bought a box of 20 gladiolus bulbs from the Green-house Nursery. The box states that if less than 90 percent of the bulbs germinate, the manufacturer will refund the price of the entire box. Suppose that the probability of germination is only 0.8. What is the probability that my wife will not get a refund on her purchase?

3.2   A physician knows that the probability that a patient will recover from a certain rare disease, without any drugs, is 0.25. To test the effectiveness of a new drug she gives it to 25 patients having this disease and decides, beforehand, to discredit the drug unless at least 10 of the patients recover. What is the probability that she will discredit the new drug on the basis of this experiment even though it raises the recovery rate (the probability that any one patient will recover) to 0.4? What is the probability that she will fail to discredit the new drug on the basis of this experiment even though the new drug is totally ineffective, that is, the recovery rate is still 0.25?

3.3   A small commuter airline uses planes with 20 seats. Its experience shows that 10 percent of individuals reserving space in a flight do not show up. If the company takes 23 reservations for each flight, what is the probability that it will be able to accommodate everyone appearing, without bumping anyone?

3.4   A multiple choice exam has N questions, each of which has k possible answers. A student knows the correct answer to n of these questions. For the remaining $N - n$ questions, he checks the answers completely at random. For each correct answer the student gets one point and for each wrong answer he gets $-a$ points (that is, a points are subtracted). Derive an expression for the expected value of the total score for the test.

If you want this value to be exactly equal to n, so that the student does not gain any points, on average, for randomizing, what should the value of a be?

3.5   If you watch American TV, you would have noticed commercials that induce people to return the "Publisher's Clearing House Sweepstakes forms." Have you wondered what the cost-benefit analysis of that is? Here is your chance to do a little bit of statistical decision theory.

The publisher of a magazine has two strategies to consider: to have no sweepstakes or to offer sweepstakes. If there are no sweepstakes, it costs α dollars to mail each subscription form to N persons. The probability that a person receiving this form will subscribe to the magazine is P_n. On each subscription received, the publisher makes a net revenue of π dollars. If a sweepstakes option is chosen, he has additional costs. Let β ($> \alpha$) be the cost per person of mailing sweepstakes coupons to N persons and S be the total cost of the prizes plus any TV promotional expenses. The probability that a person receiving the sweepstakes form will subscribe is P_s.

(a)   Derive the expected net profit for each strategy.

(b)   Describe the decision rule that you would recommend to the publisher, including when he should be indifferent between the two strategies. Carefully justify your recommendation.

3.6   A computer requires 20 memory chips to work. At any time, the probability that a given microchip is defective is 0.10. Let X be the number of chips to be ordered to ensure that 20 chips work. Derive the density function of the random variable X. On average, how many chips should you order to ensure that the computer works?

3.7   In Exercise 2.5, suppose you sell a one year term insurance of $25,000 to each, that is, you pay $25,000 if one of them dies, $50,000 to their children if both of them die, and nothing if both live through the year. If you want to make an expected profit of $200, what premium should you charge *each of them* assuming that you charge the same premium for each policy. Suppose, instead, that the premiums are different and in proportion to the respective probabilities of death. Calculate the premiums for the same $200 expected profit.

3.8 Show that for the standard normal $N(0, 1)$, $E(|X|) = \sqrt{(2/\pi)}$.

3.9 Let X be $N(\mu, \sigma^2)$. Truncate the density of X to the left at a and the right at b, that is, X takes only the values between a and b, both being finite.

    (a)   Derive the density function of this truncated normal.

    (b)   Compute the mean of the above truncated normal.

    (c)   Show that this mean is between a and b, and that if $a = \mu - c$ and $b = \mu + c$, this mean is μ.

3.10 Derive the density function of $Y = \ln X$, when X has the standard normal density function.

3.11 Let X be the standard normal $N(0,1)$. Consider the transformation $Y = X^2$. Derive the distribution function $F(Y) = P(Y \le y)$. From that, derive the density function $f(y) = F'(y)$. Next derive the characteristic function of Y. Use that to obtain its mean and variance.

3.12 When a pair of dice is thrown, let the total points be denoted by X. Construct the table of probabilities $P(X = x) = f(x)$. Suppose you are paid the square of the total score. Calculate your expected payoff.

3.13 Let the random variable X have the discrete uniform distribution that takes the values $1, 2, \ldots, n$ with equal probability. Derive the PF and CDF and graph them.

3.14 For the discrete distribution with $f(x; \theta) = k\theta^x$, $0 < \theta < 1$, and $x = 0, 1, 2, \ldots$, evaluate k in terms of θ. Derive the mth moment and deduce the mean and variance of the distribution.

3.15 For the continuous random variable with density function $f(x; \theta) = kx^{\theta-1}$, $0 < x < 1$, $\theta > 0$, obtain k in terms of θ, derive its CDF, and graph the PDF and CDF. Next compute μ'_m and deduce the mean and variance.

3.16 The PDF of a continuous random variable is $f(x; \theta) = k\theta^{-x}$, $x > 0$, $\theta > 1$. Evaluate k in terms of θ, derive the CDF, and graph it. [*Hint*: $a^b = e^{b \ln a}$.] Derive the characteristic function of this distribution and from it compute the mean and variance.

· 3.17   The continuous random variable X has PDF $f(x;\theta) = kxe^{-x/\theta}$, $x > 0$, and $\theta > 0$. Express k in terms of θ and obtain $F(x;\theta)$. Derive the characteristic function of this distribution and from it compute the mean and variance.

3.18   In Exercises 3.15 through 3.17, derive an expression for the conditional density (also known as the **truncated density**) of X given that $X > a$, that is, $f(x)/P(X > a)$.

3.19   The density function of the **gamma distribution** is given by

$$f(x) = \frac{1}{\beta^{\alpha}\,\Gamma(\alpha)}\, x^{\alpha-1}\, e^{-x/\beta}$$

with $x > 0$, $\alpha > 0$, $\beta > 0$, and $\Gamma(\alpha) = \int_0^\infty y^{\alpha-1}e^{-y}dy$ for $\alpha > 0$ is known as the **gamma function**. Derive its characteristic function. Use that to derive its mean and variance.

3.20   Derive the *lower quartile* (the 25th percentile), the median, and the *upper quartile* (the 75th percentile) for the uniform distribution on $[a, b]$.

3.21   Let $X \sim N(\mu, \sigma^2)$. Derive the density function of $Y = e^X$ which is lognormal.

4

SOME SPECIAL DISTRIBUTIONS

In the previous chapter we had presented a number of discrete and continuous random variables and described their properties. In this chapter we discuss a number of other families of distributions, their uses, and interrelations among each other. This is done separately for discrete and continuous random variables. For completeness, we have summarized the properties of the distributions already encountered. Table 4.1 at the end of this chapter has a tabular summary of these distributions and their main characteristics.

4.1 DISCRETE DISTRIBUTIONS

The Binomial Distribution

As seen before, this distribution arises when there are only two possible outcomes to an experiment, one labeled a *success* and the other labeled a *failure*. If p is the probability of a success ($q = 1 - p$) and there are n independent trials of the experiment, then we have the following.

$$f(x;\theta) = B(x;n,p) = \binom{n}{x} p^x q^{n-x} = \frac{n!}{x!\,(n-x)!} p^x q^{n-x}$$

$$x = 0, 1, \ldots, n \qquad 0 \le p \le 1 \qquad q = 1 - p$$

$$\phi(t) = (q + pe^{it})^n \qquad E(X) = np \qquad \mathrm{Var}(X) = npq$$

Simple Random Walk

Suppose the closing price of a stock moves up or down randomly by exactly one dollar for each day the Stock Exchange is open (this process is known as a **simple random walk**). The probability of an upward movement is p and that for a downward movement is $1-p$. Let X be the change in the price of the stock after n trading days. What is $f(x)$?

After n days, x can range from $-n$ to $+n$. Suppose there are k upward movements (successes) and $n-k$ downward movements (failures) in n trading days. The probability of that outcome is $\begin{bmatrix} n \\ k \end{bmatrix} p^k (1-p)^{n-k}$ and $x = k - (n-k) = 2k - n$, which implies that $k = (x+n)/2$. Hence,

$$f(x) = P(k \text{ successes}) = \begin{bmatrix} n \\ (x+n)/2 \end{bmatrix} p^{(x+n)/2} (1-p)^{n-(x+n)/2}$$

$$x = -n, \ldots, -2, -1, 0, 1, 2, \ldots, n$$

Note that $\begin{bmatrix} n \\ k \end{bmatrix} = 0$ if k is not an integer. Also, $\begin{bmatrix} n \\ (x+n)/2 \end{bmatrix} = 0$ whenever there is no path from the starting day to x in n days.

Negative Binomial Distribution

In a binomial experiment, let Y be the number of trials to get exactly k successes. To get exactly k successes, there must be $k-1$ successes in $y-1$ trials and the next outcome must be a success. Therefore,

$$f(y; k, p) = \begin{bmatrix} y-1 \\ k-1 \end{bmatrix} p^k q^{y-k} \qquad y = k, k+1, k+2, \ldots$$

Let $X = Y - k$ be the number of failures until k successes have been obtained. The density function of X is known as the **negative binomial**.

$$f(x; k, p) = \begin{bmatrix} x+k-1 \\ k-1 \end{bmatrix} p^k q^x \qquad x = 0, 1, 2, \ldots$$

Geometric Distribution

Let X be the number of the trial at which the first success occurs. The distribution of X is known as the **geometric distribution**. It has the density function

$$f(x; p) = p(1-p)^{x-1} \qquad x = 1, 2, 3, \ldots$$

Poisson Distribution

Consider the binomial distribution $B(x; n, p)$ and let $n \to \infty$, $p \to 0$ but in such a way that $np = \lambda \ (> 0)$ for all n and p. Thus the probability of a success is very small and the number of trials is large. Such a situation often arises in medical research where the probability of a disease is small and the number of patients is large. What is the distribution in this case?

$$B(n, p) = \frac{n(n-1)\cdots(n-x+1)}{x!} p^x (1-p)^{n-x}$$

$$= \frac{n(n-1)\cdots(n-x+1)}{x!} \frac{\lambda^x}{n^x} \left[1 - \frac{\lambda}{n}\right]^{n-x}$$

$$= \frac{1\left[1 - \frac{1}{n}\right]\left[1 - \frac{2}{n}\right]\cdots\left[1 - \frac{x-1}{n}\right]}{x!} \lambda^x \left[\left[1 - \frac{\lambda}{n}\right]^n\right]^{\frac{n-x}{n}}$$

Now let $n \to \infty$. We get $\lim_{n \to \infty} B(n, p) = (\lambda^x e^{-\lambda})/x!$ because $\lim_{n \to \infty} [1 - (\lambda/n)]^n = e^{-\lambda}$. Thus we get

$$f(x;\lambda) = \frac{e^{-\lambda}\lambda^x}{x!} \qquad x = 0, 1, 2, \ldots$$

which is known as the **Poisson distribution**. Let us verify that $\sum_0^\infty f(x;\lambda) = 1$.

$$\sum_0^\infty e^{-\lambda} \frac{\lambda^x}{x!} = e^{-\lambda}\left[1 + \frac{\lambda}{1!} + \frac{\lambda^2}{2!} + \cdots\right] = e^{-\lambda} e^{\lambda} = 1$$

Thus $f(x;\lambda)$ is indeed a PF.

Example 4.1

Suppose that the number of tornados hitting a certain area in a year is a random variable whose probability distribution is Poisson with $\lambda = 6$. To find the probability that fewer than 4 tornados will hit the area, we want

$$\sum_0^3 f(x \mid \lambda = 6) = \sum_0^3 \frac{e^{-6} 6^x}{x!}$$

Table B.2 of Appendix B has

$$1 - F(k-1) = \sum_k^\infty \frac{e^\lambda \lambda^x}{x!}$$

for various values of λ and k. We see from the table that $P(E) = 1 - 0.8488 = 0.1512$.

An Alternative Derivation of the Poisson Distribution

Aside from being a limit of the binomial distribution, Poisson probabilities can also be derived independently. This derivation, however, requires the use of differential equations and their solutions. Readers not interested in this may skip to the next section.

In *queuing theory*, the arrival of a new customer on the check-out line or the making of a phone call in a specific small interval is modeled as follows (it is known as a **queuing process**).

Let $f(x, t)$ be the probability of x successes in a time interval of length t, when (a) the probability of a success during the interval $(t, t+\Delta t)$ is $\alpha \Delta t + O(\Delta t)$, where $O(x)$ is a function such that $O(x)/x \to 0$ as $x \to 0$, (b) the probability of more than one success in that interval is $O(\Delta t)$, and (c) the probability of a success in that interval does not depend on what happened in previous non-overlapping intervals. It is readily seen that the queuing process approximates the above very well. To derive the explicit form of $f(x, t)$, consider $f(x, t+\Delta t)$, which is the probability of x successes in $t+\Delta t$ time. This can come about in only one of two mutually exclusive ways: the x successes might have occurred in the first t periods with no success in the last $(t, t+\Delta t)$ period, or alternatively, $x-1$ successes might have taken place in $(0, t)$, with one success in $(t, t+\Delta t)$. The probability for the former event is $f(x, t)(1-\alpha \Delta t)$ and the probability for the latter event is $f(x-1, t)\alpha \Delta t$. Hence

$$f(x, t+\Delta t) = f(x, t)(1-\alpha \Delta t) + f(x-1, t)\alpha \Delta t$$

By taking appropriate terms to the left we get

$$\frac{f(x, t+\Delta t) - f(x, t)}{\Delta t} = \alpha f(x-1, t) - \alpha f(x, t)$$

Taking the limit as $\Delta t \to 0$, we get the infinite system (one for each x) of differential equations:

$$\frac{df(x, t)}{dt} = \alpha f(x-1, t) - \alpha f(x, t)$$

Let $f(0,0) = 1$ be the initial condition for $x = 0$ and $f(x, 0) = 0$ for $x > 0$. Also $f(x, t) = 0$ for $x < 0$. For $x = 0$, the differential equation becomes

$$\frac{df(0, t)}{dt} = -\alpha f(0, t) \qquad \text{or} \qquad \frac{1}{f(0, t)}\frac{df}{dt} = -\alpha$$

The solution to this is $\ln f(0, t) = -\alpha t + C$, where C is a constant of integration. From the initial condition, $C = 0$. Thus $f(0, t) = e^{-\alpha t}$. The differential equation for $x = 1$ is

$$\frac{df(1, t)}{dt} = \alpha e^{-\alpha t} - \alpha f(1, t)$$

which can be rewritten as

$$\frac{d}{dt}\left[e^{\alpha t}f(1,\,t)\right] \;=\; \alpha$$

The solution to this is $e^{\alpha t}f(1,\,t)=\alpha t$ or $f(1,\,t)=\alpha te^{-\alpha t}$. By proceeding similarly, we get $f(x,\,t)=e^{-\alpha t}(\alpha t)^x/\,x!$. The above queuing process is thus a **Poisson process** with $\lambda=\alpha t$. This model is also appropriate when dealing with the failure of an equipment, $f(x,\,t)$ being the probability of x failures in the time $0-t$, provided that $f(x,\,t)$ satisfies the conditions (a), (b), and (c) specified earlier.

Hypergeometric Distribution

The binomial distribution is often referred to as **sampling with replacement**, which is needed to maintain the same probabilities across the trials. In other situations this may not be possible or applicable. For instance, suppose an electronic component factory ships components in lots of 100, of which a few (say 10) are defective. A quality controller draws a sample of five to test. The draw is usually **sampling without replacement**, that is, after drawing one part it is not put back to draw a second one. Thus the probability for a second draw is not the same as that of the first draw. Suppose we wish to know the probability that two of the five are defective. More generally, let there be a objects in a certain class (defective) and b objects in another class (nondefective). We draw a random sample of size n without replacement. What is the probability of observing x defectives, that is, x from class A? First, there are $\binom{a+b}{n}$ ways of drawing n objects from $a+b$ objects, which is the total number of possible outcomes. To get x from class A, there are $\binom{a}{x}$ possible ways. For *each* such outcome, there are $\binom{b}{n-x}$ possible ways of drawing from B. Thus there are $\binom{a}{x}\binom{b}{n-x}$ ways of drawing x from A and $n-x$ from B. Therefore the required probability is

$$f(x;n,\,a,\,b) \;=\; \frac{\binom{a}{x}\binom{b}{n-x}}{\binom{a+b}{n}}$$

where $x=0,\,1,\,2,\,\ldots,\,n$, $0<a$, $b\le n$, and n, a, and b are integers. This distribution is called a **hypergeometric distribution**. In the example, the probability is $\binom{10}{2}\binom{90}{3}\div\binom{100}{5}$.

4.2   CONTINUOUS DISTRIBUTIONS

Uniform Distribution on an Interval

This distribution arises when the probability density function is a constant over an interval. The density function and other characteristics of this distribution are summarized below.

$$f(x) = \frac{1}{b-a} \qquad a \le x \le b$$

$$E(X) = \frac{a+b}{2} \qquad \text{Var}(X) = \frac{(b-a)^2}{12}$$

Normal Distribution

The normal distribution $N(\mu, \sigma^2)$ is symmetric around its mean and is distributed with the bell shape in Figure 3.4. Its density function and other characteristics are given below.

$$f(x) = \frac{1}{\sigma\sqrt{2\pi}} \exp\left[-\frac{(x-\mu)^2}{2\sigma^2}\right] \qquad -\infty < x < \infty$$

$$E(X) = \mu \qquad V\,ar(X) = \sigma^2$$

$$\phi(t) = \exp\left[i\mu t - \frac{\sigma^2 t^2}{2}\right]$$

The Exponential Distribution

The distribution $f(x;\theta) = (1/\theta)e^{-x/\theta}$ for $x > 0$ and $\theta > 0$, is called the **exponential distribution**. Its CDF is $F(x, \theta) = 1 - e^{-x/\theta}$. It can be shown that the exponential distribution provides an appropriate model for the lifetime of an equipment. To see this, suppose the probability that a given equipment will fail during a small time interval Δx is $(1/\theta)\Delta x$, independent of x. Now divide 0 to x into n equal intervals with $\Delta x = x/n$. The probability that an equipment will *not* fail in *one* of these intervals is $[1 - (x/n\theta)]$. Hence it will not fail in 0 to x with probability $[1 - (x/n\theta)]^n$. As $n \to \infty$ we get $\lim_{n \to \infty} [1 - (x/n\theta)]^n = e^{-x/\theta}$ by the definition of the exponential function. Thus $F(x;\theta)$ is the probability that a given equipment will fail in the interval 0 to x.

The exponential process is closely related to the queuing process described earlier when we discussed the Poisson distribution. In a queuing process, let X be the length of the interval between two successive arrivals in a queue. $P(X > x) = P(\text{no arrivals in time interval}$

of length x). Using the derivation for equipment failure, we have this probability as $e^{-x/\theta}$. Hence $F(x) = 1 - e^{-x/\theta}$, which yields the same density function.

Practice Problem

4.1 Graph the exponential density for $\theta = 1, 2,$ and 5.

Example 4.2

Suppose that the life of a light bulb has an exponential distribution with $\theta = 400$. What is the probability that 4 out of 5 bulbs chosen independently at random have life in excess of 500 hours?

Let X be the life of a bulb. Then $p = P(X \geq 500)$ is the probability of observing that a bulb lasts for at least 500 hours.

$$f(x) = \frac{1}{\theta} e^{-x/\theta}$$

$$p = P(X \geq 500) = \int_{500}^{\infty} \frac{1}{\theta} e^{-x/\theta} dx$$

$$= [-e^{-x/\theta}]_{500}^{\infty} = e^{-500/\theta} = e^{-5/4}$$

The probability that 4 out of 5 bulbs have $X \geq 500$ is $\binom{n}{y} p^y q^{n-y}$ where $n = 5$, $y = 4$, and $p = e^{-5/4}$. Thus the required probability is $5p^4 q = 5e^{-5}[1 - e^{-5/4}]$. It will be noted that this example mixes an exponential distribution with a binomial one.

Let us now derive the mean and variance of the exponential distribution. We saw in the previous chapter that this task is made considerably easier if we first obtain the characteristic function.

$$\phi(t) = E(e^{iXt}) = \int_{-\infty}^{\infty} e^{ixt} \frac{1}{\theta} e^{-x/\theta} dx = \int_{-\infty}^{\infty} \frac{1}{\theta} e^{-x(1-i\theta t)/\theta} dx$$

Making the substitution $y = x(1 - i\theta t)$, this becomes

$$\phi(t) = \int_{-\infty}^{\infty} \frac{1}{\theta} e^{-y/\theta} \frac{dy}{1 - i\theta t} = \frac{1}{1 - i\theta t}$$

because the rest of it is the exponential density which integrates to 1.

To compute the mean and variance we differentiate the characteristic function with respect t twice and evaluate them at the origin.

$$\phi'(t) = \frac{i\theta}{(1 - i\theta t)^2} \qquad \phi''(t) = \frac{2(i\theta)^2}{(1 - i\theta t)^3}$$

Therefore, $iE(X) = \phi'(0) = i\theta$, and hence the mean of the distribution is θ. $i^2E(X^2) = \phi''(0) = 2i^2\theta^2$. Therefore, $E(X^2) = 2\theta^2$ from which it follows that the variance is $\sigma^2 = \theta^2$. This means that the exponential distribution has both the mean and the standard deviation equal to θ.

The Weibull Distribution

In some more general situations the conditions for the exponential distribution are not met. An alternative used in these cases is the **Weibull distribution**, which has the density function

$$f(x;a,b) = abx^{b-1}e^{-ax^b} \qquad x > 0 \qquad a,\, b > 0$$

This distribution is also used to model the breaking strength of certain materials. Note that when $b = 1$, this reduces to the exponential distribution.

Lognormal Distribution

A random variable X is said to have the **standard lognormal distribution** if $Z = \ln X$ has the standard normal distribution

$$f_Z(z) = \frac{1}{\sqrt{2\pi}}e^{-z^2/2} \qquad -\infty < z < \infty$$

Then, using Theorem 3.5,

$$f_X(x) = \frac{1}{\sqrt{2\pi}}\left[e^{-(\ln x)^2/2}\right] \cdot \frac{1}{x} \qquad 0 < x < \infty$$

Because $\ln X$ is defined only for positive X and most economic variables take only positive values, this distribution is very popular in economics [see Aitchison and Brown (1966), Crow and Shimizu (1988) for examples]. It has been used to model the size of firms, stock prices at the end of a trading day, income distributions, expenditure on particular commodities, and certain commodity prices.

It is useful to derive the mth moment of a lognormal random variable, that is, $\mu'_m = E(X^m)$. We have,

$$E(X^m) = \int_0^\infty x^m \frac{1}{\sqrt{2\pi}} e^{-(\ln x)^2/2} \cdot \frac{1}{x}\, dx$$

Making the substitution $z = \ln x$ we get,

$$E\,(X^m) \;=\; \int_{-\infty}^{\infty} e^{mz}\,\frac{1}{\sqrt{2\pi}}\,e^{-z^2/2}\,dz$$

$$=\; \int_{-\infty}^{\infty} \frac{1}{\sqrt{2\pi}}\,e^{-[(z-m)^2-m^2]/2}\,dz \;=\; e^{m^2/2}$$

because the rest of the integral is 1. It follows from this that,

$$E\,(X) \;=\; \sqrt{e} \qquad\qquad \mathrm{Var}\,(X) \;=\; e^2 - e$$

The lognormal distribution can be extended by letting $Y = \ln X$ have the general normal distribution $N(\mu, \sigma^2)$. Using Theorem 3.5 again we have,

$$f\,(x) \;=\; \frac{1}{\sigma\sqrt{2\pi}}\,\exp\!\left[-\frac{(\ln x - \mu)^2}{2\sigma^2}\right]\cdot\frac{1}{x}$$

Practice Problem

4.2   By proceeding as before, show that $E\,(X^m) = e^{m\mu + (m^2\sigma^2/2)}$ when $\ln X \sim N(\mu, \sigma^2)$. Deduce the mean and variance from this.

Pareto Distribution

Although the lognormal distribution is often used to model the distribution of incomes, it has been found to approximate incomes in the middle range very well but to fail in the upper tail. A more appropriate distribution for this purpose is the **Pareto distribution**, which has the density

$$f\,(x;\alpha) \;=\; \frac{\alpha}{(1+x)^{(\alpha+1)}} \qquad 0 < x < \infty \qquad \alpha > 0$$

The Pareto distribution is also commonly used in investigations involving insurance claims.

Extreme Value Distribution

For modeling extreme values such as the peak electricity demand in a day, maximum rainfall, and so on, we can use the **extreme value distribution** which, in its standard form, has the following density.

$$f\,(x) \;=\; e^{-x}\,\exp[-e^{-x}] \qquad -\infty < x < \infty$$

Logistic Distribution

The distribution such that $F(x) = 1/(1+e^{-x})$ is called the **logistic distribution**. Its density function is

$$f(x) = F'(x) = \frac{e^{-x}}{(1+e^{-x})^2} \qquad -\infty < x < \infty$$

This distribution is frequently used to model growth relations, especially those of new products which have initially slow sales, a period of very rapid increase in sales, followed by a tapering off of the growth rates.

Cauchy Distribution

The standard **Cauchy distribution** has the density function

$$f(x) = \frac{1}{\pi(1 + x^2)} \qquad -\infty < x < \infty$$

It will be seen in the next chapter that the Cauchy distribution arises when the ratio of two independent normal variates is computed. The distribution is also useful because it throws some light on the uneasy properties of certain estimators encountered in econometrics.

Practice Problems

4.3   Verify that $f(x)$ integrates to 1 for the Cauchy distribution.
[*Hint*: Use the transformation $x = \tan\theta$.]

4.4   Graph the lognormal, logistic, Pareto, and Cauchy densities.

Gamma Distribution

This distribution has the density function

$$f(x;\alpha,\beta) = \frac{1}{\beta^{\alpha}\Gamma(\alpha)} x^{\alpha-1}e^{-(x/\beta)} \qquad x > 0, \qquad \alpha,\beta > 0$$

where $\Gamma(\alpha)$ is known as the **gamma function** $\int_0^{\infty} y^{\alpha-1} e^{-y}dy$ for $\alpha > 0$. The Gamma distribution is a useful two-parameter family of distributions. In some cases, this distribution has been found to be a better approximation to the distribution of lifetime of equipments than the exponential. It can also be derived from the sum of squares of several independent normal random variables. When $\alpha = 1$, the Gamma variate reduces to the exponential case.

Theorem 4.1

For the gamma function defined as $\Gamma(\alpha) = \int_0^\infty x^{\alpha-1} e^{-x} dx,$

(a) $\Gamma(1) = 1$ *and* $\Gamma(\tfrac{1}{2}) = \sqrt{\pi}$.

(b) $\Gamma(\alpha+1) = \alpha\Gamma(\alpha)$ *and hence, if α is a positive integer, then*
$\Gamma(\alpha+1) = \alpha!$.

[The gamma function is thus a generalization of the factorial for integers.]

Proof:

(a)

$$\Gamma(1) = \int_0^\infty e^{-x} dx = [-e^{-x}]_0^\infty = 1. \qquad \Gamma(\tfrac{1}{2}) = \int_0^\infty x^{-\frac{1}{2}} e^{-x} dx$$

Let $x = y^2/2$. Then

$$\Gamma(\tfrac{1}{2}) = \int_0^\infty \frac{\sqrt{2}}{y} e^{-y^2/2} \, y \, dy$$

But $\int_{-\infty}^\infty (1/\sqrt{2\pi}) e^{-y^2/2} dy = 1$, because the integrand is $N(0,1)$. Also, by symmetry, $\int_0^\infty (1/\sqrt{2\pi}) e^{-y^2/2} dy = \tfrac{1}{2}$. Using this above we get $\Gamma(\tfrac{1}{2}) = \sqrt{\pi}$.

(b)   Integrate $\Gamma(\alpha+1)$ by parts. We get

$$\Gamma(\alpha+1) = \int_0^\infty x^\alpha e^{-x} dx$$

$$= [-x^\alpha e^{-x}]_0^\infty + \int_0^\infty \alpha x^{\alpha-1} e^{-x} dx = \alpha\Gamma(\alpha)$$

because the first term is zero. If α is a positive integer, this recursive formula can be applied repeatedly to give $\Gamma(\alpha+1) = \alpha!$.

Beta Distribution

The density function for this distribution has the form

$$f(x) = \frac{1}{B(m, n)} x^{m-1}(1-x)^{n-1} \qquad 0 < x < 1, \qquad m, n > 0$$

where the constant is

$$B(m, n) = \int_0^1 x^{m-1}(1-x)^{n-1}$$

(which is known as the **beta function**). This two-parameter family of distributions is flexible and can be used to model several different shapes. If $m = n = 1$, this reduces to the uniform distribution. The beta function can be shown to satisfy the following equation.

$$B(m, n) = \frac{\Gamma(m)\,\Gamma(n)}{\Gamma(m+n)}$$

4.3   EXTENSIONS OF DISTRIBUTIONS

More General Parametrization

In a number of situations, we have encountered density functions which are more general versions of basic functions. For instance, the standard normal $N(0,1)$ can be extended to the general case of $N(\mu, \sigma^2)$. The standard exponential density is e^{-x}, which can be extended to $(1/\theta)\,e^{-x/\theta}$. The Weibull distribution encountered in the previous chapter is a further extension of the exponential case. In this section we list a number of such extensions that make the families of densities more flexible and richer in parametrization. In later chapters, we make use of some of these extended versions of distributions in illustrating a number of results.

Exponential: $f(x;\mu,\theta) = \dfrac{1}{\theta} e^{-(x-\mu)/\theta}$ $x > \mu,\ \theta > 0$

Pareto: $f(x;\theta,x_0) = \dfrac{\theta}{x_0}\left[\dfrac{x_0}{x}\right]^{\theta+1}$ $x > x_0,\ \theta > 0$

Logistic: $f(x;\alpha,\beta) = \dfrac{(1/\beta)\,e^{-(x-\alpha)/\beta}}{\left[1 + e^{-(x-\alpha)/\beta}\right]^2}$

$$-\infty < x < \infty,\quad -\infty < \alpha < \infty,\quad \beta > 0$$

Cauchy: $f(x) = \dfrac{1}{\pi\beta\left[1 + \dfrac{(x-\alpha)}{\beta}\right]^2}$

$$-\infty < x < \infty,\quad -\infty < \alpha < \infty,\quad \beta > 0$$

Lognormal: $f(x) = \dfrac{1}{\sqrt{2\pi}\,(x-\theta)}\, e^{-[\ln(x-\theta)-\mu]^2/(2\sigma^2)}$

$$x > \theta,\quad -\infty < \mu < \infty,\quad \sigma > 0$$

Pearson family: $f(x)$ satisfies the following differential equation.

$$\frac{1}{f(x)} \frac{df}{dx} = \frac{x+a}{b_0 + b_1 x + b_2 x^2}$$

where a and the b's are constants. Several of the densities we specified earlier are special cases of the Pearson family, which covers a wide variety of shapes of probability distributions and is widely used in **curve fitting**, that is, in empirically fitting relationships to observed data (see Exercise 4.6).

Exponential family: The family of distributions that can be written in the form

$$f(x;\theta) = \exp[U(x) + \sum_{i=1}^{i=k} T_i(x)A_i(\theta) + B(\theta)]$$

is called the **exponential family** (also known as the **Koopman family**). Most of the distributions discussed earlier can be expressed in this form and hence belong to the exponential family (verify it).

The Hazard Function

Studies on reliability, instantaneous failure rates, layoffs, and so on often use a function called the **hazard function** which is closely related to the PDF and CDF of a distribution. For example, we might be interested in the conditional density that a working woman who took a leave of absence for giving birth to a child will return to work at time x given that she has not worked until that time. The hazard function is defined as

$$h(x) = \frac{f(x)}{1 - F(x)} \qquad x > \alpha$$

where $f(x)$ and $F(x)$ are the PDF and CDF of the random variable X and α is a parameter. The hazard function is thus a conditional density function.

Example 4.3

In labor economics studies, a commonly used form for $f(x)$ is the lognormal density. The standard normal density is often denoted by $\phi(x)$ [not to be confused with the characteristic function $\phi(t)$], and its CDF is denoted by $\Phi(x)$. Using this notation, the hazard function for the lognormal distribution is expressed as follows (verify it).

$$h(x;\mu,\alpha) = \frac{\dfrac{1}{(x-\alpha)}\,\phi\left[\dfrac{\ln(x-\alpha)-\mu}{\sigma}\right]}{1-\Phi\left[\dfrac{\ln(x-\alpha)-\mu}{\sigma}\right]}$$

Although we have listed a variety of distributions that are used both by statisticians and economists, the list here is not complete, nor are the descriptions detailed. Casella and Berger (1990), Fraser (1976), Mood, Graybill, and Boes (1974) have excellent summaries of the properties of several discrete and continuous random variables along with graphs.

EXERCISES

4.1   Derive the hazard functions for the exponential, logistic, and Weibull distributions presented earlier.

4.2   Verify the characteristic function, mean, and variance of as many of the densities in Table 4.1 as you can.

4.3   Let X be the exponential distribution with density function ae^{-ax} for $a > 0$. Show that $Y = X^{1/b}$ ($b > 0$) has the Weibull distribution.

4.4   Let X have the Weibull distribution. Show that $Y = X^b$ ($b > 0$) has the exponential density function.

4.5   Graph the Weibull distribution for different values of a and b.

4.6   Which of the continuous density functions in Table 4.1 can be expressed as special cases of the Pearson family?

Table 4.1   Summary of Distributions

DISTRIBUTION	$f(x)$	Domains	μ	σ^2	$\phi(t)$
DISCRETE DENSITIES					
BERNOULLI	$p^x q^{1-x}$	$x=0,1$ $0\le p\le 1$	p	pq $(q=1-p)$	$q+pe^{it}$
BINOMIAL	$\begin{bmatrix}n\\x\end{bmatrix}p^x q^{n-x}$	$x=0,1,\dots n$ $n=1,2\dots$ $0\le p\le 1$	np	npq	$(q+pe^{it})^n$
GEOMETRIC	$p(1-p)^x$	$x=0,1,2,\dots$ $0<p\le 1$	$\dfrac{1-p}{p}$	$\dfrac{1-p}{p^2}$	$\dfrac{p}{1-qe^{it}}$
HYPER-GEOMETRIC	$\dfrac{\begin{bmatrix}a\\x\end{bmatrix}\begin{bmatrix}b\\n-x\end{bmatrix}}{\begin{bmatrix}a+b\\n\end{bmatrix}}$	$x=0,1,\dots n$ $a=0,1,\dots$ $b=1,2,\dots$	$\dfrac{na}{a+b}$	$\dfrac{na}{a+b}\cdot\dfrac{b}{a+b}\cdot\dfrac{a+b-n}{a+b-1}$	--
NEGATIVE BINO-MIAL	$\begin{bmatrix}x+k-1\\k-1\end{bmatrix}p^k q^x$	$x=0,1,2,\dots$ $0<p<1$ $k>0$	kq/p	kq/p^2	$p^k(1-qe^{it})^{-k}$
POISSON	$\dfrac{e^{-\lambda}\lambda^x}{x!}$	$x=0,1,2,\dots$ $\lambda>0$	λ	λ	$\exp[\lambda(e^{it}-1)]$

(Continued)

Table 4.1   (Continued)

DISTRIBUTION	$f(x)$	Domains	μ	σ^2	$\phi(t)$
CONTINUOUS DENSITIES					
BETA	$k\,x^{m-1}(1-x)^{n-1}$	$0 \le x \le 1$ $m, n > 0$	$\dfrac{m}{m+n}$	$\dfrac{mn}{(m+n+1)(m+n)^2}$	--
CAUCHY	$\dfrac{1}{\pi(1+x^2)}$	$-\infty < x < \infty$	doesn't exist	doesn't exist	$e^{-|t|}$
CHI-SQUARE	$\dfrac{(x/2)^{n/2-1}e^{-x/2}}{2\Gamma(n/2)}$	$x > 0$ $n = 1, 2, ...$	n	$2n$	$(1-2it)^{-n/2}$
EXPONENTIAL	$\dfrac{1}{\theta}e^{-x/\theta}$	$x \ge 0$ $\theta > 0$	θ	θ^2	$(1-it\theta)^{-1}$
F	$k\,\dfrac{x^{(m-2)/2}}{[1+(m/n)x]^{(m+n)/2}}$	$x > 0$ $m, n = 1, 2, ...$	$\dfrac{n}{n-2}$ $n > 2$	$\dfrac{2n^2(m+n-2)}{m(n-2)^2(n-4)}$ $n > 4$	--

For the Beta distribution, $k = \dfrac{\Gamma(m+n)}{\Gamma(m)\Gamma(n)}$.   For the F-distribution, $k = \left[\dfrac{m}{n}\right]^{(m/2)} \dfrac{\Gamma[(m+n)/2]}{\Gamma(m/2)\Gamma(n/2)}$.

Table 4.1 (Continued)

DISTRIBUTION	$f(x)$	Domains	μ	σ^2	$\phi(t)$
GAMMA	$\dfrac{1}{\beta^\alpha \Gamma(\alpha)} x^{\alpha-1} e^{-(x/\beta)}$	$x \geq 0$ $\alpha, \beta > 0$	$\alpha\beta$	$\alpha\beta^2$	$\left[\dfrac{1}{1-i\beta t}\right]^\alpha$
GEOMETRIC	$\theta x^{\theta-1}$	$0 < x < 1$ $\theta > 0$	$\dfrac{\theta}{\theta+1}$	$\dfrac{1}{(\theta+2)(\theta+1)^2}$	$E(x^m) = \dfrac{\theta}{\theta+m}$
LOGISTIC	$\dfrac{e^{-x}}{(1+e^{-x})^2}$	$-\infty < x < \infty$	0	$\pi^2/3$	$\pi t\, cosech\,(\pi t)$
LOGNORMAL	$\dfrac{1}{x\sigma\sqrt{2\pi}} e^{-(\ln x - \mu)^2/2\sigma^2}$	$x \geq 0$ $-\infty < \mu < \infty$ $\sigma > 0$	$e^{\mu+\frac{1}{2}\sigma^2}$	$e^{2\mu}\left[e^{2\sigma^2} - e^{\sigma^2}\right]$	--
NORMAL	$\dfrac{1}{\sigma\sqrt{2\pi}} e^{-(x-\mu)^2/2\sigma^2}$	$-\infty < x < \infty$ $-\infty < \mu < \infty$ $\sigma > 0$	μ	σ^2	$e^{i\mu t - (\sigma^2 t^2/2)}$

(Continued)

Table 4.1 (Continued)

DISTRIBUTION	$f(x)$	Domains	μ	σ^2	$\phi(t)$
PARETO	$\dfrac{\theta}{x_o}\left(\dfrac{x_o}{x}\right)^{\theta+1}$	$x > x_o$ $\theta > 0$	$\dfrac{\theta x_o}{\theta-1}$ $\theta > 1$	$\dfrac{\theta x_o^2}{\theta-2} - \left[\dfrac{\theta x_o}{\theta-1}\right]^2$ $\theta > 2$	--
STUDENT'S t	$k\left[1+\dfrac{x^2}{n}\right]^{-\frac{1}{2}(n+1)}$	$-\infty < x < \infty$ $n > 0$	0 for $n > 1$	$\dfrac{n}{n-2}$ for $n > 2$	--
UNIFORM	$\dfrac{1}{(b-a)}$	$a \le x \le b$ $-\infty < a,\, b < \infty$	$\dfrac{a+b}{2}$	$\dfrac{(b-a)^2}{12}$	$\dfrac{e^{ibt}-e^{iat}}{(b-a)it}$
WEIBULL	$abx^{b-1}e^{-ax^b}$	$x > 0$ $a,\, b > 0$	$a^{-1/b}\Gamma(1+b^{-1})$	$a^{-2/b}[\Gamma(1+2b^{-1}) - \Gamma^2(1+b^{-1})]$	--

For the Student's t-distribution, $k = \dfrac{1}{\sqrt{n}}\dfrac{\Gamma[(n+1)/2]}{\Gamma(n/2)\Gamma(1/2)}$.

5

MULTIVARIATE
DISTRIBUTIONS

So far we have been concerned only with a single random variable X and functions of X. In most cases, however, the outcome of an experiment may be characterized by more than one random variable. For instance, X may be the income and Y the total expenditures of a household. Thus we may observe a pair of random variables (X, Y). If family size Z is also added, we observe (X, Y, Z). We would thus be interested in the study of multivariate distributions, which is the subject matter of this chapter. We first discuss bivariate distributions at length and then take up the general case. The material presented here is much more difficult than the previous four chapters because it makes extensive use of vectors, matrices, and determinants summarized in Appendix A. You should first review the basic properties of those concepts before proceeding to the multivariate case.

5.1   BIVARIATE DISTRIBUTIONS

Definition 5.1 (joint distribution function)

Let X and Y be two random variables. Then the function $F_{XY}(x, y)$ = $P(X \leq x$ and $Y \leq y)$ is called the **joint distribution function**.

Because the distribution function is generally represented by $F(\cdot)$ and the density function by $f(\cdot)$, the subscript XY is used to identify the fact that the random variables in question are X and Y jointly. The joint distribution function has the following properties:

1. $F_{XY}(x, \infty)$ and $F_{XY}(\infty, y)$ are univariate distribution functions, as functions of x and y respectively.

2. $F_{XY}(-\infty, y) = F_{XY}(x, -\infty) = 0$.

As before, the random variables X and Y may be discrete or continuous.

Definition 5.2 (joint density or probability function)

Discrete probability function: $f_{XY}(x, y) = P(X=x, Y=y)$

Continuous density function: $f_{XY}(x, y) = \dfrac{\partial^2 F(x, y)}{\partial x \, \partial y}$

and hence $F_{XY}(x, y) = \displaystyle\int_{-\infty}^{x} \int_{-\infty}^{y} f_{XY}(u, v) \, du \, dv$

Note that for the joint density function to exist in the continuous case, $F_{XY}(x, y)$ must have continuous cross partial derivatives. In the univariate case, if Δx is a small increment of x, then $f_X(x)\Delta x$ is the approximate probability that $x - \frac{1}{2}\Delta x < X \leq x + \frac{1}{2}\Delta x$. Similarly, in a bivariate distribution, $f_{XY}(x, y)\Delta x \Delta y$ is the approximate probability that $x - \frac{1}{2}\Delta x < X \leq x + \frac{1}{2}\Delta x$ and $y - \frac{1}{2}\Delta y < Y \leq y + \frac{1}{2}\Delta y$. The bivariate density function satisfies the conditions $f_{XY}(x, y) \geq 0$ and $\int_{-\infty}^{\infty} \int_{-\infty}^{\infty} dF(x, y) = 1$, where $dF(x, y)$ is the bivariate analog of $dF(x)$.

Example 5.1 [Ramanathan (1992), pp. 25–26]

Suppose you roll a pair of dice. Let X be the number of threes and Y be the number of fives. We would like to derive the joint density of X and Y. Note that the sample space consists of 36 points (6^2), each having a probability of 1/36, and also that X and Y can take only the values 0, 1, or 2. The joint event $(X=1, Y=1)$ occurs when the individual scores are either (3,5) or (5,3). Thus $P(X=1, Y=1) = 2/36$. By similarly enumerating the possibilities in each case we can derive the other joint probabilities. They are as given in Table 5.1 (verify this).

The probabilities are obtained as $P(X=1, Y=0) = 8/36$ and so on. We can also compute $P(X=0)$ as $P(X=0, Y=0) + P(X=0, Y=1) + P(X=0, Y=2) = 25/36$. Can we generalize this and obtain the density function of X alone? The answer is yes. Let $f_X(x) = P(X=x)$ be the density function of X. Then

$$P(X=x) = P(X=x, Y=0) + P(X=x, Y=1) + P(X=x, Y=2)$$

from the fact that the event $X=x$ can be partitioned into the three mutually exclusive events. It follows from this that $P(X=1) = 8/36 + 2/36 + 0 = 10/36$, and similarly for other Xs. It is readily seen that the numbers in the bottom margin give $f_X(x)$. Similarly, the numbers on the right-hand margin give $f_Y(y)$. These are called the **marginal distributions**.

Table 5.1 Illustration of a joint probability distribution

x y	0	1	2	$f_Y(y)$
0	16/36	8/36	1/36	25/36
1	8/36	2/36	0	10/36
2	1/36	0	0	1/36
$f_X(x)$	25/36	10/36	1/36	1

Definition 5.3 (marginal density)

If X and Y are discrete random variables, then $f_X(x) = \Sigma_y\, f_{XY}(x, y)$ is the **marginal density of** *X and $f_Y(y) = \Sigma_x\, f_{XY}(x, y)$ is the* **marginal density of** *Y. In the continuous case, $f_X(x) = \int f_{XY}(x, y)\, dy$ is the marginal density of X and $f_Y(y) = \int f_{XY}(x, y)\, dx$ is the marginal density of Y.*

Example 5.2

In Example 5.1, suppose we are given the information that a particular event has occurred; for example, that $X = 1$. We can then ask "what is the probability that $Y = 0$ *given* that $X = 1$?" To answer this, we resort to the notion of conditional probability presented in Definition 2.6. We have

$$P(Y = 0 \mid X = 1) = \frac{P(Y = 0, X = 1)}{P(X = 1)} = 8/36 + 10/36 = 0.8$$

By proceeding similarly for all cases, we obtain the **conditional distribution of Y given X.**

Definition 5.4 (conditional density)

The **conditional density of Y given $X = x$** *is defined as $f_{Y|X}(x, y) = f_{XY}(x, y)/f_X(x)$, provided $f_X(x) \neq 0$. The* **conditional density of X given $Y = y$** *is $f_{X|Y}(x, y) = f_{XY}(x, y)/f_Y(y)$ for $f_Y(y) \neq 0$. This definition holds for both discrete and continuous random variables.*

In many situations the information that the random variable X takes a particular value x may be irrelevant to the determination of the probability that another random variable Y takes a particular value y. In Chapter 2 we introduced the notion of *statistical indepen-dence* to describe this property. That concept can be extended to several random variables as well.

Definition 5.5 (statistical independence)

*The random variables X and Y are said to be **statistically independent** if and only if $f_{X|Y}(x, y) = f_X(x)$ for all values of X and Y for which $f_{XY}(x, y)$ is defined. Equivalently, $f_{Y|X}(x, y) = f_Y(y)$ and $f_{XY}(x, y) = f_X(x) f_Y(y)$.*

In other words, the condition that Y is given (that is, is known) does not alter the probability density of X, and hence the conditional density is the same as the marginal density. Note also that indepen-dence means that $f_{XY}(x, y)$ is separable in x and y.

All the above concepts are summarized in Table 5.2.

Theorem 5.1

Random variables X and Y with joint density function $f_{XY}(x, y)$ will be statistically independent if and only if $f_{XY}(x, y)$ can be writ-ten as a product of two nonnegative functions, one in X alone and another in Y alone.

Proof: Independence implies that $f_{XY}(x, y) = f_X(x) f_Y(y)$ and hence the required condition is satisfied. Conversely, suppose $f_{XY}(x, y) = g(x)h(y)$. Then, for continuous random variables,

$$f_X(x) = \int_{-\infty}^{\infty} g(x)h(y)\, dy = k_1 g(x)$$

$$f_Y(y) = \int_{-\infty}^{\infty} g(x)h(y)\, dx = k_2 h(y)$$

where k_1 and k_2 are constants independent of x and y. Hence,

$$f_{XY}(x, y) \;=\; \frac{1}{k_1 k_2}\, f_X(x) f_Y(y)$$

It is easy to show that $k_1 k_2 = 1$, which will establish the converse.

$$1 = \int_{-\infty}^{\infty} \int_{-\infty}^{\infty} g(x)h(y)\, dxdy = \left[\int_{-\infty}^{\infty} g(x)\, dx\right]\left[\int_{-\infty}^{\infty} h(y)\, dy\right] = \frac{1}{k_1 k_2}$$

Table 5.2 Bivariate distributions

Concept	Continuous	Discrete
Density Function	$f_{XY}(x,y)\Delta x\Delta y = P\left[x-\dfrac{\Delta x}{2}<X\le x+\dfrac{\Delta x}{2}, y-\dfrac{\Delta y}{2}<Y\le y+\dfrac{\Delta y}{2}\right]$	$f_{XY}(x,y) = P(X=x, Y=y)$
Distribution Function	$F_{XY}(x,y) = \int_{Y\le y}\int_{X\le x} f_{XY}(x,y)\,dxdy$	$\sum_{Y\le y}\ \sum_{X\le x} f_{XY}(x,y)$
Marginal Densities	$\int_{-\infty}^{\infty} f_{XY}(x,y)\,dx$; $\int_{-\infty}^{\infty} f_{XY}(x,y)\,dy$	$\sum_x f_{XY}(x,y)$; $\sum_y f_{XY}(x,y)$
Conditional Densities	$f(y\,|\,x) = f_{XY}(x,y)/f_X(x)$; $f(x\,|\,y) = f_{XY}(x,y)/f_Y(y)$	$\rightarrow$ *same*
Independence	$f_{XY}(x,y) = f_X(x)f_Y(y)$	$\rightarrow$ *same*

Theorem 5.2

> If X and Y are statistically independent and a, b, c, d are real constants with $a < b$ and $c < d$, then
>
> $$P(a < X < b, \ c < Y < d) \ = \ P(a < X < b) P(c < Y < d)$$

Practice Problem

5.1   Prove Theorem 5.2.

Mathematical Expectation

The concept of **mathematical expectation** is easily extended to bivariate random variables. We have

$$E_{XY}[g(X, Y)] \ = \ \int\int g(x, y) \, dF_{XY}(x, y)$$

where the subscript indicates that the integral is over the (X, Y) space.

Moments: The rth moment of X is

$$E_{XY}(X^r) \ = \ \int\int x^r dF_{XY}(x, y) \ = \ \int x^r dF_X(x) \ = \ E_X(X^r)$$

Joint moments: $E_{XY}(X^r Y^s) \ = \ \int\int x^r y^s dF(x, y)$

Theorem 5.3

> Let X and Y be independent random variables and let $u(X)$ be a function of X only and $v(Y)$ be a function of Y only. Then,
>
> $$E_{XY}[u(X) v(Y)] \ = \ E_X[u(X)] E_Y[v(Y)]$$

Practice Problem

5.2   Prove Theorem 5.3.

Covariance

The most important joint moment is the **covariance between X and Y**. It is defined as

$$\sigma_{XY} \ = \ \text{Cov}(X, Y) = E_{XY}[(X - \mu_X)(Y - \mu_Y)] = E_{XY}(XY) - \mu_X \mu_Y$$

where $\mu_X = E(X)$ and $\mu_Y = E(Y)$. In the continuous case this takes the form

$$\sigma_{XY} \ = \ \int_{-\infty}^{\infty} \int_{-\infty}^{\infty} (x - \mu_X)(y - \mu_Y) f_{XY}(x, y) \, dx \, dy$$

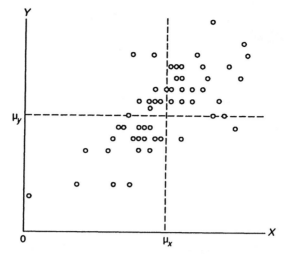

Figure 5.1 An illustration of the covariance between two random variables

and in the discrete case it is

$$\sigma_{XY} = \sum_x \sum_y (x-\mu_X)(y-\mu_Y)f_{XY}(x, y)$$

As before, the variances can be defined as $\sigma_X^2 = E[(X-\mu_X)^2] = E(X^2) - \mu_X^2$ and $\sigma_Y^2 = E[(Y-\mu_Y)^2] = E(Y^2) - \mu_Y^2$.

Just as the variance is a measure of the dispersion of a random variable around its mean, the covariance between two random variables is a measure of the joint association between them. Suppose X and Y are random variables that are positively related, so that Y increases when X increases, as illustrated in Figure 5.1 for a pair of discrete random variables. The circles represent pairs of values X and Y that are possible outcomes. The dashed lines indicate the means μ_X and μ_Y. By translating the axes to the dashed lines with origin at (μ_X, μ_Y), we can see that $X_i - \mu_X$ and $Y_i - \mu_Y$ are the distances from the new origin, for a typical outcome denoted by the suffix i. It is readily noted from the figure that the points in the first and third quadrants will make the product $(X - \mu_X)(Y - \mu_Y)$ positive, because the individual terms are either both positive or both negative. In contrast, the points in the second and fourth quadrants will make the product negative, because one of the terms is positive and the other negative. When we compute the covariance measure, which is a weighted sum of these products, the net effect is likely to be positive because there are more positive terms than the opposite. Therefore, the covariance is likely to be positive. In the case when X and Y move in opposite directions, Cov(X, Y) will be negative.

Although the covariance measure is useful in identifying the nature of the association between X and Y, it has a serious problem, namely, the numerical value is very sensitive to the units of measurement. To avoid this problem, a "normalized" covariance measure is used. This measure is called the **correlation coefficient**.

Correlation

The quantity

$$\rho_{XY} = \frac{\sigma_{XY}}{\sigma_X \sigma_Y} = \frac{\text{Cov}(X, Y)}{[\text{Var}(X)\,\text{Var}(Y)]^{1/2}}$$

is called the **correlation coefficient** between X and Y. If X and Y are positively related, that is, when X increases then so does Y, then the covariance will be positive and hence the correlation coefficient will be positive. If Cov(X, Y) = 0, then Cor(X, Y) = 0, in which case X and Y are said to be **uncorrelated**. If X and Y are independent, then $f_{XY}(x, y) = f_X(x)f_Y(y)$. Note from the definition of σ_{XY} that it is $E_{XY}[(X-\mu_X)]\,E_{XY}[(Y-\mu_Y)]$ in this case. But $E(X-\mu_X) = E(X) - \mu_X = 0$. Hence, $\sigma_{XY} = 0$ and $\rho_{XY} = 0$ *if two random variables are independent*. That the converse need not be true is seen from the counterexample given in Table 5.3 [see Ramanathan (1992), Table 2.4].

$$\text{Cov}(X, Y) = E(XY) - E(X)E(Y) \,; \quad E(X) = 8 \,; \quad E(Y) = 2$$

$$E(XY) = 6 \times 0.2 + 10 \times 0.2 + 16 \times 0.2 + 18 \times 0.2 + 30 \times 0.2 = 16$$

Therefore, Cov(X, Y) = 0. But X and Y are not independent because $P(X=6, Y=2) = 0$, but $P(X=6) = 0.4$ and $P(Y=2) = 0.2$. Hence X and Y are not independent.

Theorem 5.4: $|\rho_{XY}| \leq 1$.

Proof: Consider the random variable $(u-kv)^2$ where u and v are random variables and k is a real constant. Because $(u-kv)^2 \geq 0$, $E[(u-kv)^2] \geq 0$. But

$$E[(u-kv)^2] = E(u^2) - 2kE(uv) + k^2 E(v^2)$$

This quadratic in k must be greater than or equal to zero for all real k. To derive the necessary and sufficient condition for this, first define $a = E(u^2)$, $b = E(v^2)$, and $h = E(uv)$. The quadratic can then be written as follows.

Table 5.3 An example showing that zero covariance need not imply independence

x y	6	8	10	$f_Y(y)$
1	0.2	0.0	0.2	0.4
2	0.0	0.2	0.0	0.2
3	0.2	0.0	0.2	0.4
$f_X(x)$	0.4	0.2	0.4	1

$$bk^2 - 2hk + a = b\left[k - \frac{h}{b}\right]^2 + \frac{(ab - h^2)}{b}$$

The conditions $b > 0$ and $ab - h^2 \geq 0$ are sufficient to make the quadratic nonnegative for all real k. They are also necessary because otherwise we can choose a k for which the quadratic is negative. The required condition translates to

$$[E(uv)]^2 \leq E(u^2)E(v^2)$$

This inequality is called the **Schwartz inequality** (which has other forms also). Let $u = X - \mu_X$ and $v = Y - \mu_Y$. Then

$$[\text{Cov}(X, Y)]^2 \leq \text{Var}(X)\text{Var}(Y)$$

or $\rho_{XY}^2 \leq 1$, that is, $-1 \leq \rho_{XY} \leq 1$.

If $\rho_{XY}^2 = 1$, then $E[(u - kv)^2] = 0$ for some k, implying that $P(u = kv) = 1$, that is, $X - \mu_X = k(Y - \mu_Y)$. In other words, there is an exact linear relationship between X and Y. It should be emphasized that ρ_{XY} measures only a *linear* relationship between X and Y. It is possible to have an exact relation but a correlation less than 1, even 0. To illustrate, consider the random variable X which is distributed as Uniform $[-\theta, \theta]$ and the transformation $Y = X^2$. $\text{Cov}(X, Y) = E(X^3) - E(X)E(X^2) = 0$ because the distribution is symmetric around the origin and hence all the odd moments about the origin are zero. It follows that X and Y are uncorrelated even though there is an exact relation between them. In fact, this result holds for any distribution that is symmetric around the origin.

5.2   CONDITIONAL EXPECTATION

Definition 5.6

Let X and Y be continuous random variables and $g(Y)$ be a continuous function. Then the **conditional expectation** *(or* **conditional mean**) *of $g(Y)$* **given** *$X = x$, denoted by $E_Y|X[g(Y) \mid X]$, is given by $\int_{-\infty}^{\infty} g(y) f_Y|X(x, y)\, dy$, where $f_Y|X(x, y)$ is the conditional density of Y given X. The definition for the discrete case is analogous.*

Note that $E[g(Y) \mid X = x]$ is a function of x and is not a random variable because x is fixed. The special case of $E(Y \mid X)$ is called the **regression of Y on X** and is widely used in econometrics (more on this in Part III of the book). Theorem 5.9 has an example of conditional expectation for the normal case.

Theorem 5.5 (Law of iterated expectations)

$E_{XY}[g(Y)] = E_X[E_Y|X\{g(Y) \mid X\}]$. *That is, the unconditional expectation is the expectation of the conditional expectation.*

Proof: In the continuous random variable case,

$$E_{XY}[g(Y)] = \int\int g(y)\, f_{XY}(x, y)\, dxdy$$

But

$$f_{XY}(x, y) = f_Y|X(x, y)\, f_X(x)$$

Hence

$$E_{XY}[g(Y)] = \int [\int g(y) f_Y|X(x, y)\, dy\,]\, f_X(x)\, dx$$

The expression in the square brackets is the conditional expectation of $g(Y)$ given X. Therefore,

$$E_{XY}[g(Y)] = \int E_Y|X[g(Y) \mid X]\, f_X(x)\, dx = E_X[E_Y|X\{g(Y) \mid X\}]$$

5.3   CONDITIONAL VARIANCE

Definition 5.7

Let $\mu_{Y|X} = E(Y \mid X) = \mu^*(X)$ be the conditional mean of Y given X. Then the **conditional variance** *of Y* **given** *X is defined as $Var(Y \mid X) = E_Y|_X[(Y - \mu^*)^2 \mid X]$. This is a function of X.*

In other words, look at the subspace of $\mathcal{R}^2$ such that $X = x$ and the corresponding distribution of Y given X. The mean of this distribution is the conditional mean and its variance is the conditional variance.

Theorem 5.6

$Var_{XY}(Y) = E_X[Var(Y \mid X)] + Var_X(\mu^*)$, *that is, the variance of Y is the mean of its conditional variance plus the variance of its conditional mean.*

Proof:

$$(Y - \mu_Y)^2 = (Y - \mu^* + \mu^* - \mu_Y)^2$$
$$= (Y - \mu^*)^2 + (\mu^* - \mu_Y)^2 + 2(Y - \mu^*)(\mu^* - \mu_Y)$$
$$E_{XY}(Y - \mu_Y)^2 = E_X[E_{Y|X}(Y - \mu^*)^2 \mid X] + E_X[E_{Y|X}(\mu^* - \mu_Y)^2 \mid X]$$
$$+ 2E_X[E_{Y|X}(Y - \mu^*)(\mu^* - \mu_Y) \mid X]$$

The first term is $E_X[Var(Y \mid X)]$, the second term is $Var_X(\mu^*)$, because $E_X(\mu^*) = \mu_Y$. The third term is zero because if X is given, $\mu^* - \mu_Y$ is a constant and $E_{Y|X}(Y - \mu^*) = 0$.

Theorem 5.7: $Var(aX + bY) = a^2 Var(X) + 2ab \, Cov(X, Y) + b^2 Var(Y)$.

Proof: $E(aX + bY) = a\mu_X + b\mu_Y$.

$$Var(aX + bY) = E[a(X - \mu_X) + b(Y - \mu_Y)]^2.$$

Straightforward expansion of this gives the result.

Practice Problems

5.3   Prove that $Var(X + Y) = Var(X) + Var(Y) + 2Cov(X, Y)$.

5.4   Show that, if X and Y are independent, then $Var(X + Y) = Var(X - Y) = Var(X) + Var(Y)$.

Approximate Mean and Variance for $g(X, Y)$

As in the univariate case, occasions arise when one needs to approximate the mean and variance of a general measurable function $g(X, Y)$. A linear approximation is obtained in the neighborhood of (μ_X, μ_Y) as

$$g(X, Y) \approx g(\mu_X, \mu_Y) + \left[\frac{\partial g}{\partial X}\right]_* (X - \mu_X) + \left[\frac{\partial g}{\partial Y}\right]_* (Y - \mu_Y)$$

provided $g(\mu_X, \mu_Y)$ and the two partial derivatives exist at the mean point (μ_X, μ_Y), denoted by *. The mean and variance are

$$E[g(X, Y)] \approx g(\mu_X, \mu_Y)$$

$$\text{Var}[g(X, Y)] \approx \sigma_X^2 \left[\frac{\partial g}{\partial X}\right]_*^2 + \sigma_Y^2 \left[\frac{\partial g}{\partial Y}\right]_*^2 + 2\rho\sigma_X\sigma_Y \left[\frac{\partial g}{\partial X}\right]_* \left[\frac{\partial g}{\partial Y}\right]_*$$

Example 5.3

Consider the transformation $g(X, Y) = \sqrt{XY}$. An approximation to the mean is $\sqrt{\mu_X\mu_Y}$. To obtain the variance approximation, we need the partial derivatives

$$\frac{\partial g}{\partial X} = \frac{\sqrt{Y}}{2\sqrt{X}} \quad \text{and} \quad \frac{\partial g}{\partial Y} = \frac{\sqrt{X}}{2\sqrt{Y}}$$

An approximate variance is therefore given by

$$\text{Var}[g(X, Y)] \approx \sigma_X^2 \left[\frac{\mu_Y}{4\mu_X}\right] + \sigma_Y^2 \left[\frac{\mu_X}{4\mu_Y}\right] + \frac{1}{2}\rho\sigma_X\sigma_Y$$

Once again you should be cautioned that the approximations may be grossly in error. You should be especially careful with the variance and covariance approximations.

5.4   THE BIVARIATE NORMAL DISTRIBUTION

Definition 5.8

Let (X, Y) have the joint density

$$f_{XY}(x, y) = \frac{1}{2\pi\sigma_X\sigma_Y\sqrt{(1-\rho^2)}} \exp\left[-\frac{Q}{2(1-\rho^2)}\right]$$

where

$$Q = \left[\frac{x-\mu_X}{\sigma_X}\right]^2 - 2\rho\frac{(x-\mu_X)(y-\mu_Y)}{\sigma_X\sigma_Y} + \left[\frac{y-\mu_Y}{\sigma_Y}\right]^2$$

$$-\infty < x < \infty, \ -\infty < y < \infty, \ -\infty < \mu_X < \infty, \ \infty < \mu_Y < \infty,$$

$$\sigma_X, \sigma_Y > 0, \ \text{and} \ -1 < \rho < 1.$$

Then (X, Y) *is said to have the* **bivariate normal distribution.**
[Exercise 5.3 has a formal derivation of this.]

Practice Problems

5.5 Prove that $\int_{-\infty}^{\infty} \int_{-\infty}^{\infty} f_{XY}(x, y)\, dx dy = 1$.

5.6 Show that, $E(X) = \mu_X$, $E(Y) = \mu_Y$, $\text{Var}(X) = \sigma_X^2$, $\text{Var}(Y) = \sigma_Y^2$, and $\rho_{XY} = \rho$.

5.7 Also prove that X and Y are independent if and only if $\rho = 0$.

Theorem 5.8

If (X, Y) *is bivariate normal, then the marginal distribution of* X *is* $N(\mu_X, \sigma_X^2)$ *and that of* Y *is* $N(\mu_Y, \sigma_Y^2)$.

Proof: We have, $f_X(x) = \int_{-\infty}^{\infty} f_{XY}(x, y)\, dy$. Let $v = (y - \mu_Y)/\sigma_Y$. Then $dy = \sigma_Y\, dv$. The marginal density $f_X(x)$ is therefore

$$\int_{-\infty}^{\infty} \frac{1}{2\pi \sigma_X \sqrt{1-\rho^2}} \exp\left[-\frac{(x-\mu_X)^2}{2\sigma_X^2(1-\rho^2)} + \frac{\rho(x-\mu_X)v}{\sigma_X(1-\rho^2)} - \frac{v^2}{2(1-\rho^2)} \right] dv$$

By completing the square on v, the expression in the square brackets becomes (verify it),

$$-\frac{(x-\mu_X)^2}{2\sigma_X^2} - \frac{1}{2(1-\rho^2)}\left[v - \rho\frac{(x-\mu_X)}{\sigma_X} \right]^2$$

Let

$$w = \frac{1}{\sqrt{(1-\rho^2)}}\left[v - \rho\frac{(x-\mu_X)}{\sigma_X} \right]$$

Then $dw = \dfrac{dv}{\sqrt{1-\rho^2}}$. Thus,

$$f_X(x) = \int_{-\infty}^{\infty} \left[\frac{1}{\sigma_X\sqrt{2\pi}}\, e^{-\frac{(x-\mu_X)^2}{2\sigma_X^2}} \frac{1}{\sqrt{2\pi}} e^{-\frac{w^2}{2}} \right] dw = \frac{1}{\sigma_X\sqrt{2\pi}}\, e^{-\frac{(x-\mu_X)^2}{2\sigma_X^2}}$$

because $\int_{-\infty}^{\infty} e^{-w^2/2} dw = \sqrt{2\pi}$. X is thus a univariate normal with mean μ_X and s.d. σ_X. The result for Y follows similarly.

The converse of Theorem 5.8 need not be true, that is, if the marginal distribution of X is univariate normal, the joint density between X and Y need not be bivariate normal [see Casella and Berger (1990), Exercise 4.43].

Theorem 5.9

For a bivariate normal, the conditional density of Y given X = x is univariate normal with mean $\mu_Y + (\rho\sigma_Y/\sigma_X)(x-\mu_X)$ and variance $\sigma_Y^2(1-\rho^2)$. The conditional density of X given Y is also normal with mean $\mu_X + (\rho\sigma_X/\sigma_Y)(y-\mu_Y)$ and variance $\sigma_X^2(1-\rho^2)$.

Proof:

$$f_{Y|X}(x,y) = \frac{f_{XY}(x,y)}{f_X(x)} = \frac{1}{\sigma_Y\sqrt{2\pi(1-\rho^2)}} \exp(-Q'/2)$$

where

$$Q' = \frac{1}{1-\rho^2}\left[\frac{(y-\mu_Y)^2}{\sigma_Y^2} - 2\rho\frac{(y-\mu_Y)(x-\mu_X)}{\sigma_X\sigma_Y} + \frac{(x-\mu_X)^2}{\sigma_X^2}\right] - \frac{(x-\mu_X)^2}{\sigma_X^2}$$

which can be written as $[y-\mu_Y-(\rho\,\sigma_Y/\sigma_X)(x-\mu_X)]^2 / \sigma_Y^2(1-\rho^2)$. The theorem follows directly from this.

It was mentioned earlier that the conditional expectation $E(Y\mid X)$ is called the *regression of Y on X*. We have just seen that in the case of the bivariate normal density, this conditional expectation is of the form $\alpha + \beta X$, where α and β depend on the respective means, standard deviations, and the correlation coefficient. This is a case of a **simple linear regression** in which the conditional expectation is a linear function of X. As another example, suppose the conditional density of Y given X is exponential with the parameter $\theta = \alpha + \beta X$. Because the mean of a random variable with the exponential distribution is equal to the parameter θ (see Section 4.2), we have $E(Y\mid X) = \alpha + \beta X$. The same argument applies to the case of a Poisson density for $Y\mid X$ with the parameter $\lambda = \alpha + \beta X$, because in this case also the conditional mean is the parameter of the distribution. It will be seen in Part III that the regression model (both simple and multiple regression) plays a very important role in econometrics.

5.5   BIVARIATE TRANSFORMATIONS

In the multivariate case also we often perform transformations on the random variables and would like to know the joint density of the new sets of random variables. Here we study the bivariate case.

Theorem 5.10

Let $f_{XY}(x,y)$ be the joint density function of X and Y and $U = g(X,Y)$, $V = h(X,Y)$ be measurable transformations with the *functions g and h being continuously differentiable. Then the joint density function of U and V is given by*

$$f_{UV}(u, v) = f_{XY}[G(u, v), H(u, v)] \left| \frac{\partial(x,y)}{\partial(u, v)} \right|$$

provided the Jacobian of the transformation is nonvanishing and $x = G(u, v)$, $y = H(u, v)$ *are the inverse transformations. (Note: By Property A.33, nonvanishing Jacobian implies the existence of a local inverse at each point.)*

Proof: Essentially what is needed here is to express $\int f_{UV}(u, v) \, du \, dv$ which is the probability density in a small area in the (u, v) space in terms of the corresponding density in the (x, y) space.

Consider the inverse mapping G and H that transforms the small rectangle in B (Figure 5.2a) into the parallelogram in A. The increments in x and y can be obtained by partial differentiation.

$$dx = \frac{\partial x}{\partial u} \, du + \frac{\partial x}{\partial v} \, dv = a_{11} \, du + a_{12} \, dv \qquad (say)$$

$$dy = \frac{\partial y}{\partial u} \, du + \frac{\partial y}{\partial v} \, dv = a_{21} \, du + a_{22} \, dv$$

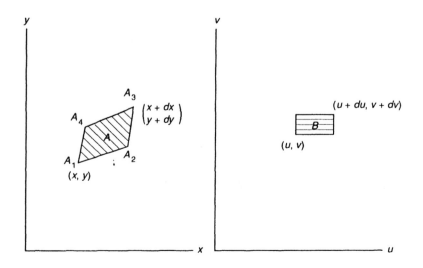

Figure 5.2a Mapping of the (X, Y) and (U, V) spaces

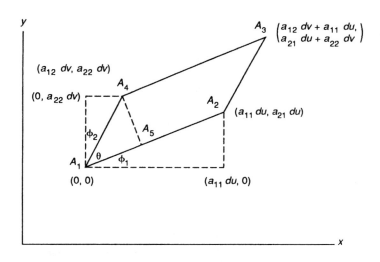

Figure 5.2b   Magnified (X, Y) space

Relative to the point A_1, A_2 has coordinates $(a_{11}\, du,\, a_{21}\, du)$ and A_4 has coordinates $(a_{12}\, dv,\, a_{22}\, dv)$, obtained alternatively by setting $du = 0$ and $dv = 0$. To evaluate the area of A, look at the magnified diagram in Figure 5.2b. (This intuitive proof goes through for other pictures of the parallelogram also.) A_3 has coordinates $(a_{11}du + a_{12}dv,\, a_{21}du + a_{22}dv)$; A_2 and A_4 have the indicated coordinates; and hence A is a parallelogram.

In Figure 5.2b, draw the perpendicular from A_4 to A_2. Also denote by θ, ϕ_1, and ϕ_2 the three angles marked in the diagram.

$$\text{Area } A = (A_1A_2)(A_4A_5) = (A_1A_2)(A_1A_4) \sin \theta$$

Because, $\theta + \phi_1 + \phi_2 = \pi/2$, $\sin \theta = \cos(\phi_1 + \phi_2) = \cos \phi_1 \cos \phi_2 - \sin \phi_1 \sin \phi_2$.

$$
\begin{aligned}
\text{Area } A &= (A_1A_2)(A_1A_4)[\cos(\phi_1 + \phi_2)] \\
&= (A_1A_2)(A_1A_4)[\cos \phi_1 \cos \phi_2 - \sin \phi_1 \sin \phi_2] \\
&= (A_1A_2 \cos \phi_1)(A_1A_4 \cos \phi_2) - (A_1A_2 \sin \phi_1)(A_1A_4 \sin \phi_2) \\
&= a_{11}\, du\, a_{22}\, dv - a_{21}\, du\, a_{12}\, dv
\end{aligned}
$$

Thus, the area of parallelogram A becomes $|J|\, du\, dv$, where J is the Jacobian of the transformation (see Appendix A) given by

$$J = \begin{bmatrix} a_{11} & a_{12} \\ a_{21} & a_{22} \end{bmatrix} = \begin{vmatrix} \dfrac{\partial x}{\partial u} & \dfrac{\partial x}{\partial v} \\ \dfrac{\partial y}{\partial u} & \dfrac{\partial y}{\partial v} \end{vmatrix} = \dfrac{\partial(x,y)}{\partial(u,v)}$$

Using the area of A in $P(A)$, we obtain

$$f_{UV}(u, v) = |J| \, f_{XY}(x, y)$$

The function $f_{XY}(x, y)$ becomes $f_{XY}[G(u, v), H(u, v)]$ by straight-forward substitution. Therefore the joint density function of (U, V) is

$$f_{XY}[G(u, v), H(u, v)] \left| \dfrac{\partial(x,y)}{\partial(u,v)} \right|$$

Note that in order to calculate the determinant of the Jacobian matrix, we can use the relation (see Property A.34)

$$\left| \dfrac{\partial(x,y)}{\partial(u,v)} \right| = \left| \dfrac{\partial(u,v)}{\partial(x,y)} \right|^{-1}$$

because the determinant of the inverse of a matrix is the reciprocal of the determinant of the matrix. The Jacobian of the inverse transformation is the inverse of the Jacobian of the direct transformation. Also note that the absolute value of J must be used because probability measures are nonnegative.

Example 5.4

Let X and Y be two independent random variables each distributed exponentially with the parameter 1. Derive the joint and marginal densities of $U = 2(X + Y)$ and $V = X/Y$. Are they independent?

The joint density of X and Y is given by $f_{XY}(x, y) = e^{-(x+y)}$. The inverse transformations are $x = uv/[2(v+1)]$ and $y = u/[2(v+1)]$. The Jacobian of the transformation is given by

$$|J| = \left| \dfrac{\partial(x, y)}{\partial(u, v)} \right| = \begin{vmatrix} \dfrac{v}{2(v+1)} & \dfrac{u}{2(v+1)^2} \\ \dfrac{1}{2(v+1)} & -\dfrac{u}{2(v+1)^2} \end{vmatrix} = -\dfrac{u}{4(v+1)^2}$$

The joint density of U and V is therefore given by

$$f_{UV}(u, v) = \frac{ue^{-u/2}}{4(v+1)^2} \qquad u, v \geq 0$$

We note that the joint density is separable in U and V. It follows from Theorem 5.1 that the two random variables are independent. Integration with respect to v gives the density function of U. We have (verify them)

$$f_U(u) = \tfrac{1}{4}u\,e^{-u/2} \qquad \text{and} \qquad f_V(v) = \frac{1}{(v+1)^2}$$

Practice Problem

5.8   Compare the densities for U and V with the ones in Table 4.1 and verify that U has a chi-square distribution. Does V have a recognizable distribution?

5.6   THE CONVOLUTION FORMULA

A frequent application of the bivariate transformation is to obtain the distribution of the sum of two random variables. Thus, given that $f_{XY}(x, y)$ is the joint density of X and Y, we wish to find the density of the random variable $X + Y$.

Theorem 5.11

If $f_{XY}(x, y)$ is the joint density function of X and Y, then the density function of $U = X + Y$ is

$$f_U(u) = \begin{cases} \int f_{XY}(u-v,\, v)\, dv & \text{continuous case} \\[2ex] \sum_v f_{XY}(u-v,\, v) & \text{discrete case} \end{cases}$$

Proof: Because the discrete case is trivial, only the continuous case is proved here. Theorem 5.10 is the relevant one here, but it requires two transformations (in general as many transformations as the number of random variables). The trick is to use a dummy transformation $V = Y$ along with the original one $U = X + Y$. The Jacobian for this pair of transformations is

$$\frac{\partial(X, Y)}{\partial(U, V)} = \begin{bmatrix} 1 & -1 \\ 0 & 1 \end{bmatrix} = 1$$

Therefore, $f_{UV}(u, v) = f_{XY}(u-v, v)$.   The density function of U is the marginal distribution of (U, V). To obtain this, we have to sum or integrate the above over the v domain. Therefore,

$$f_U(u) \ = \ \int f_{XY}(u-v, v) \, dv \qquad \text{or} \qquad \Sigma_v \, f_{XY}(u-v, v)$$

The above relationship, which gives the density function of $X+Y$ in terms of the joint density of X and Y, is called the **convolution formula**. Using this basic approach, we can get the density functions of $X-Y$, X/Y, $X/(X+Y)$, and so on.

Example 5.5

Suppose X and Y are independently distributed as Uniform $[0, T]$. We derive the density function of $U = X + Y$ using two different approaches. The first method is to apply the technique of introducing a dummy variable, as was done for Theorem 5.11.

Method 1:   If we introduce the dummy transformation $V = Y$, then $X = U - V$. The joint density of U and V is $f_{UV}(u, v) = 1/T^2$. The following inequalities hold.

$$0 \le v \le T, \quad 0 \le u \le 2T, \quad 0 \le u-v \le T, \quad u-T \le v \le u$$

The feasible area is given in Figure 5.3. For a given u, if $u \le T$, then $0 < v < u$. Therefore,

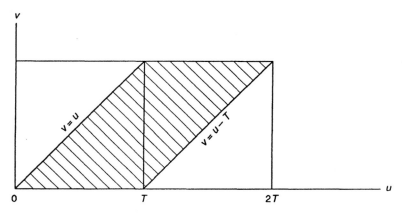

Figure 5.3   Feasible area in (U, V) space

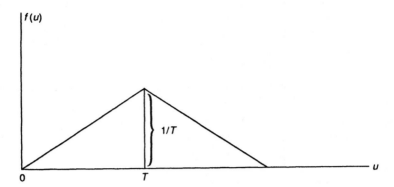

Figure 5.4   Triangular distribution

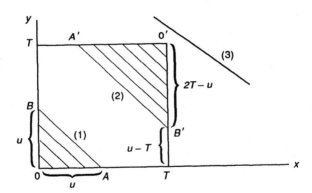

Figure 5.5   Feasible areas in (X, Y) space

$$f_U(u) = \int_0^u \frac{1}{T^2}\, dv = \frac{u}{T^2} \qquad \text{for } 0 \leq u \leq T$$

When $u > T$, we have $u - T \leq v \leq T$. In this case,

$$f_U(u) = \int_{u-T}^T \frac{1}{T^2}\, dv = \frac{2T-u}{T^2} \qquad \text{for } T \leq u \leq 2T$$

Figure 5.4 graphs this density function, known as the **triangular distribution**.

Method 2: In this method we first derive the CDF of U. $F_U(u) = P(X + Y \le u)$. $X + Y = u$ is a straight line that is one of the three lines marked in Figure 5.5. The domain for U is $[0, 2T]$, that is $0 \le u \le 2T$. If $X + Y = u \le T$, the line marked (1) is relevant. $P(X + Y \le u)$ is then the area of the shaded triangle OAB divided by T^2 [because $f_{XY}(x, y) = 1/T^2$].

$$F_U(u) = \frac{u^2}{2T^2} \text{ if } 0 \le u \le T \quad \text{and hence} \quad f_U(u) = \frac{u}{T^2}$$

If $T < X + Y \le 2T$, line (2) is in force and

$$F_U(u) = 1 - \frac{\text{Area } O'A'B'}{T^2} = 1 - \frac{1}{2T^2}(2T - u)^2$$

Therefore,

$$f_U(u) = F_U'(u) = \frac{2T - u}{T^2} \quad \text{if } T \le u \le 2T$$

If $u > 2T$, line (3) is in effect and then $F_U(u) = 1$ and hence $f_U(u) = 0$.

Practice Problems

5.9 Suppose X and Y are independent standard normal random variables. Use the approach adopted in Theorem 5.11 and show that $U = X/Y$ has the standard Cauchy distribution (see Table 4.1).

5.10 Suppose X and Y are independently and identically distributed with the density function $f_X(x) = k(x/2)^{-1/2}e^{-x/2}$, with $x > 0$, where k is a constant. Derive the density function of $X + Y$ and identify what it is from Table 4.1.

5.11 Using the same $f_X(x)$ as in Practice Problem 5.10, obtain the density function of X/Y and identify the distribution to which it corresponds.

5.12 Let $X \sim N(0, 1)$ and be independent of Y for which $f_Y(y) = k(y/2)^{-1/2}e^{-y/2}$, where k is a constant and $-\infty < x < \infty$, $y > 0$. Derive the density function of $U = X/\sqrt{Y}$.

5.7 MIXTURE DISTRIBUTIONS

In several applications of probability theory the distribution of random variables might depend on parameters or variables which themselves depend on other random variables. We saw a very simple case of this in Example 4.2. There we had a binomial random variable Y for which the probability of observing y successes out of n trials was $\binom{n}{y} p^y (1-p)^{n-y}$. But p depended on another random variable X that followed an exponential distribution with parameter θ, more specifically, $p = P(X \geq 500)$. Such a combination of distributions is called a **mixture distribution**. This might take the form $f(x;\theta)$ where θ depends on a random variable or the form $f(x \mid y)$, where Y is another random variable. In this section we explain the notion of mixing and illustrate it with an example. Casella and Berger (1990) have a number of other examples. Also, Crow and Shimuzu (1988) have an example of mixing Poisson and lognormal distributions.

Suppose $f_{X \mid Y}(x, y)$ is the conditional density of X given that $Y = y$. Let $f_Y(y)$ be the marginal density of Y. Then the joint density of X and Y is given by

$$f_{XY}(x, y) = f_{X \mid Y}(x, y) f_Y(y)$$

Our main interest, however, is in the marginal (that is unconditional) density of X. For continuous random variables, this is obtained as

$$f_X(x) = \int f_{XY}(x, y)\, dy = \int f_{X \mid Y}(x, y) f_Y(y)\, dy$$

Example 5.6

Suppose $X \mid Y$ is Poisson with

$$f_{X \mid Y}(x, y) = \frac{e^{-y} y^x}{x!} \qquad x = 0, 1, 2, \ldots$$

and Y has a gamma density with (α is a positive integer)

$$f_Y(y) = \frac{\theta^\alpha}{\Gamma(\alpha)} y^{\alpha-1} e^{-\theta y} \qquad y > 0$$

Then the joint density and the marginal density of X are given by

$$f_{XY}(x, y) = \frac{\theta^\alpha}{\Gamma(\alpha)} \frac{e^{-y(\theta+1)} y^{x+\alpha-1}}{x!}$$

$$f_X(x) = \frac{\theta^\alpha}{\Gamma(\alpha) x!} \int_0^\infty e^{-y(\theta+1)} y^{x+\alpha-1}\, dy$$

The substitution $u = y(\theta+1)$ yields the following.

$$f_X(x) = \frac{\theta^\alpha}{\Gamma(\alpha) x!} \int_0^\infty e^{-u} \frac{u^{x+\alpha-1}}{(\theta+1)^{x+\alpha-1}} \frac{du}{(\theta+1)}$$

$$f_X(x) = \frac{\theta^\alpha}{\Gamma(\alpha)\,x!}\frac{1}{(\theta+1)^{x+\alpha}}\int_0^\infty e^{-u}u^{x+\alpha-1}\,du$$

$$= \frac{\theta^\alpha}{\Gamma(\alpha)\,x!}\frac{1}{(\theta+1)^{x+\alpha}}\Gamma(x+\alpha)\left[\frac{1}{\Gamma(x+\alpha)}\int_0^\infty e^{-u}u^{x+\alpha-1}du\right]$$

The expression in square brackets is 1 because the integrand is the Gamma density (see Table 4.1). The above expression therefore simplifies to [making use of the result from Theorem 4.1 that $\Gamma(\alpha) = (\alpha-1)!$]

$$f_X(x) = \frac{(x+\alpha-1)!}{x!\,(\alpha-1)!}\left[\frac{\theta}{\theta+1}\right]^\alpha\left[\frac{1}{\theta+1}\right]^x = \binom{x+\alpha-1}{\alpha-1}\left[\frac{\theta}{\theta+1}\right]^\alpha\left[\frac{1}{\theta+1}\right]^x$$

which is the density function of the negative binomial with $p = \theta/(\theta+1)$. It follows that the mixture of Poisson and gamma distributions leads to a negative binomial distribution.

5.8   BIVARIATE CHARACTERISTIC FUNCTIONS

Definition 5.9

The **joint characteristic function of** (X, Y) is defined as

$$\phi_{XY}(t_1, t_2) = E_{XY}\left[e^{it_1X + it_2Y}\right]$$

which, for a continuous pair of random variables, is

$$\int_{-\infty}^\infty\int_{-\infty}^\infty e^{it_1X + it_2Y}f_{XY}(x, y)\,dxdy$$

We prove a number of theorems on bivariate characteristic functions that are useful in later applications.

Theorem 5.12

The random variables X and Y are statistically independent if and only if their joint characteristic function is the product of the characteristic functions of the corresponding marginal distributions, that is, $\phi_{XY}(t_1, t_2) = \phi_X(t_1)\phi_Y(t_2)$.

Proof: If X and Y are independent,

$$\phi_{XY}(t_1, t_2) = \int_{-\infty}^\infty\int_{-\infty}^\infty e^{it_1X + it_2Y}f_X(x)f_Y(y)\,dxdy$$

We can integrate separately and get $\phi_X(t_1)\phi_Y(t_2)$ on the right-hand side.

The converse is also true. Since $\phi_{XY}(t_1, t_2) = \phi_X(t_1)\phi_Y(t_2)$, by retracing the above steps we get,

$$\phi_{XY}(t_1, t_2) = \int_{-\infty}^{\infty} \int_{-\infty}^{\infty} e^{it_1 X + it_2 Y} f_X(x) f_Y(y)\, dx dy$$

Because the characteristic function is unique in the multivariate case also, the density function $f_{XY}(x, y) = f_X(x)f_Y(y)$, thus implying independence.

Theorem 5.13

The characteristic function of a bivariate normal is

$$\phi_{XY}(t_1, t_2) = \exp\left[it_1\mu_X + it_2\mu_Y - \tfrac{1}{2}(t_1^2\sigma_X^2 + t_2^2\sigma_Y^2 + 2\rho t_1 t_2 \sigma_X \sigma_Y)\right]$$

Practice Problem

5.13 Prove Theorem 5.13.

Theorem 5.13 is useful in establishing the fact that a linear combination of a bivariate normal random variable is a univariate normal random variable. This is proved below.

Theorem 5.14

If (X, Y) is bivariate normal, then $U = aX + bY$ has the univariate normal distribution with mean $a\mu_X + b\mu_Y$ and variance $a^2\sigma_X^2 + 2ab\rho\sigma_X\sigma_Y + b^2\sigma_Y^2$.

Proof: $\phi_U(t) = E_{XY}(e^{iUt}) = E_{XY}\left[e^{iaXt + ibYt}\right] = \phi_{XY}(at, bt)$

$$= \exp[it(a\mu_X + b\mu_Y) - \tfrac{1}{2}(a^2\sigma_X^2 + 2ab\rho\sigma_X\sigma_Y + b^2\sigma_Y^2)t^2]$$

But this is the characteristic function for the univariate normal.

Theorem 5.15

If $\phi_{XY}(t_1, t_2)$ is the characteristic function of (X, Y), then the characteristic functions of the marginal distributions of X and Y are respectively $\phi_{XY}(t, 0)$ and $\phi_{XY}(0, t)$.

Proof: $\phi_X(t) = E_X(e^{iXt}) = \int_{-\infty}^{\infty} e^{ixt} f_X(x)\, dx$

$$= \int_{-\infty}^{\infty} e^{ixt} \int_{-\infty}^{\infty} f_{XY}(x, y)\, dxdy = \phi_{XY}(t,\, 0)$$

Similarly for Y.

Theorem 5.16

Let X and Y be random variables with joint characteristic function $\phi_{XY}(t_1, t_2)$. Then X and Y are statistically independent if and only if $\phi_{XY}(t_1, t_2) = \phi_{XY}(t_1, 0)\,\phi_{XY}(0, t_2)$.

Proof: This theorem follows readily from Theorems 5.12 and 5.15.

Theorem 5.17

If $\phi_X(t)$ and $\phi_Y(t)$ are the characteristic functions of the two independent random variables X and Y respectively and $\phi_{X+Y}(t)$ is the characteristic function of the random variable $X + Y$, then $\phi_{X+Y}(t) = \phi_X(t)\,\phi_Y(t)$.

The proof of this theorem follows trivially by setting $t_1 = t_2$ in each step of the proof of Theorem 5.12. *The converse, however, need not be true as will be seen from Exercise 5.8.*

Theorem 5.18

If the assumptions of Theorem 5.10 hold, then

$$\phi_{UV}(t_1, t_2) = E_{XY}\left[e^{it_1 g(X, Y) + it_2 h(X, Y)}\right]$$

This theorem is far from trivial because, by definition, $\phi_{UV}(t_1, t_2) = E_{UV}[e^{it_1 U + it_2 V}]$. How do we know then that it is legitimate to take the expectation over XY?

Proof: $\phi_{UV}(t_1, t_2) = \int_U \int_V e^{it_1 u + it_2 v} f_{UV}(u, v)\, dudv$.   In proving Theorem 5.10 we showed that $f_{UV}(u, v) = f_{XY}(x, y)|J|$ and that $dxdy = |J|dudv$. With these, carry out the integration over the (X, Y) space.

$$\phi_{UV}(t_1, t_2) = \iint_{XY} e^{it_1 g(x, y) + it_2 h(x, y)} f_{XY}(x, y)\, dxdy$$

$$= E_{XY}[e^{it_1 g(X, Y) + it_2 h(X, Y)}]$$

For a better understanding of the material covered so far on bivariate distributions, it would be extremely useful to do exercises 5.1 through 5.16 at this point.

5.9   MULTIVARIATE DENSITY FUNCTIONS

All the results derived for the bivariate case can be generalized to n random variables. In the rest of the chapter, concepts and results from matrix algebra reviewed in Appendix A are used even more extensively. The level of technical difficulty is also considerably higher, but the details presented here are essential to the foundations of multivariate and multiparameter analyses of later chapters.

The joint density function of $X_1, X_2, \ldots, X_n$ will have the form $f(x_1, x_2, \ldots, x_n)$. If $P(X_i \leq x_i$ for *all* $i) = F_X(x_1, \ldots, x_n)$, then F is the CDF. If the Xs are continuous random variables,

$$f_X(x_1, x_2, \ldots, x_n) = \frac{\partial^n F_X(x_1, x_2, \ldots, x_n)}{\partial x_1 \partial x_2 \cdots \partial x_n}$$

Expected values, marginal densities, and conditional densities are all defined similarly. If $f(x_1, \ldots, x_n) = \Pi_i f_i(x_i)$, then the Xs are independent by definition.

The multivariate characteristic function is $\phi_X(t) = E_X(e^{it'X})$, where t is a column vector and X is a column vector of Xs, both $n \times 1$, and $t'X = t_1 X_1 + t_2 X_2 + \cdots + t_n X_n$ is the innerproduct. If the Xs are independent, then $\phi_X(t) = \Pi_{i=1}^{i=n} \phi_{X_i}(t_i)$ and conversely. Similar to Theorem 5.17, we have

$$\phi_{X_1+X_2+\cdots+X_n}(t) = \prod_{i=1}^{n} \phi_{X_i}(t)$$

if all the Xs are independent. The general version of Theorem 5.10 for multivariate transformations is

$$f_U(u) = f_X[G(u)] \left| \frac{\partial x}{\partial u} \right|$$

in vector notation, or more completely, let the transformations be

$$U_1 = g_1(X_1, X_2, \ldots, X_n)$$
$$U_2 = g_2(X_1, X_2, \ldots, X_n)$$
$$\cdots\cdots\cdots\cdots$$
$$U_n = g_n(X_1, X_2, \ldots, X_n)$$

and the inverse transformations be $X_i = G_i(U_1, U_2, \ldots, U_n)$. Then the joint density function $f_U(u_1, \ldots, u_n)$ is given by

$$f_X[G_1(u_1, \ldots, u_n), G_2(u_1, \ldots, u_n), \ldots G_n(u_1, \ldots, u_n)] \, |J|$$

where $|J| = \left| \dfrac{\partial(x_1, x_2, \ldots, x_n)}{\partial(u_1, u_2, \ldots, u_n)} \right|$.

The characteristic function of a marginal density can be obtained analogously to Theorem 5.15. For example,

$$\phi_{X_1, X_2}(t_1, t_2) = \phi_X(t_1, t_2, 0, 0, \ldots, 0)$$

5.10   THE MULTIVARIATE NORMAL DISTRIBUTION

In econometrics, the most widely used distribution is the multivariate normal distribution. Here we first state a few general results and then derive a variety of results for the multivariate distribution. The rest of this chapter makes considerable use of matrices and determinants (see Appendix A for a review).

Definition 5.10 (mean vector)

Let $X' = (X_1, X_2, \ldots, X_n)$ be an n-dimensional vector random variable defined in $\mathcal{R}^n$ with a density function $f_X(x)$, $E(X_i) = \mu_i$, and $\mu' = (\mu_1, \mu_2, \ldots, \mu_n)$. Then the mean of the distribution is $\mu = E(X)$, where μ and $E(X)$ are $n \times 1$ vectors, and hence $E(X-\mu) = 0$.

Definition 5.11 (covariance matrix)

The covariance between X_i and X_j is defined as $\sigma_{ij} = E[(X_i - \mu_i)(X_j - \mu_j)]$, where $\mu_i = E(X_i)$. The matrix

$$\Sigma = \begin{bmatrix} \sigma_{11} & \sigma_{12} & . & . & . & \sigma_{1n} \\ \sigma_{21} & \sigma_{22} & . & . & . & \sigma_{2n} \\ . & . & . & . & . & . \\ \sigma_{n1} & \sigma_{n2} & . & . & . & \sigma_{nn} \end{bmatrix}$$

*also denoted as Var(X), is called the **covariance matrix of X**. In matrix notation, this can be expressed as $\Sigma = E[(X-\mu)(X-\mu)']$, where $X-\mu$ is $n \times 1$. [Note: Diagonal elements are variances.]*

Several properties of μ and Σ are listed.

Property 5.1:   *If* $\underset{m \times 1}{Y} = \underset{m \times n}{A} \ \underset{n \times 1}{B} + \underset{m \times 1}{b}$ *, then* $E(Y) = A\mu + b$.

Property 5.2: Σ *is a symmetric positive semi-definite matrix.*

   Proof: If c is a constant $n \times 1$ vector,

$$\text{Var}(c'X) = E\,[(c'X - c'\mu)(c'X - c'\mu)']$$
$$= E\,[c'(X-\mu)(X-\mu)'c] = c'\Sigma c$$

   Because variance is a nonnegative number, this quadratic form is nonnegative for all c, implying that Σ is positive semi-definite. If $c'\Sigma c = 0$ for some c, then $\text{Var}(c'X) = 0$, which implies linear dependence.

Property 5.3: Σ *is positive definite if and only if it is nonsingular* [this follows from Property A.32].

Property 5.4: $\Sigma = E(XX') - \mu\mu'$.

   Proof: $E(X-\mu)(X-\mu)' = E(XX' - X\mu' - \mu X' + \mu\mu')$

$$= E(XX') - E(X)\mu' - \mu E(X') + \mu\mu' = E(XX') - \mu\mu'$$

Note that because μ is a constant, $E(X\mu') = E(X)\mu' = \mu\mu'$.

Property 5.5:   *If* $Y = AX + b$, *then the covariance matrix of* Y *is* $A\Sigma A'$.

   Proof: We have, $E(Y) = A\mu + b$. Hence,

$$\text{Var}(Y) = E\,[(AX - A\mu)(AX - A\mu)'] = E\,[A(X-\mu)(X-\mu)'A']$$
$$= AE\,[(X-\mu)(X-\mu)']A' = A\Sigma A'$$

   Note that, because A is a linear transformation matrix and the expectation is a linear operator, we can take expectations inside a matrix multiplication.

Approximate Mean and Covariance of $g(X)$

   The concept of approximating the mean and variance of measurable functions of random variables is easily extended to the multivariate case also. Let $g_i(X) = g_i(X_1, X_2, \ldots, X_n)$, for $i = 1, 2, \ldots, m$ be

m measurable functions of the random vector X. The n partial derivatives $g_{ij} = \partial g_i / \partial x_j$ for $j = 1, 2, \ldots, n$ may be arranged as an $n \times 1$ vector (call it q_i) so that

$$q_i' = \left[\frac{\partial g_i}{\partial x_1}, \frac{\partial g_i}{\partial x_2}, \ldots, \frac{\partial g_i}{\partial x_n} \right]$$

The generalization of the linear approximation to the n-variate case is

$$g_i(X) \approx g_i(\mu) + \sum_{j=1}^{j=n} g_{ij}(X_j - \mu) = g_i(\mu) + q_i'(X - \mu)$$

An approximation to the expected value is given by $E[g_i(X)] \approx g_i(\mu)$ or in vector notation, $E[g(X)] \approx g(\mu)$. Applying Property 5.5, the approximate covariance matrix is, $\text{Var}[g(X)] \approx Q \Sigma Q'$, where $Q = [\partial g_i / \partial x_j]$ is the $m \times n$ matrix of the partial derivates $\partial g_i / \partial x_j$.

We reiterate the caveat presented in earlier sections regarding the dangers of approximations. They could lead to seriously misleading implications.

The Multivariate Normal Distribution

Let $X_1, X_2, \ldots, X_n$ be n independent random variables each of which is $N(0,1)$. Then their joint density function is the product of individual density functions and is the **standard multivariate normal density**.

$$f_X(x) = f_X(x_1, x_2, \ldots, x_n) = \left[\frac{1}{\sqrt{2\pi}} \right]^n e^{-\Sigma x_i^2 / 2} = \left[\frac{1}{2\pi} \right]^{n/2} e^{-x'x/2}$$

The corresponding characteristic function is $e^{-\Sigma t_i^2 / 2}$ or $e^{-t't/2}$.

Now make the linear transformation $Y = AX + \mu$, where A and μ are constant. By assumption, $E(X) = 0$ and $E(XX') = I_n$, the identity matrix. Therefore $E(Y) = \mu$ and $E[(Y-\mu)(Y-\mu)'] = E(AXX'A') = AA' = \Sigma$ (say). Thus the vector Y has mean μ and covariance matrix Σ. Let A be nonsingular so that A^{-1} exists. The inverse transformation is $X = A^{-1}(Y-\mu)$ and $AA' = \Sigma$. To obtain the joint density function, we need the Jacobian. But because the transformation is linear, the Jacobian is A^{-1} and its determinant is $1/|A|$ (see Property A.34). The density function of Y is therefore

$$f_Y(y) = \frac{1}{(2\pi)^{n/2}} |A|^{-1} \exp \left[\frac{-(y-\mu)'(A^{-1})'A^{-1}(y-\mu)}{2} \right]$$

$$f_Y(y) = \frac{1}{(2\pi)^{n/2}|A|} \exp\left[\frac{-(y-\mu)'\Sigma^{-1}(y-\mu)}{2}\right]$$

because $\Sigma = AA'$, and hence $\Sigma^{-1} = (A')^{-1}A^{-1} = (A^{-1})'A^{-1}$ by Properties A.5 and A.6. Also, because $|A'| = |A|$ by Property A.20, $|\Sigma| = (|A|)^2$, which means that $|A| = |\Sigma|^{1/2}$. We thus have the density function of the **general multivariate normal distribution N(μ, Σ)** as

$$f_Y(y) = \frac{1}{(2\pi)^{n/2}|\Sigma|^{1/2}} \exp\left[\frac{-(y-\mu)'\Sigma^{-1}(y-\mu)}{2}\right]$$

where the mean vector of Y is μ and the covariance matrix of Y is Σ.

Characteristic Function of N(μ, Σ)

The characteristic function of Y can be derived from that of X.

$$\phi_Y(t) = E_X(e^{iY't}) = E_X[e^{i(AX+\mu)'t}] = e^{i\mu't}E_X(e^{iX'A't})$$

$$= e^{i\mu't}\ \phi_X(A't) = e^{i\mu't}\ \exp[-(A't)'(A't)/2]$$

Noting that $AA' = \Sigma$, we have $\phi_Y(t) = \exp\left[i\mu't - (t'\Sigma t/2)\right]$.

Property 5.6

If Y is multivariate normal, then $Y_1, Y_2, \ldots, Y_n$ will be independent if and only if Σ is diagonal. [This is easy to see by inspection of $\phi_Y(t)$.]

Property 5.7

A linear combination of multivariate normal random variables is also multivariate normal. More specifically, let $Y \sim N(\mu, \Sigma)$. Then $Z = AY \sim N(A\mu, A\Sigma A')$, where A is an $n \times n$ matrix.

Proof:

$$\phi_Z(t) = E(e^{iZ't}) = E(e^{iY'A't}) = \phi_Y(A't) = \exp\left[i\mu'A't - t'A\Sigma A't/2\right]$$

which is the characteristic function of $N(A\mu, A\Sigma A')$.

Property 5.8

If $Y \sim N(\mu, \Sigma)$ and Σ has rank $k < n$, then there exists a nonsingular $k \times k$ matrix A such that the $k \times 1$ matrix $X = [A^{-1}\ O](Y-\mu)$ is *a k-variate normal with zero mean and covariance matrix I_k, where O is a $k \times (n-k)$ matrix of zeros.*

Proof: If Σ has rank k, then, from Property A.17 of Appendix A, there exists a nonsingular matrix A such that AA' is a submatrix of Σ that is nonsingular. It is easy to verify from Property 5.7 that $X \sim N(0, I_k)$.

Marginal and Conditional Distributions of $N(\mu, \Sigma)$

Let $Y \sim N(\mu, \Sigma)$, and consider the following partition

$$Y = \begin{bmatrix} Y_1 \\ Y_2 \end{bmatrix} ; \quad \mu = \begin{bmatrix} \mu_1 \\ \mu_2 \end{bmatrix} ; \quad \Sigma = \begin{bmatrix} \Sigma_{11} & \Sigma_{12} \\ \Sigma_{21} & \Sigma_{22} \end{bmatrix}$$

where the n random variables are partitioned into n_1 and n_2 variates $(n_1 + n_2 = n)$.

Theorem 5.19

Given the above partition, the marginal distribution of Y_1 is $N(\mu_1, \Sigma_{11})$ and the conditional density of Y_2 given Y_1 is multivariate normal with mean $\mu_2 + \Sigma_{21}\Sigma_{11}^{-1}(Y_1-\mu_1)$ and covariance matrix $\Sigma_{22} - \Sigma_{21}\Sigma_{11}^{-1}\Sigma_{12}$.

Before proving this theorem we need two lemmas.

Lemma 5.1

The partitioned inverse of Σ can be written as follows, which is readily derived from Property A.8 of Appendix A.

$$\Sigma^{-1} = \begin{bmatrix} \Sigma_{11}^{-1} + FEF' , & -FE \\ -EF' , & E \end{bmatrix} = \begin{bmatrix} \Sigma_{11}^{-1} & 0 \\ 0 & 0 \end{bmatrix} + \begin{bmatrix} -F \\ I_{n_2} \end{bmatrix} E \begin{bmatrix} -F' , & I_{n_2} \end{bmatrix}$$

where $F = \Sigma_{11}^{-1}\Sigma_{12}$, and $E = (\Sigma_{22}-\Sigma_{21}\Sigma_{11}^{-1}\Sigma_{12})^{-1}$.

Lemma 5.2: $|\Sigma| = |\Sigma_{11}| \cdot |E^{-1}|$.

See Maddala (1977), pp. 446–447 for a proof of this.

Proof of Theorem 5.19:   Consider the quadratic form $Q = (Y-\mu)'\Sigma^{-1}(Y-\mu)$. Using Lemma 5.1 it can be written as

$$Q = [(Y_1-\mu_1)', (Y_2-\mu_2)'] \begin{bmatrix} \Sigma_{11}^{-1} & 0 \\ 0 & 0 \end{bmatrix} \begin{bmatrix} Y_1-\mu_1 \\ Y_2-\mu_2 \end{bmatrix}$$

$$+ [(Y_1-\mu_1)', (Y_2-\mu_2)'] \begin{bmatrix} -F \\ I_{n_2} \end{bmatrix} E[-F', I_{n_2}] \begin{bmatrix} Y_1-\mu_1 \\ Y_2-\mu_2 \end{bmatrix}$$

Expanding these we obtain the following:

$$Q = (Y_1-\mu_1)'\Sigma_{11}^{-1}(Y_1-\mu_1) +$$

$$[(Y_2-\mu_2) - F'(Y_1-\mu_1)]' \, E \, [Y_2-\mu_2 - F'(Y_1-\mu_1)]$$

Call the first term Q_1 and the second term Q_2. The joint density of (Y_1, Y_2) is

$$f_Y(y_1, y_2) = \frac{1}{(2\pi)^{n/2}|\Sigma|^{1/2}} \exp[-Q/2]$$

From Lemma 5.2,

$$|\Sigma|^{1/2} = |\Sigma_{11}|^{1/2}|E|^{-1/2}$$

Hence $f_Y(y_1, y_2)$ can be rewritten as follows.

$$\frac{1}{(2\pi)^{n_1/2}|\Sigma_{11}|^{1/2}} \exp(-Q_1/2) \cdot \frac{1}{(2\pi)^{n_2/2}|E|^{-1/2}} \exp(-Q_2/2)$$

$$= f_{Y_1}(y_1) \cdot f_{Y_2|Y_1}(y_1, y_2)$$

Note that the first part involves only Y_1 and Σ_{11}. If you fix Y_1 in the second part, it has the form of the multivariate normal density of Y_2 *for a given* Y_1 which integrates to 1 with respect to Y_2. The first part is therefore the marginal density of Y_1. It follows from the definition of joint probability that the second part is the conditional density of Y_2 given Y_1. It is evident that they are both multivariate normal. Hence

$$Y_1 \sim N(\mu_1, \Sigma_{11})$$

$$Y_2|Y_1 \sim N[\mu_2 + F'(Y_1-\mu_1), E^{-1}]$$

Because $F = \Sigma_{11}^{-1}\Sigma_{12}$ and $E^{-1} = \Sigma_{22} - \Sigma_{21}\Sigma_{11}^{-1}\Sigma_{12}$, we have

$$Y_2|Y_1 \sim N[\mu_2 + \Sigma_{21}\Sigma_{11}^{-1}(Y_1-\mu_1), \Sigma_{22} - \Sigma_{21}\Sigma_{11}^{-1}\Sigma_{12}]$$

The next step is to derive the distribution of the quadratic form $(Y-\mu)'\Sigma^{-1}(Y-\mu)$, which is very useful in testing hypotheses. But before being able to do that we have to study another important distribution called **chi-square**. This is done next.

5.11   THE CHI-SQUARE DISTRIBUTION

The **chi-square distribution** is a special case of the gamma distribution introduced in Chapter 4. The density function for gamma is

$$f_X(x) = \frac{1}{\beta^\alpha \Gamma(\alpha)} x^{\alpha-1} e^{-(x/\beta)} \qquad x > 0, \qquad \alpha, \beta > 0$$

where $\Gamma(\alpha)$ is the gamma function $= \int_0^\infty y^{\alpha-1} e^{-y} dy$ for $\alpha > 0$. Consider the special case when $\alpha = n/2$ and $\beta = 2$. The density function then becomes

$$f_X(x) = \frac{(x/2)^{(n/2)-1} e^{-x/2}}{2\Gamma(n/2)} \qquad x > 0$$

The above distribution is called the chi-square distribution with n **degrees of freedom** and is written as χ_n^2.

An even more special case arises when the degree of freedom is 1. It is easily shown that if $Z \sim N(0,1)$, then $Y = Z^2 \sim \chi_1^2$. Thus the square of a standard normal is chi-square with 1 d.f. To show this, first note that the distribution function of Y is

$$F_Y(y) = P(Y \le y) = P(Z^2 \le y) = P(-\sqrt{y} \le Z \le \sqrt{y})$$
$$= 2P(Z \le \sqrt{y}) = 2F_Z(\sqrt{y})$$

because of the symmetry of $N(0, 1)$ around the origin. The density function is

$$f_Y(y) = \frac{dF_Y(y)}{dy} = 2F_Z'(\sqrt{y}) \frac{1}{2\sqrt{y}}$$
$$= \frac{1}{\sqrt{y}} f_Z(\sqrt{y}) = \frac{1}{\sqrt{2\pi}} y^{-1/2} e^{-y/2}$$

This can be written as [noting from Theorem 4.1 that $\Gamma(\frac12) = \sqrt{\pi}$]

$$f_Y(y) = \frac{1}{2\sqrt{\pi}} (y/2)^{-1/2} e^{-y/2} = \frac{1}{2\Gamma(\frac12)} (y/2)^{-1/2} e^{-y/2}$$

which is the same as the density of χ_1^2.

Practice Problem

5.14 Show that the characteristic function of χ_1^2 is $(1 - 2it)^{-1/2}$.

Theorem 5.20: *The characteristic function of χ_n^2 is $(1-2it)^{-n/2}$.*

Proof: Using the density function given above, we have

$$\phi(t) = \int_0^\infty \frac{(x/2)^{(n/2)-1}e^{-x/2}}{2\Gamma(n/2)} e^{ixt}\, dx$$

$$= \int_0^\infty \frac{(x/2)^{(n/2)-1}e^{-x(1-2it)/2}}{2\Gamma(n/2)}\, dx$$

Define $y = (1-2it)x$. Thus the above integral becomes

$$\phi(t) = \left[\int_0^\infty \frac{e^{-y/2}(y/2)^{(n/2)-1}}{2\Gamma(n/2)}\, dy \right] (1-2it)^{-n/2}$$

But the expression in the square brackets is 1 because the integrand is the density for χ_1^2. Therefore $\phi(t) = (1-2it)^{-n/2}$.

Corollary

The chi-square distribution has the **additive property,** *that is, if $X \sim \chi_m^2$, $Y \sim \chi_n^2$, and X and Y are independent, then their sum $X+Y \sim \chi_{m+n}^2$. Thus the sum of independent chi-square is also chi-square with d.f. as the sum of the d.f.*

Proof: From Theorem 5.17, the characteristic function of the sum of independent random variables is the product of their individual characteristic functions. Therefore,

$$\phi_{X+Y}(t) = \phi_X(t)\phi_Y(t) = (1-2it)^{-m/2}(1-2it)^{-n/2}$$

$$= (1-2it)^{-(m+n)/2}$$

which is the characteristic function of a chi-square distribution with $m+n$ d.f.

Theorem 5.21

Let $Z_1, Z_2, \ldots, Z_n$ be independently and identically distributed as $N(0, 1)$. Then the distribution of $X = \Sigma Z_i^2$ is χ_n^2. Thus the distribution of the sum of squares of n independent standard normal random variables is χ_n^2.

Proof: Since Z_i is $N(0, 1)$, $Y_i = Z_i^2$ is χ_1^2. It follows by the additive property that $X = \Sigma\, Y_i = \Sigma\, Z_i^2$ is χ_n^2.

Theorem 5.22

If $X_i \sim N(\mu_i, \sigma_i^2)$, $i = 1, 2, \ldots, n$ and $X_1, X_2, \ldots, X_n$ are all independent, then $Y = \Sigma_{i=1}^{i=n}\, [(X_i - \mu_i)/\sigma_i]^2$ has the chi-square distribution with n d.f.

Proof: Let $Z_i = (X_i - \mu_i)/\sigma_i$. Then $Z_i \sim N(0,1)$ for all i and the Z_is are all independent. Hence by Theorem 5.21, $Y = \Sigma\, Z_i^2 \sim \chi_n^2$.

Mean and Variance of Chi-square

Recall from Chapter 3 (Section 3.6) that to get the first moment of χ_n^2 (that is, the mean), we take a derivative of $\phi(t)$ once and evaluate at $t = 0$. That will equal i times the mean.

$$\phi'(t) \;=\; \frac{-n}{2}(1 - 2it)^{(-n/2)-1}(-2i)$$

$$\phi'(0) \;=\; ni \;=\; \mu_1' i$$

Hence the mean is n. To get the second moment we need $\phi''(0)$.

$$\phi''(t) \;=\; \frac{-n}{2}\left[-\frac{n}{2}-1\right](1 - 2it)^{(-n/2)-2}(-2i)^2$$

$$\phi''(0) \;=\; i^2\mu_2' \;=\; i^2 n(n+2)$$

Hence the second moment, $\mu_2' = n(n+2)$.

$$\sigma^2 \;=\; \mu_2' - (\mu_1')^2 \;=\; n(n+2) - n^2 \;=\; 2n$$

Therefore the χ_n^2 distribution has mean n and variance $2n$.

5.12   DISTRIBUTIONS OF QUADRATIC FORMS

We are now ready to derive the distributions of quadratic forms of normal variates that will be needed in hypothesis testing.

Theorem 5.23

Let $Y \sim N_n(\mu, \Sigma)$. Then the quadratic form $(Y-\mu)'\Sigma^{-1}(Y-\mu) \sim \chi_n^2$ where $n = rank\,(\Sigma)$ and is the number of Y's.

Proof: Because Σ is a positive definite matrix, by Properties 5.3 and A.16 there exists a nonsingular matrix A such that $AA' = \Sigma$ (A is sometimes referred to as the **square root matrix** of Σ). Now define the new random variable $X = A^{-1}(Y-\mu)$. Because X is a linear combination of the Y's, X is also multivariate normal (Property 5.7, Section 5.10).

$$E(X) = A^{-1}E(Y-\mu) = 0$$

$$E(XX') = A^{-1}E[(Y-\mu)(Y-\mu)'](A^{-1})' = A^{-1}AA'(A')^{-1} = I_n$$

Note that, from Property A.6, $(A^{-1})' = (A')^{-1}$. Hence the Xs are all independent and distributed as $N(0,1)$. Thus, by Theorem 5.21, $X'X \sim \chi_n^2$. But,

$$X'X = (Y-\mu)'(A^{-1})'A^{-1}(Y-\mu) = (Y-\mu)'\Sigma^{-1}(Y-\mu)$$

Hence $(Y-\mu)'\Sigma^{-1}(Y-\mu) \sim \chi_n^2$.

Theorem 5.24

Let $X \sim N_n(0, I_n)$ and A be an $n \times n$ symmetric matrix of rank $k < n$. Then $X'AX \sim \chi_k^2$ if and only if A is idempotent.

Proof: Because A is a symmetric matrix, by Property A.30 there exists a nonsingular orthogonal matrix C (for which $CC' = I$) such that $C'AC$ is diagonal and the elements are the characteristic roots λ_i. Now let $Y = C'X$. Then $X = CY$ and $X'AX = Y'C'ACY = \Sigma_i \lambda_i Y_i^2$. Thus $X'AX \sim \chi_k^2$ if and only if k of the λ's are equal to 1 and the others are zero, that is, if and only if A is idempotent.

Theorem 5.25

Let $Y \sim N_n(0, \Sigma)$ and A be $n \times n$ and symmetric with rank $k < n$. Then $Y'AY \sim \chi_k^2$ if and only if $A\Sigma A = A$.

Proof: Let $\Sigma^{1/2}$ be the "square root" of Σ (that is, it is a nonsingular matrix D such that $DD' = \Sigma$). Define $X = \Sigma^{-1/2}Y$. Then $X \sim N_n(0, I)$. $Y'AY = X'\Sigma^{1/2}A\Sigma^{1/2}X$. From Theorem 5.24, this has a χ_k^2 distribution if and only if $\Sigma^{1/2}A\Sigma^{1/2}$ is idempotent.

$$\Sigma^{1/2}A\Sigma^{1/2} \cdot \Sigma^{1/2}A\Sigma^{1/2} = \Sigma^{1/2}A\Sigma^{1/2}$$

Pre- and post-multiplying both sides by $\Sigma^{-1/2}$ we get $A\Sigma A = A$.

Theorem 5.26

Let $X \sim N_n(0, I_n)$ and A be $n \times n$. Then the characteristic function of $X'AX$ is $|I - 2itA|^{-1/2}$.

Proof:

$$\phi_{X'AX}(t) = E_X[e^{itx'Ax}] = \frac{1}{(2\pi)^{n/2}} \int e^{itx'Ax} e^{-x'x/2} \, dx$$

$$= \frac{1}{(2\pi)^{n/2}} \int e^{-x'(I-2itA)x/2} \, dx$$

$$= |I - 2itA|^{-1/2} \left[\frac{1}{(2\pi)^{n/2}} \frac{1}{|I - 2itA|^{-1/2}} \int e^{-x'(I-2itA)x/2} \, dx \right]$$

But the expression in square brackets is the multivariate normal density with mean 0 and $\Sigma^{-1} = (I - 2itA)$. As $N(0, \Sigma)$ should integrate to 1, the above expression becomes $|I - 2itA|^{-1/2}$.

Theorem 5.27

If $X \sim N_n(0, I)$, then the quadratic forms $X'A_1X$ and $X'A_2X$ are independent if and only if $A_1 A_2 = 0$.

Proof: From Theorem 5.26, $U = X'A_1X$ has the characteristic function $|I - 2itA_1|^{-1/2}$, $V = X'A_2X$ has the characteristic function $|I - 2itA_2|^{-1/2}$, and $U + V = X'(A_1 + A_2)X$ has the characteristic function $|I - 2it(A_1+A_2)|^{-1/2}$. By Theorem 5.17, if U and V are independent, then $\phi_{U+V}(t) = \phi_U(t)\phi_V(t)$. This implies that

$$|I - 2it(A_1 + A_2)| = |I - 2itA_1| \cdot |I - 2itA_2|$$

Because $|A| \, |B| = |AB|$ when A and B are $n \times n$, the right-hand side becomes $|I - 2it(A_1+A_2) + 4i^2t^2A_1A_2|$. We note that the left and right sides will be equal if and only if $A_1A_2 = 0$. Thus the independence of U and V implies $A_1A_2 = 0$.

To prove that $A_1A_2 = 0$ implies the independence of U and V, note that (k is a constant)

$$\phi_{UV}(t_1, t_2) = E\left[e^{it_1U + it_2V}\right] = k \int e^{it_1x'A_1x + it_2x'A_2x - (x'x/2)} \, dx$$

$$= k \int e^{-x'(I - 2it_1A_1 - 2it_2A_2)x/2} \, dx$$

$$= |I - 2it_1A_1 - 2it_2A_2|^{-1/2} = \phi_U(t_1)\phi_V(t_2)$$

because $A_1A_2 = 0$. Hence U and V are independent.

Theorem 5.28

If $X \sim N_n(0, I)$ and A is an idempotent matrix of order n, then the quadratic form $X'AX$ and the linear form $b'X$ are independent if $Ab = 0$.

Proof: The quadratic form $X'AX$ can be written as $(AX)'(AX)$. Also, the covariance matrix between AX and $b'X$ is $E(AXX'b) = 0$ because $Ab = 0$. Because AX and $b'X$ are linear functions of normal variates that have zero covariance, they are also independent. It follows from from this that $b'X$ is independent of a function of AX, in particular, independent of $X'AX$.

Theorem 5.29

If $X \sim N(0, I_n)$ and A_1 and A_2 are $n \times n$ matrices of rank k_1 and k_2 respectively, then $X'A_1X$ and $X'A_2X$ are independent chi-square variates if $A_1A_2 = 0$ and A_1 and A_2 are both idempotent.

Proof: From Theorem 5.27, $A_1A_2 = 0$ and from Theorem 5.24, A_1 and A_2 must both be idempotent.

Corollary: *The necessary and sufficient conditions for Theorem 5.29 to hold when $X \sim N_n(0, \Sigma)$ are $A_1\Sigma A_1 = A_1$, $A_2\Sigma A_2 = A_2$, and $A_1\Sigma A_2 = 0$.*

Proof: Make the transformation $Y = \Sigma^{-1/2}X$. Then Y is distributed as $N(0, I_n)$. The condition $A_1\Sigma A_2 = 0$ follows from Theorem 5.29 and the other two results follow from Theorem 5.25.

5.13 MULTINOMIAL DISTRIBUTIONS

In a Bernoulli trial, an experiment has exactly two outcomes. Some experiments, however, may have more than two possible outcomes. Suppose in a given trial, one of k mutually exclusive events must occur, call them $A_1, A_2, \ldots, A_k$. For instance, we could classify an observation into one of k different classes. Let $P(A_i) = P_i$ and define the random variable Y_i, which takes the value 1 if A_i occurs and 0 otherwise. Thus $P(Y_i = 1) = P_i$. The joint density of the Y's can be written as

$$f(y_1, y_2, \ldots, y_n) = P_1^{y_1}P_2^{y_2}\cdots P_k^{y_k}$$

which is a discrete random variabled called the **multinomial distribution**.

The Multinomial Logit

A special type of the multinomial distribution, commonly used in econometric models involving qualitative responses (whether or not to strike, whether or not to purchase a car, and so forth), is the **multinomial logit** for which P_i is conditioned on one or more random variables X and the functional form is the logistic function discussed in Section 4.2. In particular,

$$P_i = \frac{e^{x_i\theta}}{\sum\limits_{j=1}^{j=k} e^{x_j\theta}}$$

If a cumulative normal distribution is used instead of the logistic, we have a **multinomial probit**. For details about the multinomial probit and logit models, you should refer to Pudney (1989), McFadden (1984), Amemiya (1981, 1984), and Maddala (1983).

EXERCISES

5.1   Let X be distributed as Uniform$(0, \theta)$, with $\theta > 0$. Compute the correlation between X and X^2 and verify that it is neither 0 nor 1.

5.2   Do the same when X has the discrete uniform density over 1, 2, ..., n.

5.3   Let X and Y be independently distributed as $N(0, 1)$. Define

$$U = \mu_x + [\sigma_x \sqrt{(1+\rho)/2}]X + [\sigma_x \sqrt{(1-\rho)/2}]Y$$
$$V = \mu_y + [\sigma_y \sqrt{(1+\rho)/2}]X - [\sigma_y \sqrt{(1-\rho)/2}]Y$$

Show that U and V have the bivariate normal distribution presented in Definition 5.8.

5.4 X and Y have the joint density function $f_{XY}(x, y)$. Let $U = \ln X$ and $V = \ln Y$ be the natural logarithms of X and Y.

(a)   Derive the joint density function of U and V.

(b)   Show that U and V will be independent *if and only if* X and Y are independent.

5.5   Let X_1 and X_2 be independently and identically distributed (abbreviated as **iid**) as $\lambda \, e^{-\lambda x}$ with $x, \lambda > 0$. Derive the joint density of $Y_1 = X_1 + X_2$ and $Y_2 = X_1/X_2$ as well as the marginal densities of Y_1 and Y_2. Are Y_1 and Y_2 independent?

5.6   Let X and Y be *iid* as $N(0, 1)$. Define the random variable $Z = XY$. Derive the characteristic function of Z. From that, derive the mean and variance of Z.

5.7   Suppose Y_1 and Y_2 are bivariate normal with means μ_1 and μ_2, and covariance matrix

$$\Sigma = \begin{bmatrix} \sigma_1^2 & \rho\sigma_1\sigma_2 \\ \rho\sigma_1\sigma_2 & \sigma_2^2 \end{bmatrix}$$

Let A be the square root matrix with

$$A = \begin{bmatrix} a & b \\ 0 & c \end{bmatrix} \qquad \text{and} \qquad \Sigma = AA'$$

Solve for a, b, and c in terms of σ_1, σ_2, and ρ, and use them to construct a pair of transformations $X_1 = g\,(Y_1, Y_2)$ and $X_2 = h\,(Y_1, Y_2)$ such that X_1 and X_2 are *iid* as $N(0, 1)$.

5.8   This exercise provides a counterexample to Theorem 5.17. Consider the following joint density function

$$f_{XY}(x, y) = \tfrac{1}{4}\,[1 + xy\,(x^2 - y^2)] \qquad -1 \le x, y \le 1$$

(a)   Show that the marginal densities of X and Y are $\tfrac{1}{2}$ and deduce that X and Y are not independent.

(b)   Show that the density function of $Z = X + Y$ is given by the triangular distribution

$$f_Z(z) = \begin{cases} \tfrac{1}{4}\,(2 + z) & -2 \le z \le 0 \\ \tfrac{1}{4}\,(2 - z) & 0 < z \le 2 \\ 0 & |z| > 2 \end{cases}$$

(c)   Next show that the characteristic functions of X and Y are $(\sin t)/t$ and that of Z is $(\sin t)^2/t^2$. Finally, prove the counterexample.

5.9   Let X and Y be chi-square with degrees of freedom m and n respectively. Derive the density function of $U = nX/(mY)$.

5.10 Let X and Y by *iid* as $N(\mu, \sigma^2)$. Define $U = X + Y$ and $V = X - Y$. Find the joint characteristic function of U and V (no integration is necessary) and use it to examine whether U and V are independent. Identify their marginal distributions.

5.11 X and Y are random variables with corresponding means μ_X and μ_Y, variances σ_X^2 and σ_Y^2, and covariance σ_{XY}. Derive the value of b for which $\text{Var}(Y - bX)$ will be minimum (b is a constant).

5.12 Let X and Y be *iid* with the density function $f(x) = \theta e^{-\theta x}$ with x and $\theta > 0$. Derive the density function of $U = X + Y$.

5.13 Derive the density function of the ratios of two independent uniform random variables defined on $[0, \theta]$, $\theta > 1$. That is, derive the density of $U = X/Y$, when X and Y are *iid* as Uniform $(0, \theta)$.

5.14 Let X and Y be two independent random variables with

$$f_X(x) = k_1 e^{-x/2} x^{(m-2)/2} \qquad x > 0$$
$$f_Y(y) = k_2 e^{-y/2} y^{(n-2)/2} \qquad y > 0$$

k_1 and k_2 are constants that depend on m and n which are positive integers.

(a) Show that $U = X + Y$ and $V = X/Y$ are independent. Derive the marginal densities of U and V (except for the constant of proportionality). Independently verify the density of U by the convolution formula.

(b) Show that $U = X + Y$ and $W = X/(X+Y)$ are independent. Derive the marginal density of W.

(c) By inspection of Table 4.1 state what the distributions of U, V, and W are. Be sure to specify the parameters of these distributions.

5.15 Let X_1 and X_2 be *iid* as $N(0, 1)$. Define $Y_1 = X_1 + X_2$, and $Y_2 = X_1^2 + X_2^2$.

(a) Show that the joint characteristic function of Y_1 and Y_2 is

$$\frac{\exp[-t_1^2/(1 - 2it_2)]}{1 - 2it_2} \qquad -\infty < t_1 < \infty, \qquad -\infty < t_2 < \frac{1}{2}$$

    (b)   Are Y_1 and Y_2 independent?

    (c)   From (a) derive the characteristic functions of the marginal distributions of Y_1 and Y_2 and identify the distributions they correspond to.

5.16 X and Y are random variables and $U = g(X)$ and $V = h(Y)$ are monotonic, measurable, continuous, and differentiable functions. Derive the joint density function of U and V. Show that U and V will be independent if and only if X and Y are independent.

5.17   Find the approximate mean and variance of $g(X, Y) = X/Y$.

5.18   Suppose $g(X, Y)$ is a measurable transformation of the random variables X and Y. Use a second-order Taylor series expansion (ignore higher order terms) and derive a second-order approximation to $E[g(X, Y)]$.

5.19   Let X_i be *iid* as Uniform $[0, 1]$, $i = 1, 2, \ldots, n$. Define $Y_1 = X_1/X_2$, $Y_2 = X_2/X_3, \ldots, Y_{n-1} = X_{n-1}/X_n$, and $Y_n = X_n$. Obtain the joint density function of the Ys and show that they are independent.

5.20   Let $X \sim \text{MVN}(\mu, \Sigma)$, with $\text{rank}(\Sigma) = n$. Consider the linear transformations $Y = a'X$ and $Z = b'X$, where a and b are constant $n \times 1$ vectors.

    (a)   Derive the joint characteristic function of Y and Z.

    (b)   Show that Y and Z are bivariate normal and derive mean and covariance matrix.

    (c)   Write down the conditional density of Y given X.

    (d)   Derive the necessary and sufficient condition for Y and Z to be independent.

    (e)   What does this condition reduce to if the X's are *iid* as $N(0, \sigma^2)$?

5.21   Let X be multivariate normal with mean vector μ and covariance matrix Σ. Σ is given to be diagonal with σ_i^2 as the ith element.

    (a)   Write down the characteristic function of a univariate normal random variable with mean μ and variance σ^2. Use this to show that the joint characteristic function of the X_i's is

$$\phi_X(t) = \exp[\Sigma\, i\mu_i t_i - (\Sigma\, \sigma_i^2 t_i^2 / 2)]$$

Let $Y = a_1 X_1 + a_2 X_2 + \cdots + a_n X_n$, where the a's are constants.

(b) Derive the characteristic function of Y and show that it corresponds to that of a univariate normal.

(c) Deduce (or derive directly) the mean and variance of Y.

(d) Write down a function of all the Xs, μ_i, and σ_i that is distributed as chi-square with n degrees of freedom. Prove (with citations of appropriate theorems) that your variable is chi-square.

5.22 Let $X_1, X_2, \ldots, X_m$ be *iid* as normal with mean μ_x and variance σ^2, and $Y_1, Y_2, \ldots, Y_n$ be *iid* as normal with mean μ_y and the same variance σ^2. Assume that X_i and Y_j are independent for all i, j. Let $\bar{X} = \Sigma X_i / m$ and $\bar{Y} = \Sigma Y_j / n$.

(a) State the distribution of $Z = \bar{X} - \bar{Y}$.

(b) Derive the distribution of

$$U = \frac{\overset{i=m}{\underset{i=1}{\Sigma}} (X_i - \bar{X})^2 + \overset{j=n}{\underset{j=1}{\Sigma}} (Y_j - \bar{Y})^2}{\sigma^2}$$

5.23 Let u be an $n \times 1$ vector of random variables each *iid* as $N(0, \sigma^2)$, so that u has the multivariate normal distribution $N(0, \sigma^2 I)$. Consider the transformation $z = \beta + (X'X)^{-1}X'u$, where X is an $n \times k$ matrix of fixed values, that is, X *is not a random variable*. β is a $k \times 1$ vector of fixed values. X has rank $k < n$. Denote the matrix $X(X'X)^{-1}X'$ by A.

(a) *Without any formal integration*, derive the characteristic function of z and show that z has a multivariate normal distribution.

(b) From that, deduce the mean and covariance matrix of z.

(c) Show that the matrix A is idempotent and has rank k. [*Hint*: use Property A.15 and the fact that trace $(AB) =$ trace (BA)]

(d)   Derive the distribution of $(z - \beta)' X'X (z - \beta)/\sigma^2$. State its mean and variance.

5.24   Let X be MVN(μ, Σ) and A be an $n \times n$ matrix of full rank.

(a)   Derive the characteristic function of $Y = AX$ from that of X.

(b)   Using that characteristic function show that Y also is MVN.

(c)   Write down the mean of Y and its covariance matrix.

(d)   What is the distribution of the quadratic form,

$$(X - \mu)' A' \Omega^{-1} A (X - \mu),$$

where $\Omega = A\Sigma A'$? State its degrees of freedom.

5.25   Let X be MVN(0, I_n). Consider $Y = AX$ and $Z = BX$, where A and B are $n \times n$ matrices of full rank. Show that $Y \sim$ MVN(0, AA') and $Z \sim$ MVN(0, BB') and that they are jointly MVN. Derive their covariance matrix in terms of A and B. State the condition under which Y and Z will be statistically independent.

Part II

STATISTICAL INFERENCE

Chapter 6 makes the transition from probability theory to statistical inference, and introduces the notion of sampling and sampling distributions. Chapter 7 examines what happens when the sample size is increased indefinitely, that is, it studies the asymptotic properties of sequences of random variables. Chapter 8 deals with the estimation of unknown parameters of a distribution and Chapter 9 is concerned with testing hypotheses on those parameters.

6

SAMPLING THEORY

The **probability model** consists of a probability space represented by the triple $(S, \mathcal{F}, P)$ and a random variable X represented by the family of density functions $f(x;\theta)$, where x is a particular value that X can take, and θ are parameters belonging to the parameter space Θ. The totality of elements about which some information is desired is called a **population**. Examples of populations are single family homes in a community, employees in a firm, television sets produced by an electronics company, and cities in a state or country. Sometimes a population is also referred to as the **parent population**.

In probability theory, we assume that the density function is either known or can be derived from some fundamental principle (as was done for the Poisson and exponential distributions). We then calculate the probabilities of various events and also obtain the measures of location and dispersion of the random variables. In reality, however, parameters of probability distributions are rarely known a priori. One can form frequency distributions based on observed data such as those in Table 2.1 and use the observed shape to identify the family of distributions that might approximate the data generating process (DGP). The actual values of the parameters underlying the identified family of distributions are unknown, however, and have to be estimated in some way.

Statistical inference is the subject that deals with the problems associated with the estimation of the unknown parameters underlying statistical distributions, measuring their precision, testing hypotheses on them, using them in generating forecasts of random variables, and so forth. Because it is expensive to measure all the elements of a population (or **statistical universe**) underlying an experiment, an investigator often resorts to measuring attributes for a small portion of the population (known as a **sample**) and draws conclusions or makes policy decisions based on the data obtained. Drawing a sample is especially important in measuring attributes such as the life of a light bulb, a television picture tube, or an automobile tire. In cases such as these, it is not feasible to measure the entire population in order to obtain average life times. Instead, a sample will be drawn from which estimates can be obtained. This leads to the question "what are the properties of the measures obtained from sample observations and what is their reliability?" In this chapter we develop the basics of **sampling theory** that will be important in statistical inference.

6.1 INDEPENDENT, DEPENDENT, AND RANDOM SAMPLES

The observations on a random variable X are denoted by x_1, $x_2, \ldots, x_n$, where n is the number of observations. These are thus realizations of an experiment repeated n times. If X represents a k–variate random variable, observations will be x_{ij} for $i = 1$, $2, \ldots, k$, $j = 1, 2, \ldots, n$. For the present, however, we will confine ourselves to univariate random variables.

It should be emphasized that, although an observation is simply a measured attribute, conceptually it can be treated as a random variable because of the uncertainty in its value. Thus, if we select a household at random and obtain its annual income in a given year, then we will treat the observed value (say x_1) as a random variable also because if we choose another household, we will obtain a different value, thus attesting to the inherent random process.

Independent Sample

Sample observations can be drawn in a variety of ways that lead to their being sometimes independent and sometimes dependent. These ideas are formalized in the rest of this section.

Definition 6.1

*The observations $x_1, x_2, \ldots, x_n$ are said to form an **independent sample** if the joint density function of the x_i's has the form*

$$f_X(x_1, x_2, \ldots, x_n) = \prod_{i=1}^{i=n} f_{X_i}(x_i ; \theta_i)$$

Because f_{X_i} might be different across i, here we are not assuming that the x's have the same distribution.

Random Sample

Definition 6.2

A **random sample** *from a population is a set of independent, identically distributed (abbreviated as* **iid***) random variables* $x_1, x_2, \ldots, x_n$, *each of which has the same distribution as X.*

It follows from the above definition that the joint density of $x_1, x_2, \ldots, x_n$ is simply the product of individual densities. Thus, if $f(x)$ is the common density function of the X_i's, then

$$f_X(x_1, x_2, \ldots, x_n) = f(x_1) f(x_2) \cdots f(x_n) = \prod_{i=1}^{i=n} f(x_i)$$

Dependent Sample

Suppose the observations obtained on the random variable X are over time. In this situation, the assumption of independence of observations between years is not likely to hold. Even if the observations are from a cross section, there may be a dependency among observations. In this case we have a **dependent sample**. The joint density $f_X(x_1, x_2, \ldots, x_n ; \theta)$ can be factored as follows.

$$f(x_n \mid x_1, x_2, \ldots, x_{n-1} ; \theta) f(x_1, x_2, \ldots, x_{n-1} ; \theta)$$

Repeated use of this decomposition yields the following expression for the joint density (for a given x_0).

$$f(x_1, x_2, \ldots, x_n ; \theta) = \prod_{i=1}^{i=n} f(x_i \mid x_1, x_2, \ldots, x_{i-1} ; \theta)$$

In the rest of this chapter we focus our attention on random samples only. Dependent samples are discussed in more detail in Part III when we consider problems relating to time series data.

6.2   SAMPLE STATISTIC

Definition 6.3

A **statistic** *is a function of the observable random variable(s) that does not contain any unknown parameters.*

The following are examples of statistics:

Sample mean:
$$\bar{x} = \frac{x_1 + x_2 + \cdots + x_n}{n}$$

Sample variance:
$$s^2 = \frac{\sum_{i=1}^{i=n}(x_i - \bar{x})^2}{n-1} = \frac{\sum x_i^2 - n\bar{x}^2}{n-1}$$

[The reason for dividing by $n-1$ will be clear later.]

Sample standard deviation: s

Sample moments:
$$m_r = \frac{1}{n}\sum_{i=1}^{i=n} x_i^r \qquad r = 1, 2, \ldots$$

Sample statistics might also involve two random variables. For example, the **sample covariance** measure between X and Y is given by

$$s_{xy} = \frac{1}{n-1}\sum_i (x_i - \bar{x})(y_i - \bar{y})$$

The **sample correlation coefficient** is given by

$$r_{xy} = \frac{s_{xy}}{s_x s_y} = \frac{\sum(x_i - \bar{x})(y_i - \bar{y})}{[\sum(x_i - \bar{x})^2]^{1/2}[\sum(y_i - \bar{y})^2]^{1/2}}$$

where s_x and s_y are the sample standard deviations of X and Y.

Theorem 6.1

If $x_1, x_2, \ldots, x_n$ is a random sample from a population with mean μ and variance σ^2 and all the c_i's are constant, then

$$Y = c_1 x_1 + c_2 x_2 + \cdots + c_n x_n = c'x$$

has the following expectation and variance.

$$E(Y) = \mu(\sum c_i) = (c_1 + c_2 + \cdots + c_n)\mu$$

$$\text{Var}(Y) = \sigma^2(c_1^2 + c_2^2 + \cdots + c_n^2) = \sigma^2 c'c$$

Corollary: $E(\bar{x}) = \mu; \quad \text{Var}(\bar{x}) = \sigma^2/n$

We have seen a more general case of this in Properties 5.1 and 5.5 of multivariate distributions (Chapter 5).

Proof: $E(Y) = E(\sum_i c_i x_i) = \sum c_i E(x_i) = \mu(\sum c_i)$

$$\text{Var}(Y) = E[Y - E(Y)]^2 = E[\sum c_i (x_i - \mu)]^2$$

$$= E[\sum c_i^2 (x_i - \mu)^2] + \sum_{i \neq j} \sum c_i c_j E[(x_i - \mu)(x_j - \mu)]$$

But because the x's are independent,

$$\text{Cov}(x_i, x_j) = E[(x_i - \mu)(x_j - \mu)] = 0$$

for all $i \neq j$. Therefore, only the first term remains and becomes $\sum c_i^2 E[(x_i - \mu)^2] = \sigma^2(\sum c_i^2)$. The corollary follows by setting $c_i = 1/n$ for all i.

Thus the expected value of the mean of a sample is the mean of the population. Furthermore, $E(m_r) = \mu_r'$.

6.3 SAMPLING DISTRIBUTIONS

Because a sample statistic is a function of random variables, it has a statistical distribution. This is called the **sampling distribution** of the statistic. If we obtain a sample of n observations and compute the statistic, we obtain a numerical value. By repeating this process we get a sequence of values of the statistic. These can be tabulated in the form of a frequency distribution as in Table 2.1, which can then be used to identify the distribution that the statistic follows.

As an example, suppose $x_1, x_2, \ldots, x_n$ are *iid* as $N(\mu, \sigma^2)$. Then because linear combinations of normal variates are also normally distributed, Theorem 6.1 implies that $\bar{x} \sim N(\mu, \sigma^2/n)$ or equivalently the standardized version $z = \sqrt{n}\,(\bar{x} - \mu)/\sigma \sim N(0,1)$. Thus the sampling distribution of the mean is normal with the same mean but a much smaller variance. As the sample size increases, the variance of the mean decreases. It will be seen in the next chapter that under quite general conditions, z will converge to a normal distribution as $n \to \infty$, *even though the parent distribution was not normal.*

Distribution of the Sample Variance of a Normal Variate

Theorem 6.2

Let $x_1, x_2, \ldots, x_n$ be a random sample from $N(0,1)$. Then

(a) $\sum_{i=1}^{i=n} (x_i - \bar{x})^2$ has the chi-square distribution with $n-1$ d.f.

(b) $\bar{x}$ has the normal distribution $(0, 1/n)$.

(c) $\bar{x}$ and $\Sigma\,(x_i - \bar{x})^2$ are statistically independent.

Proof: Let $u = n\bar{x}^2$ and $v = \Sigma\,(x_i - \bar{x})^2$. Then we have

$$u = n(\Sigma\,x_i)^2/n^2 = (1/n)(\Sigma_i x_i^2 + \Sigma_{i \neq j}\Sigma\, x_i x_j)$$

This can be expressed as the quadratic form $x'A_1 x$, where

$$A_1 = \frac{1}{n}\begin{bmatrix} 1 & 1 & . & . & . & . & . & 1 \\ 1 & 1 & . & . & . & . & . & 1 \\ . & & & & & & & \\ 1 & 1 & . & . & . & . & . & 1 \end{bmatrix}$$

Similarly,

$$v = \Sigma\,(x_i - \bar{x})^2 = \Sigma\, x_i^2 - n\bar{x}^2 = x'x - x'A_1 x = x'(I - A_1)x = x'A_2 x$$

From Theorem 5.27, u and v are independent if $A_1 A_2 = 0$. $A_1 A_2 = A_1(I - A_1) = A_1 - A_1^2$. But $A_1^2 = A_1 A_1 = (1/n^2)\,B$, where

$$B = \begin{bmatrix} n & n & . & . & . & . & . & n \\ n & n & . & . & . & . & . & n \\ . & & & & & & & . \\ n & n & . & . & . & . & . & n \end{bmatrix}$$

It is readily verified that $A_1^2 = A_1$ and hence A_1 is idempotent. Therefore $A_1 A_2 = 0$, from which it follows that u and v are independent. From Theorem 5.24, if A_1 is idempotent of rank k, then $x'A_1 x \sim \chi_k^2$. Because A_1 is symmetric and idempotent, by Appendix Property A.15, rank $(A_1) = $ trace $(A_1) = 1$. Hence $u \sim \chi_1^2$, from which it follows that $\sqrt{n}\,\bar{x} = \sqrt{u} \sim N(0,1)$ or, equivalently, $\bar{x} \sim N(0, 1/n)$. It is easy to see that $A_2 = I - A_1$ is also idempotent. Also,

$$\text{trace}\,(A_2) = \text{trace}\,(I) - \text{trace}\,(A_1) = n - 1$$

Therefore, $v = x'A_2 x \sim \chi_{n-1}^2$.

Corollary 1

If $Y_i \sim N(\mu, \sigma^2)$ and are iid, then

$$\frac{\Sigma\,[(Y_i - \bar{Y})^2]}{\sigma^2} \sim \chi_{n-1}^2 \qquad\qquad \bar{Y} \sim N(\mu, \sigma^2/n)$$

and they are independent.

This corollary follows from the result in Example 3.5 that $x_i = (Y_i - \mu)/\sigma \sim N(0,1)$.

Corollary 2

If $Y_i \sim N(\mu, \sigma^2)$ for all i and are independent, then $E(s^2) = \sigma^2$, where $s^2 = [\Sigma (Y_i - \bar{Y})^2]/(n-1)$. In other words, the expected value of the sample variance is equal to the population variance.

Proof: Since $\Sigma (x_i - \bar{x})^2 \sim \chi^2_{n-1}$ when $x_i \sim N(0,1)$, setting $x_i = (Y_i - \mu)/\sigma$ we have, $[\Sigma (Y_i - \bar{Y})^2]/\sigma^2 \sim \chi^2_{n-1}$. Because the mean of a χ^2_m is m (see Table 4.1),

$$E\left[\frac{\Sigma (Y_i - \bar{Y})^2}{\sigma^2}\right] = n - 1$$

which implies that $E(s^2) = \sigma^2$. Also $[(n-1)s^2]/\sigma^2 \sim \chi^2_{n-1}$. The need to divide by $n-1$ is now evident. The variance of χ^2_m is $2m$ (see Table 4.1), and therefore

$$\text{Var}\left[\frac{(n-1)s^2}{\sigma^2}\right] = 2(n-1) \text{ and hence } \text{Var}(s^2) = \frac{2\sigma^4}{n-1}$$

[See pp. 323–324 of Fraser (1976) for an alternative proof of Theorem 6.2.]

The Student's *t*-distribution

In econometrics, the most frequently encountered sampling distributions are the *t*-distribution and the *F*-distribution. The first is discussed here and the second in the next section.

Let Z be a standard normal variate and U be a chi-square with n d.f. and let them be independent. We shall now derive the distribution of $t = Z/(\sqrt{U/n})$, that is, the ratio of a standard normal to the square root of χ^2_n. The joint density of Z and U is (see Table 4.1)

$$f_{ZU}(z,u) = k_1 e^{-z^2/2} u^{(n/2)-1} e^{-u/2} \qquad -\infty < Z < \infty \qquad U > 0$$

where k_1 is a constant independent of z and u. Let $t = Z/\sqrt{(V/n)}$ and $V = U$. The joint density of t and V is given by

$$f_{tV}(t,v) = k_2 e^{-t^2 v/(2n)} v^{(n/2)-1} e^{-v/2} \sqrt{v}$$

because the Jacobian of the transformation is $\partial z / \partial t = \sqrt{(v/n)}$. To get the marginal density of t, integrate $f_{tV}(t,v)$ over the range of v. Thus,

$$g(t) = \int_0^\infty k_2 e^{-v[1+(t^2/n)]/2} \, v^{(n-1)/2} \, dv$$

Now make the transformation $w = v[1+(t^2/n)]$. We have

$$g(t) = \int_0^\infty k_2 e^{-w/2} \, w^{(n-1)/2} \left[1+\frac{t^2}{n}\right]^{-(n+1)/2} dw$$

The integrand is proportional to the density of χ^2, which integrates to a constant independent of t. Therefore the density function of t is of the form

$$g(t) = k_3 \left[1+\frac{t^2}{n}\right]^{-(n+1)/2} \qquad -\infty < t < \infty$$

Note that like the normal distribution, t is also symmetric to the origin. The constant k_3 is evaluated from the condition

$$\int_{-\infty}^\infty k_3 \left[1+\frac{t^2}{n}\right]^{-n+1/2} dt = 1$$

It can be shown that

$$k_3 = \frac{\Gamma[(n+1)/2]}{\Gamma(1/2)\Gamma(n/2)} \cdot \frac{1}{\sqrt{n}}$$

where Γ is the gamma function encountered in Theorem 4.1 (Section 4.2). The parameter n is the number of degrees of freedom as in the chi-square case. Thus the ratio of a $N(0,1)$ to the square root of χ^2_n has the t-distribution with n d.f. with the following density function.

$$f_t(x) = \frac{\Gamma[(n+1)/2]}{\Gamma(\frac{1}{2})\Gamma(n/2)} \cdot \frac{1}{\sqrt{n}} \left[1+\frac{x^2}{n}\right]^{-(n+1)/2} \qquad -\infty < t < \infty$$

Theorem 6.3

Let $x_1, x_2, \ldots, x_n$ be a random sample from $N(\mu, \sigma^2)$. Then $t = (\bar{x}-\mu)/(s/\sqrt{n})$ has the Student's t-distribution with $n-1$ d.f., where s is the sample standard deviation obtained from $s^2 = \Sigma(x_i - \bar{x})^2/(n-1)$.

Proof: We know that $(x_i - \mu)/\sigma \sim N(0,1)$. From Theorem 6.2, Corollary 1,

$$\frac{\Sigma\,(x_i - \bar{x}\,)^2}{\sigma^2} = \frac{(n-1)s^2}{\sigma^2} \sim \chi^2_{n-1}$$

and $(\bar{x}-\mu)/(\sigma/\sqrt{n}) \sim N(0,1)$. Also, they are independent. Hence

$$t = \frac{\bar{x}-\mu}{\sigma/\sqrt{n}} \div \left[\frac{(n-1)s^2}{\sigma^2(n-1)}\right]^{1/2} \sim t_{n-1}$$

which simplifies to

$$t = \frac{\bar{x}-\mu}{s/\sqrt{n}} \sim t_{n-1}$$

Note that $(\bar{x}-\mu)/(\sigma/\sqrt{n}) \sim N(0,1)$, but if we replace σ by the sample standard deviation s, then $t = (\bar{x}-\mu)/(s/\sqrt{n})$ is distributed as t_{n-1}. This will be used in Chapter 9 to test hypotheses on μ.

The F-distribution

The t-distribution is closely related to the F-distribution, which is developed here. Let U be χ^2_m, V be χ^2_n, and U be independent of V. We shall derive the distribution of

$$F = \frac{U/m}{V/n} = \frac{nU}{mV}$$

The joint density of U and V is (see Table 4.1)

$$f_{UV}(u,v) = k_1 u^{(m-2)/2} v^{(n-2)/2} e^{-(u+v)/2} \qquad u,v > 0$$

Therefore, the joint density of F and $W = V$ is (using $u = Fmw/n$)

$$f_{FW}(F,w) = k_2 (Fw)^{(m-2)/2} w^{(n-2)/2}\, w\, \exp\left[-[(Fmw/n)+w]/2\right]$$

noting that the Jacobian is $\partial u/\partial F = mw/n$. Constants k_1 and k_2 are independent of u, v, and F. The marginal distribution of F is obtained by integrating over w. Thus (replacing F by x)

$$f_F(x) = k_2 x^{(m-2)/2} \int_0^\infty w^{(m+n-2)/2} \exp\left[-[1+(xm/n)]w/2\right] dw$$

Now let $[1+(xm/n)]w = z$. We get

$$f_F(x) = \frac{k_2 x^{(m-2)/2}}{\left[1+\dfrac{mx}{n}\right]^{(m+n)/2}} \int_0^\infty z^{(m+n-2)/2} e^{-w/2} dz = k_3 \frac{x^{(m-2)/2}}{\left[1+\dfrac{mx}{n}\right]^{(m+n)/2}}$$

The constant k_3 is obtained, as usual, by integrating $f_F(x)$ over x and setting it equal to 1. It can be shown that

$$k_3 = \frac{\Gamma[(m+n)/2]}{\Gamma(m/2)\Gamma(n/2)} \left[\frac{m}{n}\right]^{m/2}$$

The two parameters of the F-distribution are m and n, which are also referred to as the degrees of freedom. The degrees of freedom of the numerator is always quoted first, that is, $F_{m,n}$ is the F–distribution with d.f. m and n with the following density function.

$$f_F(x) = \frac{\Gamma[(m+n)/2]}{\Gamma(m/2)\Gamma(n/2)} \left[\frac{m}{n}\right]^{m/2} \frac{x^{(m-2)/2}}{\left[1+\dfrac{mx}{n}\right]^{(m+n)/2}}$$

Theorem 6.4

Let $x_1, x_2, \ldots, x_n$ be a random sample from $N(\mu, \sigma^2)$. Then $F = (\bar{x}-\mu)^2/(s^2/n)$ has the F-distribution with d.f. $(1, n-1)$ and $\sqrt{F}$ has the t-distribution with d.f. $n-1$.

Proof: We know that

$$\frac{(\bar{x}-\mu)}{(\sigma/\sqrt{n})} \sim N(0,1) \qquad \frac{[(n-1)s^2]}{\sigma^2} \sim \chi^2_{n-1}$$

and they are independent. Hence $(\bar{x}-\mu)^2/(\sigma^2/n) \sim \chi^2_1$. Therefore,

$$F = \frac{(\bar{x}-\mu)^2}{(\sigma^2/n)} + \frac{s^2}{\sigma^2} \sim F_{1,n-1}$$

This simplifies to $F = (\bar{x}-\mu)^2/(s^2/n) \sim F_{1,n-1}$. From Theorem 6.3 we have, $\sqrt{F} = (\bar{x}-\mu)/(s/\sqrt{n}) \sim t_{n-1}$.

Thus, if t is a student's t with $n-1$ d.f., then $F = t^2$ has an F–distribution with d.f. 1 and $n-1$. The converse is also true.

Theorem 6.5

Let $x_1, x_2, \ldots, x_m$ be a random sample from $N(\mu_1, \sigma_1^2)$ and $y_1, y_2, \ldots, y_n$ be a random sample from $N(\mu_2, \sigma_2^2)$, and let $\bar{x}, \bar{y}$ be the sample means and s_1^2, s_2^2 the sample variances. Also let x_i and y_j be independent for all i and j. Then

$$F = \frac{s_1^2/\sigma_1^2}{s_2^2/\sigma_2^2} \sim F_{m-1,n-1}$$

Proof: The proof is trivial because we know from Corollary 1 of Theorem 6.2 that $(m-1)s_1^2/\sigma_1^2 \sim \chi_{m-1}^2$ and $(n-1)s_2^2/\sigma_2^2 \sim \chi_{n-1}^2$. By the definition of the F-distribution it follows that

$$F = \frac{s_1^2/\sigma_1^2}{s_2^2/\sigma_2^2} \sim F_{m-1,n-1}$$

The F-distribution is commonly used in testing the hypothesis $\sigma_1^2 = \sigma_2^2$, that is, that two independent normal populations have equal variances.

Sampling Distribution of the Binomial Mean

Let $x_1, x_2, \ldots, x_n$ be a random sample from a Bernoulli population with probability of success p, $x_i = 1$ with probability p, and $x_i = 0$ with probability $1-p$. The density function for a single trial may be written as $f(x, p) = p^x(1-p)^{1-x}$. The joint density function of $x_1, x_2, \ldots, x_n$ is

$$f_X(x_1, x_2, \ldots, x_n) = \prod_{i=1}^{n} p^{x_i}(1-p)^{1-x_i} = p^{\Sigma x_i}(1-p)^{n-\Sigma x_i}$$

Let $k = \Sigma_{i=1}^{i=n} x_i$. Note that k is the number of successes in n trials, the density for which is

$$f_k(k) = \binom{n}{k}p^k(1-p)^{n-k}$$

The density function for the mean $U = \bar{x} = k/n$ is easy to obtain.

$$f_U(u) = P(U = u) = P(\Sigma x_i = k) = \binom{n}{k}p^k(1-p)^{n-k}$$

Expressing k as nu, we get

$$f_U(u) = \binom{n}{nu}p^{nu}(1-p)^{n-nu}$$

$$E(\bar{x}) = \frac{1}{n}\Sigma E(x_i) = p$$

$$\text{Var}(\bar{x}) = \frac{1}{n}\text{Var}(x_i) = p(1-p)/n$$

Sampling Distribution of the Poisson Mean

Let $x_1, x_2, \ldots, x_n$ be a random sample from Poisson (λ). Then

$$f(x_1, x_2, \ldots, x_n) = \frac{e^{-n\lambda}\lambda^{\Sigma x_i}}{\prod_{i=1}^{n} x_i!} \qquad x_i = 0, 1, 2, \ldots$$

The distribution of the sample sum Σx_i is easy to obtain from the characteristic function (see Table 4.1).

$$\phi_{x_i}(t) \;=\; \exp(\lambda e^{it} - \lambda)$$

Hence $\phi_{\Sigma x_i}(t) = \Pi_{i=1}^{i=n} \phi_{x_i}(t) = \exp(n\lambda e^{it} - n\lambda)$. This is the characteristic function of Poisson $(n\lambda)$. Hence

$$f_{\Sigma x_i}(\Sigma x_i) \;=\; \frac{e^{-n\lambda}(n\lambda)^{\Sigma x_i}}{(\Sigma x_i)!}$$

As $\bar{x} = \Sigma x_i / n$, its density function is

$$f(\bar{x}) \;=\; \frac{e^{-n\lambda}(n\lambda)^{n\bar{x}}}{(n\bar{x})!}$$

$E(\bar{x}) = [\Sigma E(x_i)]/n = \lambda$ and $\text{Var}(\bar{x}) = [\text{Var}(x_i)]/n = \lambda/n$.

Order Statistics and Their Sampling Distributions

In Chapter 3 we discussed the median, the quartile, and other quantiles. The median is that value of the random variable on either side of which 50 percent of the population lies. To identify the **sample median** one would arrange the observations in increasing order (call them $Y_1, Y_2, \ldots, Y_n$), and identify the median in the sample. The Y's are called **order statistics**, which play important roles in some estimation and hypothesis testing problems. For instance, an electric utility company would like to know the peak electricity consumption for the following day and the probability that the peak will exceed a designated level. Because the peak is the maximum for the day (Y_n in our notation) and is important for the determination of which power plants to turn on, one would be interested in the probability distribution of the peak demand. In this section we discuss the distributions of selected order statistics, namely, the maximum and the minimum. For a discussion of other order statistics such as the median and the range, or of Y_r, refer to Mood, Graybill, and Boes (1974, pp. 251–256) and Lindgren (1976, pp. 217–221).

Sampling Distribution of the Maximum

Denoting the maximum value among the X_i's by Y_n, we have

$$F_{Y_n}(y) = P(Y_n \leq y) = P(X_i \leq y, \text{ for all } i) = \prod_{i=1}^{i=n} F_{X_i}(y)$$

We have thus expressed the CDF of Y_n in terms of the marginal distributions of the X's. If the X's are identically distributed, then

$$F_{Y_n}(y) = [F_X(y)]^n$$

In the case of continuous random variables, we have

$$f_{Y_n}(y) = \frac{d}{dy}F_{Y_n}(y) = n[F_X(y)]^{n-1}f_X(y)$$

Sampling Distribution of the Minimum

By proceeding similarly, we can derive the following for the minimum.

$$F_{Y_1}(y) = 1 - \prod_{i=1}^{i=n}[1 - F_{X_i}(y)]$$

If the X's are also identically distributed,

$$F_{Y_1}(y) = 1 - [1 - F_X(y)]^n$$

In the continuous case we have,

$$f_{Y_1}(y) = n[1 - F_X(y)]^{n-1}f_X(y)$$

6.4 MONTE CARLO SIMULATIONS OF DATA

In the previous section we derived analytically the statistical distributions of a number of sample statistics that are commonly used in estimation and hypothesis testing. In many practical situations, however, such an analytic derivation is impossible to obtain. In these cases, a researcher often resorts to a **Monte Carlo simulation** of the experiment. This consists of a number of steps; (1) assuming an underlying probability distribution with preselected parameters, obtain computer-generated random numbers to simulate a synthetic data set, (2) use the simulated data to estimate the parameters of the distribution, and (3) replicate the experiment numerous times (often over 500 or 1000 times) by varying the initial parameters and study the distribution of the fitted parameters around the known surrogate parameters.

The crucial step in conducting a Monte Carlo study is the generation of random numbers. Computer-generated algorithms exist for obtaining random numbers [see Press *et al* (1988), Chapter 7]. The starting point of all these is the generation of uniform deviates that lie in the range (0,1). From these we can obtain the random observations for the uniform distribution on any real interval. To illustrate, suppose a computer program is used to generate uniform random observations in the interval $(0,1)$ denoted by $x_1, x_2, \ldots, x_n$. These can be converted to a Uniform(a, b) by the transformation $y_i = a + (b - a)x_i$. To obtain random deviates from the standard exponential distribution with the parameter 1, the appropriate transformation is $y_i = -\ln(x_i)$.

Because the cumulative distribution function $F(\cdot)$ of a random variable (Y) is always between 0 and 1, the natural way of obtaining random numbers for any distribution with a known CDF is to first generate random numbers on Uniform$(0,1)$ (denoted by x_i), then find the value y_i that has the fraction x_i of probability area to the left of it in the distribution for Y. The value y_i is the desired random observation from the distribution corresponding to F. This technique is commonly used to generate random numbers from the standard normal $N(0, 1)$ which can be transformed to any $N(\mu, \sigma^2)$. The book by Press *et al* (1988) has the methods for converting a bivariate uniform density to a standard bivariate normal (see Exercise 6.9) as well as the algorithms for obtaining random numbers from the Poisson, gamma, and binomial distributions.

EXERCISES

6.1   Consider the random sample $x_1, x_2, \ldots, x_n$ from the exponential distribution, $f(x) = \theta e^{-x\theta}$, with x and $\theta > 0$. Show that the distribution of their sum, $Y = x_1 + x_2 + \cdots + x_n$ is a gamma. To prove that, carry out the following steps.

   (a)   Show that the characteristic function of X is $\theta/(\theta - it)$.

   (b)   From that, derive the characteristic function of Y.

   (c)   Using the density function of gamma given in Table 4.1, derive its characteristic function.

   (d)   Compare the characteristic functions in (c) and (b) and establish that Y has the gamma distribution. What are the parameters of that distribution?

6.2   Let $x_1, x_2, \ldots, x_m$ be a random sample from a normal distribution with mean μ_x and variance σ^2, and $y_1, y_2, \ldots, y_n$ be a random sample from a normal distribution with mean μ_y and the *same variance* σ^2. Also, x_i and y_j are independent for all

i, j. Let $\bar{x}, \bar{y}$ be the sample means and s_x^2, s_y^2 the sample variances.

(a) Write down the distribution of $z = \bar{x} - \bar{y}$, its mean, and variance.

(b) Derive the distribution of

$$u = \frac{(m-1)s_x^2 + (n-1)s_y^2}{\sigma^2}$$

Justify your answer with appropriate theorems.

(c) Using z and u, construct a random variable that has the t-distribution. State its degrees of freedom.

6.3 Let x_i $(i = 1, 2, \ldots, n)$ be a random sample with $E(x_i) = \mu$ and $\text{Var}(x_i) = \sigma^2$. Consider the statistic $u(x) = \Sigma w_i x_i$, where the w's are constants.

(a) Derive the condition for $E[u(x)]$ to be equal to μ.

(b) Derive the variance of $u(x)$.

(c) Suppose we choose w_i so that $\text{Var}[u(x)]$ is minimum subject to the condition that $E[u(x)] = \mu$. Derive the optimum value of w_i and the corresponding $u(x)$.

6.4 Let $x_1, x_2, \ldots, x_n$ be a random sample from a population that is Uniform $(0, \theta)$, $0 < x_i < \theta$ for all i. Denote by Y_n the largest of the observations. Compute the CDF of Y_n and then show that its density function is

$$f_{Y_n}(y) = \frac{ny^{n-1}}{\theta^n} \qquad 0 < y < \theta$$

6.5 Consider the random sample x_i $(i = 1, 2, \ldots, n)$ from the exponential distribution with density function $\theta e^{-\theta x}$, $x, \theta > 0$. Derive the CDF and PDF of $Y_n = \max(x_1, x_2, \ldots, x_n)$.

6.6 Verify the CDF and PDF for the smallest observation in a random sample.

6.7 Let x_i $(i = 1, 2, \ldots, n)$ be a random sample from a distribution with density function $f_X(x)$ and let $Y_1, Y_2, \ldots, Y_n$ be the corresponding order statistics. For the order statistic of order r (denoted by Y_r), show that

$$F_{Y_r}(y) = \sum_{i=r}^{i=n} \binom{n}{i} [F_X(y)]^i [1 - F_X(y)]^{n-i}$$

[Refer to Mood, Graybill, and Boes (1974), pp. 252–256 for help.]

6.8   Verify that you can generate random numbers from the standard exponential distribution by first generating observations $x_1, x_2, \ldots, x_n$ for a random variable (X) distributed as Uniform$(0,1)$ and then using the transformation $y = -\ln(x)$.

6.9   Press *et al* (1988) have shown how a bivariate uniform random variable can be transformed to a bivariate normal in order to generate a pair of bivariate normal random numbers. To see this, let X_1 and X_2 be *iid* as Uniform$(0,1)$. Consider the transformations

$$Y_1 = \sqrt{-2\ln(X_1)}\,\cos(2\pi X_2)$$
$$Y_2 = \sqrt{-2\ln(X_1)}\,\sin(2\pi X_2)$$

(a)   Verify that the inverse transformations are

$$X_1 = \exp[-(Y_1^2 + Y_2^2)/2]$$
$$X_2 = \frac{1}{2\pi}\arctan(Y_2/Y_1)$$

(b)   Show that the Jacobian of the transformation is

$$-\left[\frac{1}{\sqrt{2\pi}}e^{-Y_1^2/2}\right]\left[\frac{1}{\sqrt{2\pi}}e^{-Y_2^2/2}\right]$$

(c)   Derive the joint density function of Y_1 and Y_2 and verify that each Y_i is *iid* as $N(0, 1)$. Describe how the property just derived can be exploited to generate random numbers from a bivariate normal distribution.

7

ASYMPTOTIC
DISTRIBUTION THEORY

In Chapter 6 we derived the exact distributions of several statistics based on a random sample of observations. In many situations it is not possible to derive such exact distributions. In testing hypotheses, however, we are constantly called upon to compute statistics based on observations and to draw inferences about hypotheses from the numerical values of these statistics. As many such statistics are complicated functions of observations, exact distributions are often not available. The problem disappears in most cases, however, if the sample size is large because we can then derive approximate distributions. Hence the need for large sample or asymptotic distribution theory. This chapter presents an introduction to asymptotic distribution theory. For an excellent treatment of the subject refer to Lukacs (1968), Chung (1974), and White (1984). Other useful references are Dhrymes (1970), Fisz (1963), Fraser (1976), Gnedenko and Kolmogorov (1954), Loeve (1955), Rao (1965), Spanos (1989), and Wilks (1962).

7.1   DIFFERENT TYPES OF CONVERGENCE

Probability theory uses several different types of convergence. Before discussing that, however, let us review the notions of convergence from real analysis, some of which have already been used in earlier chapters.

Definition 7.1 (limit of a sequence)

*Suppose $a_1, a_2, \ldots, a_n$ constitute a sequence of real numbers. If there exists a real number a such that for every real $\varepsilon > 0$, there exists an integer $N(\varepsilon)$ with the property that for all $n > N(\varepsilon)$), we have $\mid a_n - a \mid < \varepsilon$, then we say that a is the **limit** of the sequence $\{a_n\}$ and write $\lim_{n \to \infty} a_n = a$.*

Intuitively, if a_n lies in an ε neighborhood of a $(a - \varepsilon, a + \varepsilon)$ for all $n > N(\varepsilon)$, then a is said to be the limit of the sequence $\{a_n\}$. Examples where limits exist are

$$\lim_{n \to \infty} [1 + (1/n)] = 1 \quad \text{and} \quad \lim_{n \to \infty} [1 + (a/n)]^n = e^a$$

The sequences $\sum_{i=1}^{i=n} (i^2)$ and $(-1)^n$ do not have limits.

The notion of convergence is easily extended to that of a function $f(x)$.

Definition 7.2 (limit of a function)

The function $f(x)$ has the limit A at the point x_0, if for every $\varepsilon > 0$ there exists a $\delta(\varepsilon) > 0$ such that $|f(x) - A| < \varepsilon$ whenever $0 < |x - x_0| < \delta(\varepsilon)$.

With this background we are ready to explore the different types of stochastic convergence.

Definition 7.3 (convergence in distribution)

*Given a sequence of random variables X_n whose CDF is $F_n(x)$, and a CDF $F_X(x)$ corresponding to the random variable X, we say that X_n **converges in distribution** to X, and write $X_n \xrightarrow{d} X$, if $\lim_{n \to \infty} F_n(x) = F_X(x)$ at all points x at which $F_X(x)$ is continuous.*

Intuitively, convergence in distribution occurs when the distribution of X_n comes closer and closer to that of X as n is increased indefinitely. Thus, $F_X(x)$ can be taken to be an approximation to the distribution of X_n when n is large. An example of convergence of this type is the Poisson distribution that was derived in Section 4.1 as the limit of a sequence of binomial distributions with some restrictions. We will see another example of convergence in distribution when we discuss the central limit theorem in Section 7.5.

Definition 7.4 (convergence in probability)

*The sequence of random variables X_n is said to **converge in probability** to the real number x if $\lim_{n \to \infty} P[|X_n - x| > \varepsilon] = 0$ for each $\varepsilon > 0$. Thus it becomes less and less likely that the random variable $X_n - x$ lies outside the interval $(-\varepsilon, +\varepsilon)$. Equivalent definitions are given below.*

(a) $\lim\limits_{n \to \infty} P[|X_n - x| < \varepsilon] = 1, \quad \varepsilon > 0.$

(b)   *Given $\varepsilon > 0$ and $\delta > 0$, there exists $N(\varepsilon, \delta)$ such that $P[|X_n - x| > \varepsilon] < \delta$, for all $n > N$.*

(c) $P[|X_n - x| < \varepsilon] > 1 - \delta,$ for all $n > N,$ that is, $P[|X_{N+1} - x| < \varepsilon] > 1 - \delta, \; P[|X_{N+2} - x| < \varepsilon] > 1 - \delta,$ *and so on.*

We write, $X_n \xrightarrow{P} x$ *or* $\text{plim } X_n = x.$

The sequence of random variables X_n is said to converge in probability to the random variable X if the sequence of their differences $(X_n - X)$ converges in probability to 0.

As an example of convergence in probability to a real number, consider the random sample $x_1, x_2, \ldots, x_n$ from the distribution for the random variable X with mean μ and variance σ^2. Let $Y_n = \bar{x}_n$ be the sample mean. We know that $E(Y_n) = \mu$ and $\text{Var}(Y_n) = \sigma^2/n$. As n goes to infinity, the variance of Y_n becomes smaller and smaller and ultimately becomes 0. This means that Y_n comes closer and closer to the real number μ and ultimately converges to it. Thus, $\text{plim } \bar{x}_n = \mu$. This result, known as the **weak law of large numbers**, is formally proved in Section 7.3 in more general situations and is illustrated in Figure 7.1. As n increases, the distribution of $\bar{x}_n$ is more and more "tightly packed" around μ. Ultimately, the distribution collapses to the single point μ.

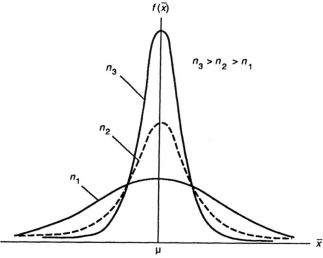

Figure 7.1   Convergence in probability of the sample mean

Definition 7.5 [convergence in mean(r)]

The sequence of random variables X_n is said to **converge in mean of order (r) to X** *$(r \geq 1)$, and designated $X_n \overset{(r)}{\to} X$, if $E[|X_n - X|^r]$ exists and $\lim_{n \to \infty} E[|X_n - X|^r] = 0$, that is, if the rth moment of the difference tends to zero. The most commonly used version is* **mean-squared convergence,** *which is when $r = 2$. Thus $X_n \overset{m.s.}{\to} X$ in this case.*

The sample mean $Y_n = \bar{x}_n$ used earlier is readily seen to converge also in mean square to μ. This follows from the fact that $E[(Y_n - \mu)^2] = \text{Var}(Y_n) = \sigma^2/n$, which tends to zero as $n \to \infty$.

Definition 7.6 (almost sure convergence)

The sequence of random variables X_n is said to **converge almost surely** *to the real number x, and is written as $X_n \overset{a.s.}{\to} x$, if $P[\lim X_n = x] = 1$. In other words, the sequence X_n may not converge everywhere to x, but the points where it does not converge form a set of measure zero in the probability sense. More formally, given $\varepsilon, \delta > 0$, there exists N such that*

$$P[|X_{N+1} - x| < \varepsilon, |X_{N+2} - x| < \varepsilon, ...] > 1 - \delta$$

that is, the probability of these events jointly occurring can be made arbitrarily close to 1.

X_n is said to converge almost surely to the random variable X if $(X_n - X) \overset{a.s.}{\to} 0$.

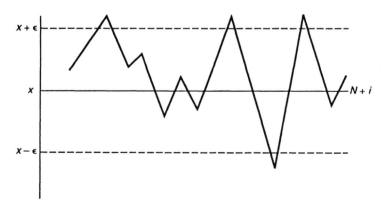

Figure 7.2   Illustration of almost sure convergence

Figure 7.2 illustrates almost sure convergence. We note that for some values of i, X_{N+i} lies outside the boundary $(x - \varepsilon, x + \varepsilon)$. Convergence in probability requires that these "outliers" have probability measure zero.

7.2 RELATIONSHIPS AMONG MODES OF CONVERGENCE

Before examining how the different types of stochastic convergence are related, we discuss a number of useful theorems.

Theorem 7.1

If $X_n \xrightarrow{d} X$ and $Y_n \xrightarrow{p} c$ $(\neq 0)$, where c is a constant, (a) $(X_n + Y_n)$ $\xrightarrow{d} (X + c)$, and (b) $(X_n / Y_n) \xrightarrow{d} (X / c)$.

Proof: Although the theorem seems trivial, its proof is not. Only (a) is proved below. [Prove (b) as an exercise.]

Let $a - c$ be a continuity point of $F_X(x)$. The event $Z_n = (X_n + Y_n) \leq a$ can be expressed as the union of the disjoint events (for $\varepsilon > 0$)

S_1: $X_n + Y_n \leq a$ and $|Y_n - c| \leq \varepsilon$

S_2: $X_n + Y_n \leq a$ and $|Y_n - c| > \varepsilon$

Hence $F_{Z_n}(a) = P(X_n + Y_n \leq a) = P(S_1) + P(S_2)$.

Step 1: To prove that $\lim_{n \to \infty} F_{Z_n}(a) \leq F_X(a - c + \varepsilon)$.

Define S_3: $|Y_n - c| > \varepsilon$.

Since $Y_n \xrightarrow{p} c$, we have $P(S_3) \to 0$. Also, $S_2 \subset S_3$, which implies that $P(S_2) \leq P(S_3)$. Because the latter tends to zero, $P(S_2)$ also tends to zero. Hence

$$\lim_{n \to \infty} F_{Z_n}(a) = \lim_{n \to \infty} P(S_1)$$

Define S_4: $(X_n + Y_n) \leq a$ and $Y_n > c - \varepsilon$.

S_4 implies that $X_n \leq a - c + \varepsilon$. Therefore,

$$P(S_4) \leq P(X_n \leq a - c + \varepsilon) = F_{X_n}(a - c + \varepsilon)$$

Because $S_1 \subset S_4$, it follows that

$$P(S_1) \le P(S_4) \le F_{X_n}(a - c + \varepsilon)$$

Taking the limit as $n \to \infty$, we establish Step 1.

Step 2:   To prove that $F_X(a - c - \varepsilon) \le \lim_{n \to \infty} F_{Z_n}(a)$.

$$P(X_n \le a - c - \varepsilon) = P(X_n \le a - c - \varepsilon, |Y_n - c| > \varepsilon) +$$
$$P(X_n \le a - c - \varepsilon, |Y_n - c| \le \varepsilon)$$

As before, the first term $\le P(|Y_n - c| > \varepsilon)$, which tends to 0 as $n \to \infty$. The second term $\le P(X_n \le a - c - \varepsilon, Y_n \le c + \varepsilon) \le P(X_n + Y_n \le a) = F_{Z_n}(a)$. Combining the two terms and taking the limit as $n \to \infty$, we get

$$F_X(a - c - \varepsilon) = \lim_{n \to \infty} P(X_n \le a - c - \varepsilon) \le \lim_{n \to \infty} F_{Z_n}(a)$$

This establishes Step 2. Putting the two steps together we get

$$F_X(a - c - \varepsilon) \le \lim_{n \to \infty} F_{Z_n}(a) \le F_X(a - c + \varepsilon)$$

Letting $\varepsilon \to 0$, because $a - c$ is a continuity point of $F_X(x)$, we get $F_{Z_n}(a) \to F_X(a-c)$ as $n \to \infty$. But $F_X(a-c)$ is the distribution function of $X + c$, that is,

$$F_{X+c}(a) = P(X + c \le a) = P(X \le a - c) = F_X(a - c)$$

Therefore, $(X_n + Y_n) \overset{d}{\to} (X + c)$.

We now state without proof a number of results relating to convergence in probability. The procedure is similar to that in Theorem 7.1 and hence the steps are quite tedious.

Theorem 7.2

If $X_n \overset{P}{\to} X$ and $Y_n \overset{P}{\to} Y$, then (a) $(X_n + Y_n) \overset{P}{\to} (X+Y)$, (b) $(X_nY_n) \overset{P}{\to} XY$, and (c) if Y_n and $Y \ne 0$, then $(X_n/Y_n) \overset{P}{\to} X/Y$.

Theorem 7.3

If $g(\cdot)$ is a continuous function, then $X_n \overset{P}{\to} X$ implies that $g(X_n) \overset{P}{\to} g(X)$. In other words, convergence in probability is preserved under continuous transformations.

Proof: By the continuity of $g(\cdot)$ we have, for given ε and $\delta > 0$,

$$|X_n - X| < \delta \quad => \quad |g(X_n) - g(X)| < \varepsilon$$

Therefore,

$$P[|X_n - X| < \delta] \leq P[|g(X_n) - g(X)| < \varepsilon]$$

Because the left-hand side converges to 1, so does the right-hand side.

Theorem 7.4

Convergence in probability implies convergence in distribution, that is, $X_n \overset{p}{\to} X => X_n \overset{d}{\to} X$, but the converse need not be true.

Proof: By definition, $X_n \overset{p}{\to} X => P[|X_n - X| > \varepsilon] \to 0$ as $n \to \infty$. Let $Y_n = X_n - X$. Then $\lim_{n \to \infty} P[|Y_n| > \varepsilon] = 0$. This implies that $Y_n \overset{p}{\to} 0$. Also, $X_n = Y_n + X$. Hence, by Theorem 7.1,

$$X_n \overset{d}{\to} (0 + X) = X$$

The theorem can also be proved directly without resorting to Theorem 7.1. By Definition 7.4b, convergence in probability implies that

$$P[X_n < x - \varepsilon] + P[X_n > x + \varepsilon] < \delta$$

which means that each of the terms must be less than δ. The following inequality therefore holds.

$$F_n(x - \varepsilon) - \delta \leq 0 \leq F_X(x) \leq 1 \leq F_n(x + \varepsilon) + \delta$$

Letting ε and δ tend to zero, we have $\lim_{n \to \infty} F_n(x) = F_X(x)$.

That the converse need not be true is illustrated with a simple example. Let X_n be a sequence of *iid* random variables with CDF $F_X(x)$. Then $X_n \overset{d}{\to} X$ trivially. But we cannot state whether X_n converges in probability or of order (r) to anything.

As a more specific example, suppose the sample space is $S = [0, 1]$ and X_n is a sequence of random variables defined as follows:

$$X_n(s) = \begin{cases} 1 & 0 \leq s \leq \frac{1}{2} \\ 0 & \frac{1}{2} < s \leq 1 \end{cases}$$

with a probability of one-half for each case. The corresponding sequence of distribution functions is (see Figure 7.3)

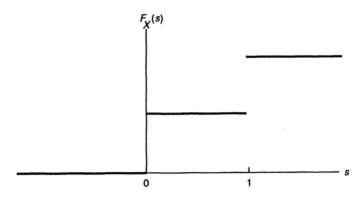

Figure 7.3 CDF for the specified sequence of random variables

$$F_{X_n}(s) = \begin{cases} 0 & s < 0 \\ 1/2 & 0 \le s \le \tfrac{1}{2} \\ 1 & s > \tfrac{1}{2} \end{cases}$$

Next suppose the random variable X is defined as

$$X(s) = \begin{cases} 0 & 0 \le s \le \tfrac{1}{2} \\ 1 & \tfrac{1}{2} < s \le 1 \end{cases}$$

with a probability of one-half for each case. It is easily verified that the corresponding CDF is the same as that for X_n, which means that $X_n \overset{d}{\to} X$ trivially. However, $X_n - X$ is always equal to 1 and hence X_n does not converge in probability to X.

Theorem 7.5

Convergence in mean of order r implies convergence in mean of an order less than r, that is, $X_n \overset{(r)}{\to} X \implies X_n \overset{(s)}{\to} X$ $(r > s)$, but the converse need not be true.

Proof: Because $[E(|Z|^r)]^{1/r}$ is a nondecreasing function of r [for proof see Lukacs (1968), Theorem 1.2.5, p. 12], we have for $r > s$

$$E(|X_n - X|^s) \le E(|X_n - X|^r)$$

It is readily apparent that if the right-hand side of the inequality goes to zero as $n \to \infty$, then so does the left-hand side, thus establishing the theorem.

The converse need not be true because even if $X_n \overset{(s)}{\to} X$, there is no assurance that higher order moments exist.

Theorem 7.6

Convergence in mean of order $r \geq 1$ implies convergence in probability, but the converse need not be true.

Proof: The proof is trivial and follows directly from Theorem 3.8. Letting $h(X) = |X_n - X|^r$ in Theorem 3.8, we see that, for $\varepsilon > 0$,

$$P[|X_n - X| > \varepsilon] \leq \frac{E[|X_n - X|^r]}{\varepsilon^r}$$

Because $X_n \overset{(r)}{\to} X$, the right-hand side has a zero limit. Therefore $X_n \overset{p}{\to} X$.

To give a counterexample for the converse, consider X_n which takes the value 0 with probability $1 - (1/n)$ and the value n with probability $1/n$. $X_n \overset{p}{\to} 0$ obviously, but $E(X_n^2) = n \to \infty$ as $n \to \infty$. Therefore convergence in probability need not imply convergence in mean square.

Theorem 7.7

Almost sure convergence implies convergence in probability, but the converse need not be true.

Proof: Let A be the set $\{|X_{N+1} - X| < \varepsilon, |X_{N+2} - X| < \varepsilon, \ldots\}$ and B_i be the set $\{|X_{N+i} - X| < \varepsilon\}$, for $i = 1, 2, \ldots$ Because $A \subset B_i$, $P(B_i) \geq P(A)$ for $i = 1, 2, \ldots$ If X_n converges almost surely to X, then $P(A) > 1 - \delta$, and hence $P(B_i) > 1 - \delta$ for $i = 1, 2, \ldots$, thus implying that X_n converges in probability also to X.

We readily see that almost sure convergence is the strongest form of convergence. For a counterexample showing that convergence in probability need not imply convergence almost surely, see Lukacs (1968, pp. 32–35). Lukacs also has examples showing that, in general, there exists no implication between convergence in mean of order r and almost sure convergence. However, there are special cases in which convergence in probability and almost sure convergence are equivalent. Figure 7.4 is a diagrammatic representation of the relationships among the different modes of convergence.

We now state several results regarding almost sure convergence. The proofs are in the Appendix of Fraser's book (1976, pp. 557–858).

Theorem 7.8

(1) $X_n \overset{a.s.}{\to} x$ *if and only if* $P[\underset{j=n}{\overset{\infty}{\sup}} |X_j - X| > \varepsilon] \to 0$ *as* $n \to \infty$ *for any* $\varepsilon > 0$.

(2)   *If* $\overset{\infty}{\underset{n=1}{\Sigma}} P(|X_n - X| > \varepsilon)$ *is finite for each* $\varepsilon > 0$, *then* $X_n \overset{a.s.}{\to} X$.

(3)   *If* $\Sigma_{n=1}^{\infty} E[|X_n - X|^r]$ *is finite for some* $r > 0$, *then* $X_n \overset{a.s.}{\to} X$.

(4)   *If the function* $g(\cdot)$ *is continuous at* X *and* $X_n \overset{a.s.}{\to} X$, *then* $g(X_n) \overset{a.s.}{\to} g(X)$. [This is proved in White (1984), p. 17.]

7.3   THE WEAK LAW OF LARGE NUMBERS

When discussing convergence in probability it was shown that as the sample size becomes large, the sample mean approaches the population mean. This is called the **weak law of large numbers (WLLN)**, which holds under a variety of different assumptions.

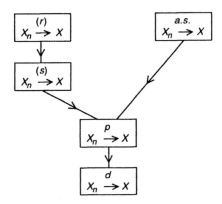

Figure 7.4   Relationships among modes of convergence

Theorem 7.9 (Khinchin's theorem)

Let $\{X_n, \ n \geq 1\}$ be a sequence of iid random variables with the finite mean μ, and let $\bar{X}_n = (\Sigma_{i=1}^{i=n} X_i)/n$. Then

$$\lim_{n \to \infty} P[\,|\bar{X}_n - \mu| > \varepsilon\,] = 0$$

or equivalently,

$$\lim_{n \to \infty} P[\,|\bar{X}_n - \mu| \leq \varepsilon\,] = 1$$

In other words, plim $\bar{X}_n = \mu$.

Proof: Expand the characteristic function $\phi_{X_i}(t)$ around the origin, using the Taylor series.

$$\phi_{X_i}(t) = 1 + \phi'(0)t + R(t)$$

where $R(t)$ is a remainder that goes to zero as t goes to zero. But $\phi'(0) = \mu i$. Therefore $\phi_{X_i}(t) = 1 + i\mu t + R(t)$. The characteristic function of X_i/n is $1 + (i\mu t/n) + R(t/n)$ and hence that of $\bar{X}_n$ is

$$\phi_{\bar{X}_n}(t) = \left[1 + \frac{i\mu t}{n} + R(t/n)\right]^n$$

Using Lemma 7.1 proved below, it is readily seen that $\lim_{n \to \infty} \phi_{\bar{X}_n}(t) = e^{i\mu t}$, which is the characteristic function of a random variable with all its density at μ. Hence the theorem.

Lemma 7.1

Let $b(x)$ be a function such that $\lim_{x \to 0} b(x)/x = 0$. Then

$$\lim_{x \to 0} \left[1 + ax + b(x)\right]^{1/x} = e^a$$

Proof: Let $y = ax + b(x) = x[a + c(x)]$, where $c(x)$ is a function that converges to 0 as $x \to 0$. Then

$$\left[1 + ax + b(x)\right]^{1/x} = \left[(1+y)^{1/y}\right]^{a + c(x)} = (1+y)^{a/y}\left[(1+y)^{1/y}\right]^{c(x)}$$

As $x \to 0$ then $y \to 0$ and the first term becomes e^a. The exponent $c(x)$ of the second term goes to zero and hence it becomes $e^0 = 1$. Hence the limit is e^a.

WLLN can be established even if we relax the assumption that the elements of the sequence $\{X_n\}$ are identically distributed, provided an alternative condition holds.

Theorem 7.10 (Chebyshev's Theorem)

Let $\{X_n\}$ be a sequence of independent random variables with means μ_n and variances σ_n^2, and let $\bar{\mu}_n = (\Sigma \mu_i)/n$. If the variances are bounded above, that is, $\sigma_n^2 < c < \infty$, then $(\bar{X}_n - \bar{\mu}_n) \overset{p}{\rightarrow} 0$.

Proof: $\text{Var}(\bar{X}_n) = 1/(n^2) \Sigma \sigma_i^2 \leq c/n$. By Chebyshev's inequality,

$$P[|\bar{X}_n - \bar{\mu}_n| \geq \varepsilon] \leq \frac{\text{Var}(\bar{X}_n)}{\varepsilon^2} \leq \frac{c}{n\varepsilon^2}$$

Because the right-hand side of Chebyshev's inequality converges to 0, so does the left-hand side, thus establishing the theorem.

When dealing with a sequence of observations over time, the independence assumption is questionable. WLLN will still hold, however, provided $\text{Var}(\bar{X}_n)$ tends to zero as n tends to infinity.

Theorem 7.11 (Markov's Theorem)

Let $\{X_n\}$ be a sequence of random variables with means μ_n, and let $\bar{X}_n$ and $\bar{\mu}_n$ be as defined in Theorems 7.9 and 7.10. If $\text{Var}(\bar{X}_n) \rightarrow 0$ as $n \rightarrow \infty$, then $(\bar{X}_n - \bar{\mu}_n) \overset{p}{\rightarrow} 0$.

Proof: By Chebyshev's inequality,

$$P[|\bar{X}_n - \bar{\mu}_n| \geq \varepsilon] \leq \frac{\text{Var}(\bar{X}_n)}{\varepsilon^2}$$

Because the right-hand side converges to zero, by assumption, so does the left-hand side, thus implying the required convergence in probability.

Theorems 7.9, 7.10, and 7.11 specified alternative conditions under which the sample mean of a sequence of random variables with zero expectations converge in probability to zero. Those conditions are, however, only sufficient but not necessary. Kolmogorov derived the necessary and sufficient condition for the WLLN to hold *without any assumption about independence or identical distribution.*

Theorem 7.12   (Kolmogorov's theorem on WLLN)

Given the definitions of $\overline{X}_n$ and $\overline{\mu}_n$ in Theorems 7.9 and 7.10, let $Z_n = \overline{X}_n - \overline{\mu}_n$. A necessary and sufficient condition for the WLLN to hold, that is, for $Z_n \overset{P}{\to} 0$, is that $\lim_{n \to \infty} E[Z_n^2/(1 + Z_n^2)] = 0$.

Proof: Let $F_n(z)$ denote the distribution function of Z_n. To prove sufficiency, first note that

$$P[\,|Z_n| \geq \varepsilon] = \int_{|z| \geq \varepsilon} dF_n(z)$$

$|z| \geq \varepsilon$ implies that

$$\frac{z^2}{1 + z^2} \geq \frac{\varepsilon^2}{1 + \varepsilon^2}$$

from which it follows that

$$P[\,|Z_n| \geq \varepsilon] = \int_{|z| \geq \varepsilon} dF_n(z) \leq \frac{1 + \varepsilon^2}{\varepsilon^2} \int_{|z| \geq \varepsilon} \frac{z^2}{1 + z^2} dF_n(z)$$

$$\leq \frac{1 + \varepsilon^2}{\varepsilon^2} \int \frac{z^2}{1 + z^2} dF_n(z) = \frac{1 + \varepsilon^2}{\varepsilon^2} E\left[\frac{Z_n^2}{1 + Z_n^2}\right]$$

The condition that the expected value of the expression in the square brackets converges to zero as n goes to infinity is therefore sufficient to establish the theorem. For necessity, note that

$$P[\,|Z_n| \geq \varepsilon] = \int_{|z| \geq \varepsilon} dF_n(z) \geq \int_{|z| \geq \varepsilon} \frac{z^2}{1 + z^2} dF_n(z)$$

$$= \int \frac{z^2}{1 + z^2} dF_n(z) - \int_{|z| < \varepsilon} \frac{z^2}{1 + z^2} dF_n(z)$$

Because the first integral is $E[Z_n^2/(1 + Z_n^2)]$ and the second integral is less than ε^2, we have

$$P[\,|Z_n| \geq \varepsilon] \geq E[Z_n^2/(1 + Z_n^2)] - \varepsilon^2$$

Therefore

$$0 \leq E\left[\frac{Z_n^2}{1 + Z_n^2}\right] \leq \varepsilon^2 + P[\,|Z_n| \geq \varepsilon]$$

Finally, letting $n \to \infty$ and $\varepsilon \to 0$, the necessity of the condition is established.

7.4   THE STRONG LAW OF LARGE NUMBERS

The WLLN stated that under certain conditions the sample mean converges in probability to the population mean. We can in fact derive a stronger result, namely, that the sample mean converges almost surely to the population mean. This is the **strong law of large numbers (SLLN)**. As in the case of the WLLN, the result holds under different restrictions on the dependence of observations, their heterogeneity, and moments.

As before, let $X_1, X_2, \ldots, X_n$ be a sequence of random variables with $E(X_i) = \mu_i < \infty$, and $\bar{X}_n = (\Sigma X_i)/n$, and $\bar{\mu}_n = (\Sigma \mu_i)/n$. Then, under certain conditions we can show that $(\bar{X}_n - \bar{\mu}_n) \overset{a.s.}{\to} 0$. The conditions parallel those for WLLN and are stated here in a number of theorems.

Theorem 7.13

If the X_i's are iid, then $(\bar{X}_n - \bar{\mu}_n) \overset{a.s.}{\to} 0$. [See Fraser (1976), pp. 560–563, for proof.]

It follows from this theorem that for $(\bar{X}_n - \bar{\mu}_n)$ to converge in probability to zero but not almost surely, the sequence $\{X_n\}$ must be either not independent or not identically distributed (or both).

Theorem 7.14   (Kolmogorov's theorem on SLLN)

If the X_i's are independent with finite variances, and if $\Sigma_{n=1}^{\infty} \mathrm{Var}(X_n)/n^2 < \infty$, then $(\bar{X}_n - \bar{\mu}_n) \overset{a.s.}{\to} 0$. [See Lukacs (1968), p. 87 or Fisz (1963), p. 223 for proof.]

In the case of an *iid* sequence of random variables, Kolmogorov derived a necessary and sufficient condition for almost sure convergence of the sample mean.

Theorem 7.15

If the X_i's are iid, then a necessary and sufficient condition for $(\bar{X}_n - \bar{\mu}_n) \overset{a.s.}{\to} 0$ is that $E|X_i - \mu_i| < \infty$ for all i. [See Rao (1973), p. 115 for proof.]

7.5   THE CENTRAL LIMIT THEOREM

Perhaps the most important theorem in large sample theory is the **central limit theorem**, which states that, under quite general (and intuitively reasonable) conditions, the mean of a sequence of random variables (such as the sample mean, for example) converges to a normal distribution *even though the parent distribution is not normal.* Thus, even if we did not know the statistical distribution of the population from which a sample is drawn, by having a large sample we can approximate quite well the distribution of the sample mean by the normal distribution. This enormously simplifies the statistical inference in such cases. In this section, we first present some results regarding the convergence of integrals and then prove the central limit theorem.

Theorem 7.16

Suppose X_n ($n \geq 1$) is a sequence of random variables with CDF $F_n(x)$, and $X_n \xrightarrow{d} X$ with the CDF $F_X(x)$. Then for any bounded continuous function $g(x)$,

$$\lim_{n \to \infty} \int_{-\infty}^{\infty} g(x)\, dF_n(x) = \int_{-\infty}^{\infty} g(x)\, dF(x)$$

In the case of continuous random variables, this takes the form

$$\lim_{n \to \infty} \int_{-\infty}^{\infty} g(x) f_n(x)\, dx = \int_{-\infty}^{\infty} g(x) f(x)\, dx$$

Thus we can take the limit inside the integral. See Fraser (1976, p. 559) for a proof of this and the following theorem on characteristic functions.

Theorem 7.17   (continuity theorem)

Suppose X_n ($n \geq 1$) is a sequence of random variables with CDF $F_n(x)$, and $X_n \xrightarrow{d} X$ with the CDF $F_X(x)$. Let the corresponding characteristic functions be $\phi_n(t)$ and $\phi(t)$. Then $\lim_{n \to \infty} \phi_n(t) = \phi(t)$. Conversely, if $\phi_n(t) \to \phi(t)$, which is continuous at $t = 0$, then $\phi(t)$ is a characteristic function and the corresponding distribution functions are such that $F_n(x) \to F(x)$.

Proof of this theorem may be obtained from Dhrymes (1970, p. 94) or Fraser (1976, p. 559). The need for continuity at $t = 0$ is illustrated with the following example. Let $X_n \sim N(0,n)$. Then $\phi_n(t) = e^{-nt^2/2} \to 1$ if $t = 0$ and 0 if $t \neq 0$. Thus $\lim \phi_n(t)$ exists but is not continuous at $t = 0$. Also F_n does not converge to F. Hence this is a counterexample.

Example 7.1

We know that for the binomial distribution $\phi_n(t) = (q + pe^{it})^n$. Let p be such that $np = \lambda$. Then

$$\phi_n(t) = \left[1 + \frac{\lambda}{n}(e^{it}-1)\right]^n$$

$$\lim_{n \to \infty} \phi_n(t) = \exp[\lambda(e^{it}-1)]$$

But this is the characteristic function of the Poisson distribution. Thus the binomial converges to the Poisson when $n \to \infty$ (for $p = \lambda/n$).

Theorem 7.18 (central limit theorem)

Let $X_1, X_2, \ldots, X_n$ be a sequence of random variables, S_n be their sum $\Sigma_{i=1}^{i=n} X_i$, and $\bar{X}_n$ be their mean S_n/n. Define the standardized mean

$$Z_n = \frac{\bar{X}_n - E(\bar{X}_n)}{\sqrt{Var(\bar{X}_n)}} = \frac{S_n - E(S_n)}{\sqrt{Var(S_n)}}$$

Then, under a variety of alternative assumptions (stated below), $Z_n \xrightarrow{d} N(0,1)$.

de Moivre's Theorem: X_i's are independent Bernoulli variates. [This was the first case established historically.]

Lindberg-Levy Theorem: X_i's are independent and identically distributed with $Var(X_i) = \sigma^2 < \infty$.

[The next two theorems retain the independence but assume different means and variances. For dependent observations see White (1984), p. 115 and 124.]

Liapounov Theorem: X_i's are independent with $E(X_i) = \mu_i$,
Var $(X_i) = \sigma_i^2$, $E\left[|X_i - \mu_i|^{2+\delta}\right] = \rho_i < \infty$ $(\delta > 0)$, and

$$\lim_{n \to \infty} \frac{(\Sigma_{i=1}^{i=n} \rho_i)^2}{(\Sigma_{i=1}^{i=n} \sigma_i^2)^{2+\delta}} = 0$$

Lindberg-Feller Theorem: X_i's are independent, $E(X_i) = \mu_i$,
Var $(X_i) = \sigma_i^2$, $s_n^2 = \Sigma_{i=1}^{i=n} \sigma_i^2$, $S_n = \Sigma_{i=1}^{i=n} X_i$, and for $\varepsilon > 0$,

$$\lim_{n \to \infty} \frac{1}{s_n^2} \sum_{i=1}^{n} \left[\int_{|x - \mu_i| > \varepsilon S_n} (x - \mu_i)^2 \, dF_i(x) \right] = 0$$

Proof: We prove only the Lindberg-Levy version of the theorem. The proof of Liapounov Theorem may be found in Loeve (1955, p. 276), and that of Lindberg-Feller Theorem in Chung (1968, p. 187).

The standardized Z_n can be written as: $Z_n = \Sigma_{i=1}^{i=n} u_i$, where $u_i = (X_i - \mu)/(\sigma \sqrt{n})$ are all *iid*. Because of the *iid* property, $\phi_{Z_n}(t) = [\phi_{u_i}(t)]^n$. Let $\phi(t)$ be the characteristic function of $(X_i - \mu)/\sigma$. Taylor expansion of $\phi(t)$ gives

$$\phi(t) = \phi(0) + \phi'(0) t + \phi''(0) \frac{t^2}{2} + R(t^2)$$

where $R(\cdot)$ is a remainder that goes to zero as t goes to zero. We have,

$$\phi(0) = 1, \qquad \phi'(0) = i \, E[(X_i - \mu_i)/\sigma] = 0$$

$$\phi''(0) = i^2 \text{Var}[(X_i - \mu_i)/\sigma] = -1$$

Therefore, $\phi_{u_i}(t) = \phi(t/\sqrt{n}) = 1 - (t^2/2n) + R(t^2/n)$. Hence,

$$\phi_{Z_n}(t) = [\phi(t/\sqrt{n})]^n = \left[1 - \frac{t^2}{2n} + R(t^2/n) \right]^n$$

By Lemma 7.1, $\lim_{n \to \infty} \phi_{Z_n}(t) = e^{-t^2/2}$ which is the characteristic function of the standard normal. By the uniqueness and continuity theorems for characteristic functions, $Z_n \xrightarrow{d} N(0,1)$.

7.6   MULTIVARIATE CENTRAL LIMIT THEOREM

In this section we extend the central limit theorem to the case of a random sample of observations on a general k-dimensional random vector. Before proving that, however, we need the following lemma.

Lemma 7.2

Let X be a random vector with $E(X) = \mu$ and covariance matrix Σ. If every nontrivial linear combination of the elements of X is normally distributed, then $X \sim MVN(\mu, \Sigma)$.

Proof: Define $Y = c'X$, where c is any $k \times 1$ vector of constants. Let the scalar random variable Y have mean v and variance σ^2. We have,

$$v = c'\mu \quad \text{and} \quad \sigma^2 = c'\Sigma c$$

By the assumption of the theorem, $Y \sim N(v, \sigma^2)$ and hence its characteristic function is (see Example 3.15),

$$E(e^{itc'X}) = \exp[itv - \tfrac{1}{2}\sigma^2 t^2] = \exp[itc'\mu - \tfrac{1}{2}t^2 c'\Sigma c]$$

Define the $k \times 1$ vector $s = tc$ and substitute for it in the above equation.

$$E(e^{is'X}) = \exp[is'\mu - \tfrac{1}{2}s'\Sigma s]$$

But this can be identified (see Section 5.10) as the characteristic function of the multivariate random variable X with mean vector μ and covariance matrix Σ (using the vector s instead of the usual t). This establishes the lemma.

Theorem 7.19

Let X_n $(n \geq 1)$ be a sequence of k-variate random variables, independently and identically distributed with the mean vector μ and covariance matrix Σ, and let $\overline{X}_n = (\Sigma_{i=1}^{i=n} X_i)/n$. Then

$$Z_n = \sqrt{n}(\overline{X}_n - \mu) \xrightarrow{d} N_k(0, \Sigma)$$

Proof: Consider a $k \times 1$ vector of constants and the linear combination $Y_n = c'(X_n - \mu)$. Y_n is a sequence of *iid* scalar random variables with mean 0 and variance $c'\Sigma c$. Therefore by the central limit theorem, $W_n = \sqrt{n}\,\bar{Y}_n \overset{d}{\to} N(0, c'\Sigma c)$. This means that the characteristic function has the property

$$\lim_{n \to \infty} \phi_{W_n}(t) = \exp[-t^2 c'\Sigma c / 2]$$

The characteristic function of Z_n is given by

$$\phi_{Z_n}(s) = E(e^{is'Z_n})$$

Choosing $s = tc$ in the above and noting that $W_n = \sqrt{n}\,\bar{Y}_n = c'Z_n$, we get

$$\phi_{Z_n}(tc) = E(e^{itc'Z_n}) = E(e^{itW_n}) \to \exp[-t^2 c'\Sigma c / 2]$$

which is the characteristic function of $N(0, c'\Sigma c)$. Thus the limit of every linear combination of the X_n's is a normal distribution. Therefore by Lemma 7.2, $Z_n \overset{d}{\to} N(0, \Sigma)$.

EXERCISES

7.1   Prove that the conditions given in Theorems 7.10 and 7.11 imply the Kolmogorov's condition in Theorem 7.12.

7.2   Let $X_n \sim \text{Bin}(n, p)$. Show that $(X_n - np)/\sqrt{npq} \overset{d}{\to} N(0, 1)$.

7.3   Let X have the Poisson distribution with parameter λ.

(a)   Show that its characteristic function is $\exp[\lambda(e^{it}-1)]$ and that its mean and variance are both λ.

(b)   Let $x_1, x_2, \ldots, x_n$ be a random sample, and let $S_n = \Sigma_{i=1}^{i=n} x_i$. Derive the characteristic function of S_n and identify its distribution, its mean, and its variance (which will depend on n and λ).

(c)   Construct Z_n, a function of S_n (and of n and λ) such that $Z_n \overset{d}{\to} N(0,1)$. Justify, using the central limit theorem, your choice of Z_n as asymptotically normal.

(d)   Derive the characteristic function of Z_n and show that it converges to $e^{-t^2/2}$, which is the characteristic function of $N(0,1)$.

7.4   Let X_n be a Poisson random variable with parameter n. Its characteristic function is (verify it)

$$\phi_n(t) = \exp\left[n\,(e^{it}-1)\right]$$

(a)   From this, derive the characteristic function of $Y_n = (X_n - n)/\sqrt{n}$.

(b)   Use it to show that $Y_n \xrightarrow{d} N(0,1)$.

7.5   Let $x_1, x_2, \ldots, x_n$ be a random sample from a distribution with $E(\ln x_i) = 0$ and $\mathrm{Var}(\ln x_i) = 1$. Let

$$Y_n = (x_1 x_2 \cdots x_n)^{1/\sqrt{n}}$$

Show that Y_n converges to a lognormal distribution.

7.6   Consider the random sample $x_1, x_2, \ldots, x_n$ from the exponential distribution, $f(x) = \theta e^{-x\theta}$, with x and $\theta > 0$. Show that the distribution of their sum, $Y = x_1 + x_2 + \cdots + x_n$ is a gamma. To show that, carry out the following steps.

(a)   Show that the characteristic function of X is $\theta/(\theta - it)$.

(b)   From that, derive the characteristic function of Y.

(c)   Using the density function of gamma given in Table 4.1, derive the characteristic function of gamma.

(d)   Compare the characteristic functions in (c) and (b) and establish that Y has the gamma distribution. What are the parameters of that distribution?

7.7   Let $X_1, X_2, \ldots, X_n$ be iid as $N(0,1)$ and $u_n = (\Sigma_{i=1}^{i=n} X_i^2)/n$. Show directly from the characteristic function of u_n that $u_n \xrightarrow{d} 1$. Verify this by the weak law of large numbers.

7.8   Suppose x_i $(i = 1, 2, \ldots, n)$ is a random sample from a distribution with mean μ and standard deviation σ. Under the conditions of the central limit theorem, show that $\sqrt{n}\,[g(\bar{x}) - g(\mu)]$ has the limiting $N[0, g'(\mu)\sigma]$ distribution, where $g(\cdot)$ is continuously differentiable in the neighborhood of μ.

7.9   Let $x_1, x_2, \ldots, x_n$ be a random sample from $N(\mu, \sigma^2)$. Let $\bar{x}$ be the sample mean and $s_n^2 = [\Sigma(x_i - \bar{x})^2]/(n-1)$ be the sample variance. Using characteristic functions, show that $s_n^2 \xrightarrow{d} \sigma^2$. Verify this by the law of large numbers. Also, show that $(n - 1)s_n^2/n \xrightarrow{P} \sigma^2$. State any theorems used in proving this.

7.10 Let x_i be a random sample from $N(\mu, \sigma^2)$ for $i = 1, 2, \ldots, n$. $\bar{x}$ is the sample mean and $s_n^2 = [\Sigma(x_i - \bar{x})^2]/(n - 1)$ is the sample variance. Define $Z_n = \sqrt{n}\ (\bar{x} - \mu)/s_n$. Show that $Z_n \xrightarrow{d} N(0,1)$.

7.11 Let $x_i \sim N(\mu, \sigma^2)$, for $i = 1, \ldots, n$. Define $Z_n = \sqrt{n}\ (e^{\bar{x}} - e^\mu)$, where $\bar{x}$ is the mean.

(a) Using a Taylor expansion, show that

$$e^{\bar{x}} = e^\mu + e^{x_0}(\bar{x} - \mu)$$

whenever $|x_0 - \mu| \le |\bar{x} - \mu|$.

(b) Citing appropriate limit theorems show that $Z_n \xrightarrow{d} N(0, \sigma^2 e^{2\mu})$.

7.12 Let Y be a random variable distributed as χ^2 with n degrees of freedom and let $Z_n = (Y - n)/\sqrt{(2n)}$. *Without using characteristic functions*, show that Z_n converges in distribution to $N(0,1)$.

7.13 Let x_i $(i = 1, 2, \ldots, n)$ be a random sample of size n from a population with density function $f(x; \theta) = \theta e^{-x\theta}$, with $x, \theta > 0$. The mean and variance are, respectively, $1/\theta$ and $1/\theta^2$. Construct a random variable that depends on the sample mean and θ and converges in distribution to $N(0,1)$. Justify your claim with appropriate theorem(s).

7.14 Let X and Y be bivariate normal with means μ_x, μ_y, standard deviations σ_x, σ_y, and correlation coefficient ρ. A random sample of size n is drawn giving the observations $x_1, y_1; x_2, y_2; \ldots; x_n, y_n$. The following sample statistics are then obtained.

$$m_{xx} = \Sigma(x_i - \bar{x})^2/n$$

$$m_{yy} = \Sigma(y_i - \bar{y})^2/n$$

$$m_{xy} = \Sigma(x_i - \bar{x})(y_i - \bar{y})/n$$

$$r_{xy}^2 = m_{xy}^2/(m_{xx}m_{yy})$$

Citing appropriate limiting theorems, show that

$$m_{xx} \xrightarrow{P} \sigma_x^2 \qquad m_{yy} \xrightarrow{P} \sigma_y^2 \qquad r_{xy} \xrightarrow{P} \rho$$

7.15   Let X be a k-dimensional random variable with $\mu_i = E(X_i)$, $v_{ij} = E(X_i X_j)$, and $\sigma_{ij} = E[(X_i-\mu_i)(X_j-\mu_j)]$. A random sample of size T is drawn from that population. The observations are arranged in a $T \times k$ matrix X. Define

$$\overline{X}_i = \frac{1}{T} \sum_{t=1}^{T} X_{ti}$$

$$s_{ij} = \frac{1}{T} \sum_{t=1}^{T} (X_{ti} - \overline{X}_i)(X_{tj} - \overline{X}_j) = \frac{1}{T} \sum_t X_{ti} X_{tj} - \overline{X}_i \overline{X}_j$$

Consider the $k \times k$ cross product matrix $X'X/T$. Using appropriate limiting theorems prove the following:

$$\overline{X}_i \xrightarrow{P} \mu_i \qquad \frac{1}{T} X'X \xrightarrow{P} V \qquad S \xrightarrow{P} \Sigma$$

where $V, S,$ and Σ are appropriate matrices. (Note: to prove a matrix convergence, show that a typical element converges to the corresponding limit.)

7.16   The conditional density of y given X is $N(X\beta, \sigma^2 I)$, where y is an $n \times 1$ vector of observations (*iid*), X is an $n \times k$ matrix of rank k $(< n)$, and β is a $k \times 1$ vector of fixed values. You are also given that $u = y - X\beta$, $\text{plim}(X'X/n) = \Sigma$, a non-singular matrix, and $\text{plim}(X'u/n) = \alpha$. Derive the probability limit of $z_n = (X'X)^{-1}X'y$. What are the conditions under which this plim will equal β? Carefully justify your steps.

8

ESTIMATION

One of the major tasks of an empirical statistician/econometrician is to find a probability model that best approximates the data generating process (DGP). The model is given by the triple $(\mathcal{R}, \mathcal{B}, F)$, where $\mathcal{R}$ is the real line, $\mathcal{B}$ is the σ-field of Borel sets introduced in Chapter 3, and F is the cumulative distribution function. An investigator usually has a pretty good idea of the random variables (X) that characterize the DGP. Based on frequency distributions or some fundamental principles, a probability density function $f(x;\theta)$ is typically formulated. A sample of observations on the random variables is then obtained. In earlier chapters we discussed a variety of ways of characterizing distributions and we derived statistical distributions of a number of statistics based on the sample observations, for both small and large samples. All the topics discussed so far were intended to prepare ourselves for the two main problems encountered in statistics, namely, the estimation of unknown parameters and the testing of hypotheses. The first problem is addressed in detail in this chapter and the second is discussed in Chapter 9.

We assume that the family of density functions $f(x;\theta)$ is known but that the values of the parameters θ are unknown. The first step in estimation is to obtain observations on one or more random variables. Let $x_1, x_2, \ldots, x_n$ be a set of observations (usually a random sample). Initially we restrict our attention to a single random variable X and a single parameter θ. The extension to multivariate and multiparameter cases is discussed later. The observations are used to construct estimates of θ. The formula for obtaining the estimate of a parameter is referred to as an **estimator** and the numerical value associated with it is called an **estimate**. Thus, an estimator is any function of $x_1, x_2, \ldots, x_n$. (Note that this is only another term for a *statistic*.)

Suppose we know from past observation that the income of a person is approximately normally distributed with mean μ and s.d. σ, both unknown. The problem of estimation is simply one of obtaining a sample of observations on incomes and then deriving estimators for

163

μ and σ. An obvious estimator of μ is $\bar{x}$, the sample mean. An alternative estimator is found by taking the largest income in the sample and the smallest income and averaging them. Which is a "better" estimate of μ? To answer this we have to develop criteria for estimators, which is done in the next two sections. Initially, only single parameters are considered. This is extended in Section 8.9 to the multiparameter case.

8.1   SMALL SAMPLE CRITERIA FOR ESTIMATORS

In this section we list a number of desirable properties for samples of finite size, in particular for small samples. The standard notation for an unknown parameter is θ and an estimator of θ is denoted by $\hat{\theta}$. We shall denote the parameter space by Θ. A function $g(\theta)$ is called **estimable** if there exists a statistic $u(x)$ such that $E[u(x)] = g(\theta)$.

Unbiasedness

An estimator $\hat{\theta}$ is called an **unbiased estimator** of θ if $E(\hat{\theta}) = \theta$, and $E(\hat{\theta}) - \theta = b(\theta)$, if it is nonzero, is called the **bias**. Note that $\hat{\theta}$ is a random variable because it is a function of random variables. Unbiasedness means that the population mean of the distribution of $\hat{\theta}$ is θ. This is obviously a desirable property. An example of an unbiased estimator is the sample mean $\bar{x}$ from a random sample of observations from $f(x; \theta)$ with $E(X) = \theta$. Because $\bar{x}$ has the property $E(\bar{x}) = \theta$, it is readily seen to be an unbiased estimator of θ.

Recall from Corollary 2 of Theorem 6.2 that the sample variance $s^2 = [\Sigma(x_i - \bar{x})^2]/(n-1)$ has the property that $E(s^2) = \sigma^2$. Thus, s^2 is an unbiased estimator of σ^2, but $[\Sigma(x_i - \bar{x})^2]/n = (n-1)s^2/n$ is not. This is the reason for dividing the sum of squares by $n-1$ rather than by n.

Practice Problems

8.1   Calculate the bias in $\hat{\sigma}^2 = [\Sigma(x_i - \bar{x})^2]/n$, the estimator of σ^2, and examine what happens as $n \to \infty$.

8.2   Show that if $\hat{\theta}$ is an unbiased estimator of θ, then $\hat{\theta}^2$ cannot be an unbiased estimator of θ^2. Is the sign of the bias positive or negative?

Although, the mean of a sample is an unbiased estimator of $\theta = E(x_i)$, it is easy to verify that the average of the smallest and largest observations is also an unbiased estimator of $E(x_i)$. Therefore unbiased estimators are not unique.

Practice Problem

8.3 Consider the estimator $\hat{\theta} = \Sigma_i\, w_i x_i$, where the w's are fixed and nonrandom ($\hat{\theta}$ is known as a **linear estimator**), and let $\theta = E(x_i)$. Derive the condition under which $\hat{\theta}$ will be an unbiased estimator of θ.

It is seen from the practice problem that there exist an infinite number of unbiased estimators and hence we need additional criteria to judge whether a particular estimator is "good."

Mean Squared Error

A commonly used measure of the adequacy of an estimator is $E[(\hat{\theta} - \theta)^2]$, which is called the **mean squared error (MSE)**. It is a measure of how close $\hat{\theta}$ is, on average, to the true θ. It can also be written as follows:

$$\text{MSE} = E[(\hat{\theta} - \theta)^2] = E[\hat{\theta} - E(\hat{\theta}) + E(\hat{\theta}) - \theta]^2$$

$E(\hat{\theta}) - \theta = b(\theta)$ is the bias in $\hat{\theta}$. Because it is independent of the observations, $b(\theta)$ is nonrandom. Hence,

$$\text{MSE} = E[\hat{\theta} - E(\hat{\theta})]^2 + b^2(\theta) + 2b(\theta)E[\hat{\theta} - E(\hat{\theta})] = \text{Var}(\hat{\theta}) + b^2(\theta)$$

because $E[\hat{\theta} - E(\hat{\theta})] = 0$. Therefore, MSE is the sum of the variance and the square of the bias.

If $\hat{\theta}$ is unbiased, $b(\theta) = 0$, and hence MSE = $\text{Var}(\hat{\theta})$. Thus, in the case of unbiased estimators we can compare the respective variances. The estimator with the smaller variance is "better" because, on average, it will be closer to the true θ. For an estimator to be "good" in the general case, we would like MSE to be small. This leads to the concept of *relative efficiency*.

Relative Efficiency

Let $\hat{\theta}_1$ and $\hat{\theta}_2$ be two alternative estimators of θ. Then the ratio of the respective MSEs, $E[(\hat{\theta}_1 - \theta)^2] \,/\, E[(\hat{\theta}_2 - \theta)^2]$, is called the

relative efficiency of $\hat{\theta}_1$ with respect to $\hat{\theta}_2$. If $\hat{\theta}_1$ and $\hat{\theta}_2$ are both unbiased, this ratio reduces to $\text{Var}(\hat{\theta}_1)/\text{Var}(\hat{\theta}_2)$. As a smaller MSE or variance means a greater precision of estimates, another desirable characteristic of an estimator is a high degree of efficiency.

If you are comparing two unbiased estimates $\hat{\theta}_1$ and $\hat{\theta}_2$, then the one with the smaller variance is better because it is "closer" to the true value. If either of them is biased, there may be a trade-off. One of them may be biased but may have a much smaller variance. The MSE is the one often used to decide whether the trade-off is worth it, lower values indicating greater efficiency. This is especially useful when forecasts from different models are compared for accuracy.

Practice Problem

8.4   Derive the variance and the mean square error of $\hat{\theta}$ in Practice Problem 8.3.

Example 8.1

Let x_i, $i = 1, 2, \ldots, n$, be a random sample from the uniform distribution on $(0, \theta)$, $0 < x_i < \theta$, for all i, with the density function $f(x) = 1/\theta$. Denote by z_n the largest observation.

$$F_{z_n}(z) = P(z_n \leq z) = P(\text{all the } x\text{'s are } \leq z)$$

$$= [P(X \leq z)]^n = \left[\int_0^z \frac{1}{\theta}\, dx\right]^n = \frac{z^n}{\theta^n}$$

Therefore the density function of z_n is

$$f_{z_n}(z) = \frac{nz^{n-1}}{\theta^n} \qquad 0 < z < \theta$$

For this distribution, $E(\bar{x}) = E(X) = \theta/2$, and hence $\hat{\theta}_n = 2\,\bar{x}$ is an unbiased estimator of θ.

$$E(z_n) = \int_0^\theta \frac{nz^n}{\theta^n}\, dx = \frac{n}{n+1}\theta$$

Hence $\theta_n^* = (n+1)z_n/n$ is also an unbiased estimator of θ. Which of these is more efficient?

$$E(X^2) = \int_0^\theta \frac{x^2}{\theta}\, dx = \frac{\theta^2}{3}$$

Therefore, $\text{Var}(X) = \theta^2/12$ and $\text{Var}(\bar{x}) = \theta^2/(12n)$. It follows that $\text{Var}(\hat{\theta}_n) = \theta^2/(3n)$.

$$E\left(z_n^2\right) = \int_0^\theta \frac{nz^{n+1}}{\theta^n} \, dz = \frac{n}{n+2} \theta^2$$

The variance of z_n can be computed as

$$\text{Var}\left(z_n\right) = \frac{n\theta^2}{(n+2)(n+1)^2}$$

Therefore, $\text{Var}\left(\theta_n^*\right) = \theta^2/[n\,(n+2)]$. Because the sample size is usually larger than 1, we have $\text{Var}\left(\theta_n^*\right) < \text{Var}\left(\hat{\theta}_n\right)$. Therefore, θ_n^* is more efficient than $\hat{\theta}_n$.

UMVU Estimators

An estimator $\hat{\theta}$ of θ is called a **uniformly minimum variance unbiased (UMVU) estimator** if $E\left(\hat{\theta}\right) = \theta$ and for any other unbiased estimator θ^*, $\text{Var}\left(\hat{\theta}\right) \le \text{Var}\left(\theta^*\right)$ for every θ. Thus, among the class of unbiased estimators, a UMVU estimator has the smallest variance.

Practice Problem

8.5 In Practice Problem 8.3, derive the UMVU estimator of θ by choosing the w_i's appropriately.

Sufficiency

In many cases, we may be able to identify a statistic $\hat{\theta}$ that summarizes all the information about θ that the entire sample observations contain. If such a statistic can be found, it would be more convenient to work with rather than the individual observations. Thus $\hat{\theta}$ is "sufficient" for θ. This notion is formalized as follows.

Definition 8.1

Let $\hat{\theta}$ be a sample statistic and θ^* any other statistic not a function of $\hat{\theta}$. Also, let $f(x;\theta)$ be the density function. $\hat{\theta}$ is said to be a **sufficient statistic** *for θ if and only if the conditional density of θ^* given $\hat{\theta}$ is independent of θ, for every choice of θ^*. Equivalently, the conditional density of the sample given $\hat{\theta}$, that is, $f(x_1, x_2, \ldots, x_n \mid \hat{\theta})$, is independent of θ.*

A sufficient statistic thus reduces the information contained in the individual observations to a single statistic. In Section 8.3 we provide a general method of testing whether a particular statistic is sufficient.

Example 8.2

From the Poisson distribution with $f(x; \theta) = e^{-\theta}\theta^x/x!$ for $x = 0$, 1, 2, . . . draw a random sample of size 2, that is, x_1 and x_2. The joint density function is

$$f(x_1, x_2; \theta) = \frac{e^{-2\theta}\theta^{x_1 + x_2}}{x_1! x_2!}$$

Now consider the two alternative statistics, $T_1 = x_1 + x_2$, and $T_2 = x_1^2 + x_2^2$. We will presently show that T_1 is a sufficient statistic but that T_2 is not. First compute the unconditional probability of T_1 (say for $T_1 = 4$). $T_1 = 4$ only for the sample points $(0, 4), (1, 3), (2, 2), (3, 1),$ and $(4, 0)$. We have

$$P(T_1 = 4) = \frac{e^{-2\theta}\theta^4}{0! \, 4!} + \frac{e^{-2\theta}\theta^4}{1! \, 3!} + \cdots = k_1 e^{-2\theta}\theta^4 \qquad (8.1)$$

where k_1 is a constant independent of θ.

$$P(X = x \mid T_1 = 4) = \frac{P(X = x \text{ and } T_1 = 4)}{P(T_1 = 4)} \qquad (8.2)$$

We need the numerator for a pair of x's such that $T_1 = x_1 + x_2 = 4$ and $X = x_1$, $X = x_2$. This is given by

$$P(X = x \text{ and } T_1 = 4) = \frac{e^{-2\theta}\theta^4}{x_1! x_2!} \qquad (8.3)$$

Hence

$$P(X = x \mid T_1 = 4) = \left[\frac{e^{-2\theta}\theta^4}{x_1! x_2!}\right] \div \left[k_1 e^{-2\theta}\theta^4\right] = \frac{1}{k_1 x_1! x_2!}$$

which is independent of θ. Thus, the conditional probability of X given $T_1 = 4$ does not depend on θ. It is easy to verify that this would work for any given T_1, not just 4. Let us now show that knowing T_2 is not adequate. Suppose we know that $T_2 = 25$. First compute $P(T_2 = 25)$, that is, $P(x_1^2 + x_2^2 = 25)$. The sample points are now $(0, 5), (3, 4), (4, 3),$ and $(5, 0)$.

$$P(x_1^2 + x_2^2 = 25) = \frac{e^{-2\theta}\theta^5}{0! \, 5!} + \frac{e^{-2\theta}\theta^7}{3! \, 4!} + \frac{e^{-2\theta}\theta^7}{4! \, 3!} + \frac{e^{-2\theta}\theta^5}{5! \, 0!} \qquad (8.4)$$

$$P(X = x \text{ and } T_2 = 25) = \frac{e^{-2\theta}\theta^{x_1 + x_2}}{x_1! x_2!}$$

or specifically for $x_1 = 3$ and $x_2 = 4$, we have

$$P\ [(3,\ 4)\ \text{and}\ T_2 = 25] = \frac{e^{-2\theta}\theta^7}{3!\ 4!} \tag{8.5}$$

The conditional probability we need is $P(X = x \mid T_2 = 25)$, which is obtained by dividing equation (8.5) by (8.4). It is easily seen that this conditional probability will have θ in it. This is because θ^7 in the numerator cannot be factored out as was done for T_1. Thus T_2 is not a sufficient statistic.

Sufficient statistics are, however, not unique (in the above example, $\bar{x}$ is also a sufficient statistic) and hence it would be useful to obtain a statistic that provides the maximum reduction. This idea leads to the concept of *minimal sufficiency*.

Minimal Sufficiency

$\hat{\theta}$ is **minimal sufficient** if, for any other sufficient statistic θ^* we can find a function $h(\cdot)$ so that $\hat{\theta} = h(\theta^*)$. This concept is, however, not used much in econometrics. Lehmann and Scheffè (1950, Theorem 6.3) provide a method of finding a minimal sufficient statistic. Further details may be found in that paper and from Casella and Berger (1990, Section 6.1.2).

Completeness

The concept of **completeness** is relevant not to estimators but to density functions. A family of densities $f(x;\theta)$ is said to be complete if $E[h(x)] = 0$ implies that $h(x) = 0$ for all θ, where $h(x)$ is a continuous function. Thus the only unbiased estimator of 0 is zero. The usefulness of this concept comes from the fact that if $f(x;\theta)$ is complete, then two functions of x with the same expectation must be identical because their difference has zero expectation. Thus, if $f(x;\theta)$ is complete, then $E[h_1(x)] = E[h_2(x)]$ implies that $h_1(x) = h_2(x)$. For more details on completeness, see Casella and Berger (1990, Section 6.1.4).

8.2   LARGE SAMPLE PROPERTIES OF ESTIMATORS

In addition to the criteria listed in the previous section, we may want to stipulate desirable properties that are applicable for large samples only. In this section we list a number of desirable **asymptotic** properties.

Asymptotic Unbiasedness

If an estimator has the property that $\mathrm{Var}(\hat{\theta}_n)$ and $\sqrt{n}\,(\hat{\theta}_n - \theta)$ tend to zero as the sample size increases, then it is said to be **asymptotically unbiased**. Note that this is a stronger requirement than to say that $E(\hat{\theta}_n - \theta)$ tends to zero. The estimate $\hat{\sigma}^2$ given in Practice Problem 8.1 is an example of an asymptotically unbiased estimator.

Consistency

Another desirable property of $\hat{\theta}$ is that as sample size n increases, $\hat{\theta}$ must approach the true θ. This property is called **consistency**. Three types of consistency measures can be identified.

Simple Consistency: Let $\hat{\theta}_1$, $\hat{\theta}_2$, ..., $\hat{\theta}_n$ be a sequence of estimators of θ. This sequence is a **simple consistent estimator** of θ if, for every $\varepsilon > 0$,

$$\lim_{n \to \infty} P(\,|\hat{\theta}_n - \theta\,| < \varepsilon) = 1 \qquad \theta \in \Theta$$

Thus $\hat{\theta}_n$ is a simple consistent estimator if $\mathrm{Plim}\,\hat{\theta}_n = \theta$.

Squared-error Consistency: The sequence $(\hat{\theta}_n)$ is a **squared-error consistent estimator** of θ if

$$\lim_{n \to \infty} E[(\hat{\theta}_n - \theta)^2] = 0$$

Strong Consistency: $\hat{\theta}_n$ is said to be **strongly consistent** if $P[\lim_{n \to \infty} \hat{\theta}_n = \theta] = 1$.

You will readily recognize the first as convergence in probability, $(\hat{\theta}_n \xrightarrow{p} \theta)$, the second as convergence in mean square $(\hat{\theta}_n \xrightarrow{m.s.} \theta)$, and the third as almost sure convergence $(\hat{\theta}_n \xrightarrow{a.s.} \theta)$. We have shown that m.s. and almost sure convergence are stronger than convergence in probability. Hence, simple consistency is weaker than squared-error and almost sure consistency. Also note that since $\mathrm{MSE} = \mathrm{Var}(\hat{\theta}_n) + b_n^2(\theta)$, squared-error consistency implies that both the bias and variance approach zero. Thus if $\mathrm{Var}(\hat{\theta}_n) \to 0$ and $b_n(\theta) \to 0$ (or $= 0$), then $\mathrm{Plim}\,\hat{\theta}_n = \theta$. Figure 8.1 illustrates a biased but consistent estimator.

In Example 8.1, the variances of both the unbiased estimators, $\hat{\theta}_n$ and θ_n^*, tend to zero as the sample size increases. Therefore they are both mean square consistent and hence by Theorem 7.6 are also simple consistent.

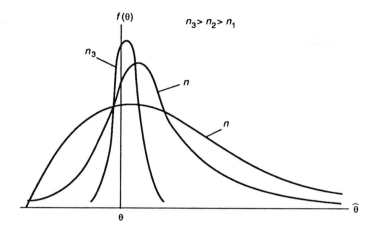

Figure 8.1   An example of a biased but consistent estimator

Does unbiasedness imply consistency? Absolutely not, as is seen from the following trivial counterexample. Let $x_1, x_2, \ldots, x_n$ be a random sample with $E(x_i) = \mu$. The first observation x_1 is an unbiased estimator of μ, but letting $n \to \infty$ is not going to make Plim $x_1 = \mu$.

The sample mean $\bar{x}_n$ from *iid* observations with $E(X) = \mu$ has the property that $E(\bar{x}_n) = \mu$ and $\text{Var}(\bar{x}_n) = \sigma^2/n \to 0$. Hence $\bar{x}_n$ is both unbiased and consistent. As stated before, from Corollary 2 of Theorem 6.2, $s^2 = [\Sigma(x_i - \bar{x})^2]/(n-1)$ has the property that $E(s^2) = \sigma^2$. Thus, s^2 is an unbiased estimator of σ^2, but $\hat{\sigma}^2 = [\Sigma(x_i - \bar{x})^2]/n = (n-1)s^2/n$ is not. It is easy to show, however, that because of the strong law of large numbers, they are both strongly consistent estimators.

$$\frac{1}{n}\Sigma(x_i - \bar{x})^2 = \frac{1}{n}\Sigma x_i^2 - \bar{x}^2$$

By the strong law, $\bar{x}$ converges almost surely to μ and $(\Sigma x_i^2)/n$ converges almost surely to $E(x_i^2)$. Therefore,

$$\hat{\sigma}^2 = \frac{1}{n}\Sigma(x_i - \bar{x})^2 \overset{a.s.}{\to} E(x_i^2) - \mu^2 = \sigma^2$$

Because $s^2 = n\hat{\sigma}^2/(n-1)$, its probability limit is also σ^2. Thus both estimators are strongly consistent.

Asymptotic Efficiency

No one estimator can be most efficient for all values of θ. Some are good for certain values of θ and others are more efficient in some other range of θ. For instance, let $\hat{\theta} = \theta_0$ be fixed regardless of observations. If $\theta = \theta_0$ or near it, then $\hat{\theta}$ is a pretty good estimate; but when θ is far from θ_0, it is a very poor estimate. When we deal with consistent estimators, however, the range of values for θ for which one estimate is more efficient than another shrinks as the sample size increases. This suggests a notion of **asymptotic efficiency** that is defined below.

Definition 8.2

*Let $\hat{\theta}_n$ be a consistent estimator of θ. $\hat{\theta}_n$ is said to be **asymptotically efficient** if there is no other consistent estimator θ_n^* for which* $\lim\limits_{n \to \infty} \left[E[(\hat{\theta}_n - \theta)^2] / E[(\theta_n^* - \theta)^2] \right] > 1$ *for all θ in some open interval.* [see Taylor and Mann (1972) for the definition of *lim*.]

Best Asymptotic Normality

The sequence of estimators $(\hat{\theta}_n)$ is a **best asymptotically normal (BAN)** estimator if all the following conditions are satisfied.

(a) $\hat{\theta}_n \xrightarrow{p} \theta$ for every $\theta \in \Theta$, that is, $\hat{\theta}_n$ is consistent.

(b)   The distribution of $\sqrt{n}(\hat{\theta}_n - \theta) \xrightarrow{d} N[0, \sigma^2(\theta)]$, where $\sigma^2(\theta) = \lim n \, \text{Var}(\hat{\theta}_n).]$

(c)   There is no other sequence (θ_n^*) that satisfies (a) and (b) and is such that $\sigma^2(\theta) > \sigma^{*2}(\theta)$ for every θ in some open interval. $[\sigma^{*2}(\theta) = \lim n \, \text{Var}(\theta_n^*)]$

It will be seen in Section 8.12 that BAN estimators do exist.

8.3   THE LIKELIHOOD FUNCTION

Let $x_1, x_2, \ldots, x_n$ be a sample of observations on X. Then the **likelihood function** of the parameter θ is defined as

$$L(\theta; x) = f_X(x_1, x_2, \ldots, x_n; \theta)$$

where $f(\cdot)$ is the joint density function of the x_i's. In the case of a random sample with *iid* observations, the likelihood function reduces to

$$L(\theta; x) = \prod_{i=1}^{n} f_X(x_i; \theta)$$

Although the expression for the likelihood function is identical to that of the joint density function, their interpretations are quite different. In the case of a discrete random variable, the density function gives the probability of observing the values x_1, x_2, and so on, *given the value of the parameter* θ. The likelihood function on the other hand is a function of θ *for a given set of observations*. As the object here is to obtain estimates of unknown parameters based on sample observations, the likelihood function is more relevant in the context of estimation.

Theorem 8.1 (factorization theorem)

A statistic $\hat{\theta}$ *will be sufficient for* θ *if and only if the likelihood function* $L(\theta; x)$ *can be factored as*

$$L(\theta; x) = h(\hat{\theta}, \theta)g(x_1, \ldots, x_n)$$

In other words, L is the product of two functions, one independent of $\hat{\theta}$ *and* θ *and the other involving only* $\hat{\theta}$ *and* θ.

Proof: This will be proved only for the discrete case. The proof for the continuous case is cumbersome [see Lehmann (1959), pp. 17–19].

To prove that the condition is sufficient, we assume that $L(\theta; x) = h(\hat{\theta}, \theta)g(x)$ and then show that $f(x \mid \hat{\theta})$ is independent of θ.

$$f(x \mid \hat{\theta} = \theta_0) = \frac{P(X = x \text{ and } \hat{\theta} = \theta_0)}{P(\hat{\theta} = \theta_0)}$$

If $\hat{\theta} \neq \theta_0$, the numerator is zero, otherwise it is equal to $L(\theta; x)$ because $\hat{\theta} = \theta_0$ is then a certainty. Thus the numerator is $h(\theta_0, \theta)g(x)$. The denominator is

$$\sum_{\hat{\theta} = \theta_0} P(X = x) = \sum_{\hat{\theta} = \theta_0} h(\hat{\theta}, \theta)g(x) = h(\theta_0, \theta) \sum_{\hat{\theta} = \theta_0} g(x)$$

Therefore $f(x \mid \hat{\theta} = \theta_0)$ is the ratio in which $h(\theta_0, \theta)$ is common and cancels out. The rest is independent of θ and hence $\hat{\theta}$ is a sufficient statistic for θ.

To prove necessity we assume that $f(x\,|\,\hat{\theta}) = c(x;\hat{\theta})$ is independent of θ. $P(X=x) = P(X=x$ and $\hat{\theta} = \theta_0)$ when $\hat{\theta} = \theta_0$, and 0 otherwise. From the definition of conditional probability,

$$P(X=x \text{ and } \hat{\theta} = \theta_0) = P(X=x\,|\,\hat{\theta} = \theta_0)\,P(\hat{\theta} = \theta_0)$$
$$= c(x;\hat{\theta})\,P(\hat{\theta} = \theta_0)$$

Thus, $L(\theta;x) = c(x;\hat{\theta})\,h(\hat{\theta},\theta)$, which is the desired factorization because $c(x;\hat{\theta})$ is independent of θ and $h(\hat{\theta},\theta)$ depends on x only through $\hat{\theta}$.

This theorem is very useful in finding sufficient statistics. All we have to do is set up the likelihood function and by inspection find a $\hat{\theta}$ that factorizes it.

Example 8.3

Let $f(x;\theta) = (1/\theta)e^{-x/\theta}$ with $x > 0$. Then

$$L(\theta;x) = \frac{1}{\theta^n}\,e^{-(\Sigma x_i)/\theta}$$

Define $\hat{\theta} = (\Sigma x_i)$. Then $L = (1/\theta^n)e^{-\hat{\theta}/\theta}$, which satisfies the factorization criterion. [Note that $\bar{x}$ is also a sufficient statistic.]

Practice Problem

8.6   Let x_i $(i = 1, 2, \ldots, n)$ be a random sample from $N(0, \theta)$. Derive the likelihood function and identify a statistic that is sufficient for θ.

The usefulness of sufficient statistics lies in their ability to help us find unbiased estimators that have the lowest variance. This is expressed in the next two theorems established by C.R. Rao and D. Blackwell.

Theorem 8.2 (Rao-Blackwell theorem)

Let x_i, $i = 1, 2, \ldots, n$, be a random sample from the distribution of X with the density function $f(x;\theta)$, and let the statistic $\hat{\theta} = \hat{\theta}(x_1, x_2, \ldots, x_n)$ be sufficient for θ. Also, let the statistic $T = T(x_1, x_2, \ldots, x_n)$, not a function of $\hat{\theta}$, be an unbiased estimator of $u(\theta)$. Denote by $v(\hat{\theta})$, the conditional expectation of T given $\hat{\theta}$. Then

(a) $E[v(\hat{\theta})] = u(\theta)$, *that is, $v(\hat{\theta})$ is also an unbiased estimator of $u(\theta)$.*

(b) $\text{Var}[v(\hat{\theta})] \le \text{Var}(T)$, *that is, the newly constructed unbiased estimator of $u(\theta)$ has a smaller (no larger) variance.*

(c) $v(\hat{\theta})$ *is independent of θ and hence can be computed from the sample.*

Before proving this theorem, let us discuss its usefulness. Our ultimate objective is to get the "best" estimator of θ [or of a function $u(\theta)$ of it]. By inspection of the likelihood function, using the factorization criterion, we may get a sufficient statistic $\hat{\theta}$. Also suppose we find by some procedure an unbiased estimator $T(x)$ of $u(\theta)$. The theorem then gives us a procedure by which we can get another unbiased estimator with a smaller variance. This procedure is called **Rao-Blackwellization**. Thus, this conditioning of a statistic by a sufficient statistic improves the statistic in the sense that the variance is generally reduced.

Proof: Part (c) is easy to establish. Denote the joint density of T and $\hat{\theta}$ by $g(T, \hat{\theta}; \theta)$ and the conditional density of T given $\hat{\theta}$ by $P(T|\hat{\theta})$. Because $\hat{\theta}$ is a sufficient statistic, $P(T|\hat{\theta})$ is independent of θ and therefore $E(T|\hat{\theta})$, which is $v(\hat{\theta})$, is also independent of θ, proving part (c). Part (a) is also easy to show. By the law of iterated expectations (Theorem 5.5), $E(T) = E_{\hat{\theta}} E[T|\hat{\theta}]$. But by assumption, $E(T) = u(\theta)$ and $E[T|\hat{\theta}] = v(\hat{\theta})$. Hence $E_{\hat{\theta}}[v(\hat{\theta})] = u(\theta)$, that is, $v(\hat{\theta})$ is unbiased. To prove part (b), recall Theorem 5.6. We have

$$\text{Var}(T) = E_{\hat{\theta}}[\text{Var}(T|\hat{\theta})] + \text{Var}[v(\hat{\theta})] \ge \text{Var}[v(\hat{\theta})]$$

Example 8.4

We illustrate Rao-Blackwellization using the uniform distribution given in Example 8.1. The likelihood function is $L(\theta; x) = 1/\theta^n$. It follows trivially that z_n, the largest observation, is a sufficient statistic for θ. In Example 8.1 we had two unbiased estimators of θ, $\hat{\theta}_n = 2\bar{x}$ and $\theta_n^* = (n+1)z_n/n$. Define the ordered series of observations as $x(i)$, that is,

$$x(1) \le x(2) \le \cdots \le x(n) = z_n$$

Because $0 \le x(i) \le z_n$ for $i = 1, 2, \ldots, n-1$, we have

$$f_{x(i)|z_n}(u) = \frac{1}{z_n} \qquad [\text{analogous to } f(x) = 1/\theta]$$

Therefore,

$$E[x(i)|z_n] = \int_0^{z_n} \frac{u}{z_n}\, du = \frac{z_n}{2}$$

Let x^* be the mean of the $n-1$ x's that are not the largest value z_n, that is, $x^* = [\Sigma_{i=1}^{n-1} x(i)]/(n-1)$. Then,

$$\bar{x} = \frac{1}{n}\Sigma x_i = \frac{1}{n}\Sigma x(i) = \frac{z_n + \overset{n-1}{\underset{i=1}{\Sigma}} x(i)}{n} = \frac{z_n + (n-1)x^*}{n}$$

It follows from the previous derivation that

$$E(\bar{x} \mid z_n) = E\left[\frac{z_n + (n-1)x^*}{n} \mid z_n\right]$$

$$= E\left[\frac{z_n}{n} \mid z_n\right] + \frac{n-1}{n} E\left[x^* \mid z_n\right]$$

$$= \frac{z_n}{n} + \frac{n-1}{n}\frac{z_n}{2} = \frac{n+1}{2}n\, z_n$$

The Rao-Blackwell theorem says that if $\hat{\theta}$ is sufficient, then $E(T|\hat{\theta}) = v(\hat{\theta})$ is more efficient than T. To apply this, we need to identify $\hat{\theta}$ and $T(x)$. We have shown that z_n is sufficient for θ and that $\hat{\theta}_n = 2\bar{x}$ is an unbiased estimator of θ. By the theorem, $E[\hat{\theta}_n |z_n]$ will be more efficient than $\hat{\theta}_n$.

$$E[\hat{\theta} |z_n] = 2E[\bar{x}|z_n] = \frac{n+1}{n} z_n = \theta_n^*$$

Therefore, θ_n^* is the Rao-Blackwellized improvement over $\hat{\theta}_n$.

Although the Rao-Blackwell theorem is useful in reducing the variance of an estimator, there may be several such estimators, all of them unbiased and all of them functions of $\hat{\theta}$ only. It will therefore be useful to narrow the search further. The following theorem accomplishes this. It states that if the density of a sufficient statistic is *complete*, then there is a unique unbiased estimator that depends on the sufficient statistic.

Theorem 8.3

As before, let $x_i \sim f(x_i; \theta)$ and let $\hat{\theta}$ be sufficient. Also let the density of $\hat{\theta}$ be complete. If a function of $\hat{\theta}$, say $v(\hat{\theta})$, exists such that $E[v(\hat{\theta})] = u(\theta)$, then $v(\hat{\theta})$ is the UMVU estimator of $u(\theta)$.

Proof: Let v^* be another unbiased estimator of θ. Then $E(v^*)$ $= u(\theta) = E(v)$, that is, $E[v - v^*] = 0$. But since the density of $\hat{\theta}$ is complete, $E[v - v^*] = 0$ implies that $v - v^* = 0$, that is, $v^* = v$. Hence $v(\hat{\theta})$ is unique. The minimum variance follows from part (b) of Theorem 8.2.

To summarize, suppose that by using some method, we find an unbiased estimator of θ. We then Rao-Blackwellize it with respect to a sufficient statistic of θ and check whether the density of the sufficient statistic is complete. If it is, our search has ended and we have the UMVU estimator of θ. In the next section, we discuss a method of obtaining estimates that may be unbiased or can be made unbiased.

8.4   THE PRINCIPLE OF MAXIMUM LIKELIHOOD

In a statistical experiment, the observations $x_1, x_2, \ldots, x_n$ come from some underlying DGP. An investigator typically attempts to approximate the DGP by a family of probability density functions $f(x; \theta)$ and then tries to infer the value of θ based on the observations. Consider two possible choices for θ, say θ_1 and θ_2. The probability of observing the sample $x_1, x_2, \ldots, x_n$ is $L(\theta_1; x)$ if θ_1 is true and $L(\theta_2; x)$ if θ_2 is true. If $L(\theta_1; x) > L(\theta_2; x)$, then $\theta = \theta_1$ gives a higher joint probability of the actual realization, namely, that x_1, $x_2, \ldots, x_n$ was observed. Fisher (1922, 1925) argued that in this case we should choose θ_1 over θ_2. This is based on the notion that an event occurs because "it is most likely to happen." The generalization of this idea is the **principle of maximum likelihood**. Simply stated, it says that among all the possible values θ can assume, the one it is *most likely* to assume is the one that maximizes the likelihood function given by

$$L(\theta; x_1, x_2, \ldots, x_n) = f(x_1, x_2, \ldots, x_n; \theta)$$

Definition 8.3

Let

$$L(\theta; x) = f(x_1, x_2, \ldots, x_n; \theta)$$

be the likelihood function of the sample. Then if the statistic $\hat{\theta} = u(x_1, \ldots, x_n)$ maximizes $L(\theta; x)$ for $\theta \in \Theta$, then $\hat{\theta}$ is said to be the **maximum likelihood estimator (MLE)** *of θ. If x_1, x_2, $x_3, \ldots, x_n$ are iid, then*

$$L(\theta; x) = \prod_{i=1}^{n} f(x_i; \theta)$$

Thus the MLE procedure selects the density function from the family $f(x; \theta)$ that makes the actual realization most probable.

If the likelihood function is differentiable with respect to θ, the first-order condition for maximizing the likelihood function is $dL/d\theta = 0$, if θ is the only parameter, or $\partial L/\partial\theta = 0$, if θ is one of several parameters. Because the logarithmic transformation is monotonic, $\partial L/\partial\theta = 0$ if and only if $\partial(\ln L)/\partial\theta = 0$. The log-likelihood ($\ln L$) is more convenient because, for *iid* observations,

$$\frac{\partial \ln L}{\partial \theta} = 0 = \sum_{i=1}^{n} \left[\frac{\partial \ln f(x_i; \theta)}{\partial \theta} \right]$$

which is often more convenient to solve.

Example 8.5

Let $x_i \sim N(\mu, \sigma^2)$. Find the MLEs of μ and σ^2.

$$L(\theta; x) = \prod_{i=1}^{n} f(x_i; \theta) = \left[\frac{1}{\sigma\sqrt{2\pi}} \right]^n e^{-\Sigma(x_i - \mu)^2/(2\sigma^2)}$$

The log-likelihood is

$$\ln L = -n \ln \sigma - n \ln \sqrt{2\pi} - \frac{1}{2\sigma^2} \sum_i (x_i - \mu)^2$$

$$\frac{\partial \ln L}{\partial \mu} = 0 = \frac{1}{2\sigma^2} 2\sum_i (x_i - \mu)$$

Therefore $n\mu = \Sigma x_i$, or $\hat{\mu} = \bar{x}$.

$$\frac{\partial \ln L}{\partial \sigma} = -\frac{n}{\sigma} + \frac{\Sigma(x_i - \mu)^2}{\sigma^3} = 0$$

Hence $\sigma^2 = [\Sigma(x_i - \mu)^2]/n$. As we do not know μ, use the MLE of $\mu (= \bar{x})$ and get $\hat{\sigma}^2 = [\Sigma(x_i - \bar{x})^2]/n$. Thus the MLE of μ is $\bar{x}$ and that of $\sigma^2 = [\Sigma(x_i - \bar{x})^2/n$. Note that $E(\bar{x}) = \mu$, but $E(\hat{\sigma}^2) \neq \sigma^2$. Therefore *the maximum likelihood method need not give unbiased estimates.* Here we can estimate μ without knowing σ^2, but not vice versa. [Verify the sufficient conditions for maximization.]

Example 8.6

Let $x_1, x_2, \ldots, x_n$ be drawn from the uniform distribution with density $f(x; a, b) = 1/(b - a)$, for $a \leq x \leq b$. The likelihood function is $L(a, b; x) = 1/(b - a)^n$. Note that L is large when $b - a$ is small, which means that we should make b as small as possible and a as large as possible. Thus, MLE of a is the smallest value of the x's and the MLE of b is the largest value of the x's. Because L is monotonic in a and b, $\partial L / \partial \theta \neq 0$.

8.5   LOWER BOUNDS FOR VARIANCES OF ESTIMATORS

We stated earlier that under certain conditions UMVU estimators may exist. We now examine what that minimum variance is and whether it is attainable. Before doing that, however, we list a number of regularity conditions that are needed for many of the results to hold.

Let $f(x; \theta)$ be the density function from which a random sample $x_1, x_2, \ldots, x_n$ is drawn. Then $\ln L = \Sigma_{i=1}^{n} \ln f(x_i; \theta)$.

Regularity Conditions

(1) θ lies in an open interval Θ of the real line (which may include $\pm \infty$).

(2)   For all $\theta \in \Theta$, $\partial L / \partial \theta$ and $\partial^2 L / \partial \theta^2$ exist and are continuous.

(3) $\int L(\theta; x)\, dx$ can be differentiated under the integral sign, that is, $\int [\partial L(\theta; x)/\partial \theta]\, dx$ exists.

(4) $E\left[\{\partial \ln L / \partial \theta\}^2 \right]$ and $E\left[\partial^2 \ln L / \partial \theta^2 \right] < \infty$ for all $\theta \in \Theta$.

(5) $\int T(x_1, x_2, \ldots, x_n) L(\theta; x)\, dx$ can be differentiated under the integral sign with respect to θ, that is, $\int T(x_1, x_2, \ldots, x_n)[\partial L(\theta; x)/\partial \theta]\, dx$ exists.

The Score and Information of the Sample

Define

$$S(\theta; x) = \frac{\partial(\ln L)}{\partial \theta} \quad \left[= \sum_{i=1}^{n} \frac{\partial \ln f(x_i; \theta)}{\partial \theta} \quad \text{for a random sample} \right]$$

$S(\theta;x)$ is called the **score function** of the sample. Several properties of the score function are easy to obtain. We know that, for continuous random variables, $\int L(\theta;x)\,dx = 1$. Differentiate both sides partially with respect to θ. Then

$$0 = \int \frac{\partial L}{\partial \theta}\,dx = \int \frac{1}{L}\frac{\partial L}{\partial \theta} L(\theta;x)\,dx = \int \frac{\partial \ln L}{\partial \theta} L\,dx = E\left[\frac{\partial \ln L}{\partial \theta}\right]$$

Hence,

$$E(S) = E\left[\frac{\partial \ln L}{\partial \theta}\right] = 0$$

that is, the score function has zero expectation. Now differentiate the above integral once more with respect to θ.

$$0 = \int \frac{\partial \ln L}{\partial \theta}\frac{\partial L}{\partial \theta}\,dx + \int \frac{\partial^2 \ln L}{\partial \theta^2} L\,dx$$

$$= \int \left[\frac{\partial \ln L}{\partial \theta}\right]^2 L(\theta;x)\,dx + \int \frac{\partial^2 \ln L}{\partial \theta^2} L\,dx$$

Therefore

$$E\left[\left\{\frac{\partial \ln L}{\partial \theta}\right\}^2\right] = -E\left[\frac{\partial^2 \ln L}{\partial \theta^2}\right] = I(\theta)$$

$I(\theta)$ is called the **information (or Fisher information) in the sample**. It follows from this that

$$\mathrm{Var}(S) = \mathrm{Var}\left[\frac{\partial \ln L}{\partial \theta}\right] = E\left[\left\{\frac{\partial \ln L}{\partial \theta}\right\}^2\right] = I(\theta)$$

Thus the score function has zero mean and variance equal to the information.

Theorem 8.4 (Cramer-Rao inequality)

Let $x_i \sim f(x;\theta)$ and $T = T(x_1, x_2, \ldots, x_n)$ be a statistic such that $E(T) = u(\theta)$, differentiable in θ. Also assume the regularity conditions given earlier. Let $u(\theta) - \theta = b(\theta)$, the bias in T. Then

$$\mathrm{Var}(T) \geq \frac{[u'(\theta)]^2}{I(\theta)} = \frac{[1 + b'(\theta)]^2}{I(\theta)}$$

where $I(\theta)$ is Fisher's information.

Corollary: If T is an unbiased estimator of θ, then $\mathrm{Var}(T) \geq 1/I(\theta)$.

The above inequality is called the **Cramer-Rao inequality** and the lower bound for Var (T) is called the **Cramer-Rao lower bound (CRLB)**.

Proof: For any T and S, $\text{Cov}^2(T, S) \le \text{Var}(T) \text{Var}(S)$, by Theorem 5.4. Let $T = T(x_1, x_2, \ldots, x_n)$ be the statistic given above for which $E(T) = u(\theta)$ and let $S = S(\theta; x)$ be the score function specified earlier. We showed that $E(S) = 0$ and $\text{Var}(S) = I(\theta)$. $\text{Cov}(T, S) = E(ST)$ because $E(S) = 0$. Also,

$$u(\theta) = E(T) = \int T L \, dx$$

Differentiate both sides with respect to θ.

$$u'(\theta) = \int T \frac{\partial L}{\partial \theta} \, dx = \int T \left[\frac{1}{L} \frac{\partial L}{\partial \theta} \right] L \, dx$$

$$= \int TSL \, dx = E(TS) = \text{Cov}(T, S)$$

Using all these, we get

$$[u'(\theta)]^2 = \text{Cov}^2(T, S) \le I(\theta) \text{Var}(T)$$

Therefore

$$\text{Var}(T) \ge \frac{[u'(\theta)]^2}{I(\theta)} = \frac{[1 + b'(\theta)]^2}{I(\theta)}$$

because $u(\theta) = \theta + b(\theta)$. If $E(T) = \theta$, then $b(\theta) = 0$ and hence the corollary follows.

Example 8.7

Consider the random sample x_i from the exponential distribution with parameter θ for which $E(x_i) = \theta$.

$$f(x; \theta) = \frac{1}{\theta} e^{-x/\theta} \qquad x > 0 \quad \theta > 0$$

$$L(\theta; x) = \frac{1}{\theta^n} e^{-(\Sigma x_i)/\theta}$$

$$\ln L(\theta; x) = -n \ln \theta - \frac{\Sigma x_i}{\theta}$$

$$S(\theta; x) = \frac{\partial \ln L}{\partial \theta} = -\frac{n}{\theta} + \frac{\Sigma x_i}{\theta^2}$$

$$\frac{\partial^2 \ln L}{\partial \theta^2} = \frac{n}{\theta^2} - \frac{2 \Sigma x_i}{\theta^3}$$

$$I(\theta) = E \left[-\frac{\partial^2 \ln L}{\partial \theta^2} \right] = -\frac{n}{\theta^2} + \frac{2n\theta}{\theta^3} = \frac{n}{\theta^2}$$

The MLE of θ is obtained from the equation $S(\theta; x) = 0$, which gives $\hat{\theta} = \bar{x}$, for which $\text{Var}(\hat{\theta}) = \theta^2 / n$. The CRLB is $1/I(\theta)$ $= \theta^2 / n$, which is the same as the variance of $\hat{\theta}$. Therefore CRLB for $\hat{\theta}$ is attained in this example.

Example 8.8

The exponential distribution used in the previous example has the property that $\text{Var}(X) = \theta^2$ and has the characteristic function $\phi(t) = (1 - it\theta)^{-1}$ (see Table 4.1). This means that $E(X^2) = 2\theta^2$ (verify it). For simplicity, let us draw a single observation x from this distribution. Choose $T(x)$ to be x^2. Then $u(\theta) = E(T) = 2\theta^2$ and $u'(\theta) = 4\theta$. The CRLB for T is therefore given by $16\theta^4$. From the characteristic function we can derive the fourth moment of X as $E(X^4) = 24\theta^4$ (verify it). Therefore,

$$\text{Var}(T) = E(X^4) - [E(X^2)]^2 = 24\theta^4 - 4\theta^4 = 20\theta^4$$

which is larger than the CR lower bound.

We see from the two examples that the CRLB is attained in some cases but not in others. An interesting question that arises is whether there exists an estimator T for which the bound is indeed attained. This is explored in the next section.

8.6   THE EXPONENTIAL FAMILY OF DISTRIBUTIONS

In Section 4.3 we introduced a family of distributions known as the **exponential family** (also known as the **Koopman family**). Here we show that there is a close relationship between the exponential family and the CRLB. For the sake of simplicity of exposition, we assume that the sample size is 1, that is, that a single observation x is drawn on the random variable X. In order for the CR bound to be attained, we need $\text{Cov}^2(T, S) = \text{Var}(T) \text{Var}(S)$. From the proof of Theorem 5.4 we see that the above is possible if and only if T is a linear function of S, that is, $T = \alpha(\theta)S + \beta(\theta)$. Since $E(S) = 0$, we have, $E(T) = \beta(\theta) = u(\theta)$. Hence, solving for S in terms of T ($L = f$ here because $n = 1$),

$$\frac{\partial \ln f(x;\theta)}{\partial \theta} = S(\theta;x) = \frac{T(x) - u(\theta)}{\alpha(\theta)}$$

Integrating partially with respect to θ, this takes the form

$$\ln f = U(x) + T(x)A(\theta) + B(\theta)$$

or

$$f(x;\theta) = \exp[U(x) + T(x)A(\theta) + B(\theta)]$$

where $A(\theta)$ and $B(\theta)$ are derived from $\alpha(\theta)$ and $u(\theta)$.

The family of distributions that can be written in the above form will be recognized as the exponential family. Thus for the CR bound to be attained, the distribution must belong to the exponential family and $T(x)$ must be the appropriate statistic. Most of the well-known distributions we have seen so far belong to this family and therefore it is possible to find an appropriate statistic for which the CRLB is attained. It is readily noted that unless a statistic $T(x)$ takes the above form, the CRLB will not be attained.

The exponential family has another interesting property. The likelihood function is

$$L(\theta;x) = \exp\left[\sum_i U(x_i) + A(\theta)\sum_i T(x_i) + nB(\theta)\right]$$

Let $\hat{\theta} = \sum T(x_i)$ be a statistic. Then $L(\theta;x)$ can be factored so that $\hat{\theta}$ becomes a sufficient statistic for θ.

Example 8.9

Consider the distribution $N(\mu, \sigma_0^2)$ with a known variance equal to σ_0^2. [Later we will look at multiparameter cases.]

$$L(\theta;x) = \left[\frac{1}{\sigma_0\sqrt{2\pi}}\right]^n \exp\left[-\frac{\sum (x_i - \mu)^2}{2\sigma_0^2}\right]$$

$$= \left[\frac{1}{\sigma_0\sqrt{2\pi}}\right]^n \exp\left[\frac{-\sum x_i^2}{2\sigma_0^2} - \frac{n\mu^2}{2\sigma_0^2} + \frac{\mu\sum x_i}{\sigma_0^2}\right]$$

We readily see by inspection that $\bar{x}$ is a sufficient statistic for μ. Since $E(\bar{x}) = \mu$, the CR inequality says

$$\text{Var}(\bar{x}) \geq \frac{1}{I(\mu)} \quad \text{where} \quad I(\mu) = -E\left[\frac{\partial^2 \ln L}{\partial \mu^2}\right]$$

$$\ln L = -n \ln \sigma_0 - n \ln \sqrt{2\pi} - [\sum (x_i - \mu)^2]/(2\sigma_0^2)$$

$$\frac{\partial^2 \ln L}{\partial \mu^2} = -\frac{n}{\sigma_0^2} \qquad \text{and hence} \qquad I(\mu) = \frac{n}{\sigma_0^2}$$

$$\frac{1}{I(\mu)} = \frac{\sigma_0^2}{n}$$

We know that $\text{Var}(\bar{x}) = \sigma_0^2/n$. Thus the CR bound is attained.

Practice Problem

8.7   Redo the above for the distribution $N(\mu_0, \theta)$, where μ_0 is a known constant.

8.7   SMALL SAMPLE PROPERTIES OF MAXIMUM LIKELIHOOD ESTIMATORS

In this section we establish several small sample properties of maximum likelihood estimators.

Theorem 8.5

If $T = T(x_1, x_2, \ldots, x_n)$ is a sufficient statistic for θ and there exists a unique MLE of θ, then the solution $\hat{\theta}$ is a function of T.

Proof: Because $T(x)$ is sufficient, $L(\theta; x) = g(T, \theta)h(x)$. Hence $\ln L = \ln g(T, \theta) + \ln h(x)$. Thus maximizing L is equivalent to maximizing $\ln g(T, \theta)$. But as $g(T, \theta)$ depends only on T and θ, the solution $\hat{\theta}$ must depend only on T. [The uniqueness assumption is needed because a counterexample can be given otherwise (see p. 244 of Lindgren, 1976).]

Theorem 8.6

If there exists an unbiased estimator T of θ that attains the CR lower bound, then T will be the MLE of θ.

Proof: We saw earlier that if the CR bound is attained by T, then T must be a linear function of the score $S(\theta; x)$. Thus $T = \alpha(\theta)S + \theta$, because $E(T)$ must be θ [note that $E(S) = 0$]. Thus

$$S(\theta; x) = \frac{\partial \ln L(\theta; x)}{\partial \theta} = \frac{T - \theta}{\alpha(\theta)} \qquad [\alpha(\theta) \neq 0]$$

For MLE, $S = 0$, which implies that $T = \hat{\theta}$.

Thus the maximum likelihood estimators have a number of important small sample properties. First, they are functions of sufficient statistics and hence contain all the information in the sample. Second, if the CR lower bound is attainable (or equivalently if the distribution belongs to the exponential family) by an unbiased estimator of θ, then it also maximizes the likelihood function. Finally, $\hat{\theta}$ can also be shown to be minimal sufficient, that is, if θ^* is any other sufficient statistic, then there exists a function $h(\cdot)$ such that $\hat{\theta} = h(\theta^*)$.

8.8 ASYMPTOTIC PROPERTIES OF MAXIMUM LIKELIHOOD ESTIMATORS

The MLE of a parameter also possesses several large sample properties. First, it can be shown to be consistent (that is, converges in probability to θ). Secondly, it is also a BAN estimator of θ (Section 8.2). Thus the MLE of θ possesses all the following asymptotic properties; (a) consistent, (b) asymptotically efficient, and (c) asymptotically normally distributed. Combined with the small sample properties of MLE of θ discussed above, it is clear that it possesses many of the desirable properties listed in Sections 8.1 and 8.2.

The proof of existence, uniqueness, consistency, asymptotic efficiency, and asymptotic normality of maximum likelihood estimators is postponed until the multiparameter case is treated.

8.9 JOINT ESTIMATION OF SEVERAL PARAMETERS

We now extend the results to the case of simultaneous estimation of several unknown parameters such as, for example, the mean and s.d. of a normal distribution. It is assumed that regularity conditions similar to those specified earlier also hold in the multiparameter case. The likelihood function is now

$$L(\theta; x) = f(x_1, x_2, \ldots, x_n; \theta_1, \theta_2, \ldots, \theta_k)$$

In the case of an *iid* random sample, the log-likelihood reduces to the following.

$$\ln L = \sum_{i=1}^{n} \ln f(x_i; \theta_1, \theta_2, \ldots, \theta_k)$$

To simplify the notation, let θ be the $k \times 1$ column vector of parameters. The maximum likelihood procedure now gives the k **likelihood equations**

$$0 = \frac{\partial \ln L}{\partial \theta_j} = \sum_{i=1}^{n} \frac{\partial \ln f(x_i; \theta)}{\partial \theta_j} \qquad j = 1, 2, \ldots, k$$

Let $S(\theta; x)$ be the column vector $(\partial \ln L / \partial \theta_j)$, called the **score vector** and S' the corresponding row vector given by

$$S'(x; \theta) = \left[\frac{\partial \ln L}{\partial \theta_1}, \frac{\partial \ln L}{\partial \theta_2}, \ldots, \frac{\partial \ln L}{\partial \theta_k} \right]$$

The maximum likelihood estimator of θ is therefore given as the solution to the vector equation $S = 0$.

Example 8.10

Suppose the random variable Y is generated by the process (known as an **autoregressive process**) $Y_t = \beta Y_{t-1} + u_t$, where u_t is *iid* as $N(0, \sigma^2)$, t denotes time ($t = 2, \ldots, T$), and β is an unknown constant in (0, 1). At time t, Y_{t-1} is known and is hence not random. Note that Y_{t-1} is defined only for $t = 2$ onward and hence the number of observations is $T - 1$. Let us derive the maximum likelihood estimators of β and σ^2. The joint density function of the u's is given by

$$f(u_2, u_3, \ldots, u_T) = \left[\frac{1}{\sigma\sqrt{2\pi}} \right]^{T-1} \exp\left[-\frac{1}{2\sigma^2} \sum_{t=2}^{t=T} u_t^2 \right]$$

The log-likelihood is given by

$$\ln L = -(T-1)\ln \sigma - \frac{T-1}{2}\ln(2\pi) - \frac{1}{2\sigma^2}\sum_{t=2}^{t=T}(Y_t - \beta Y_{t-1})^2$$

The likelihood equations are given by

$$\frac{\partial \ln L}{\partial \beta} = \frac{1}{\sigma^2}\sum_{t=2}^{t=T}(Y_t - \beta Y_{t-1})Y_{t-1} = 0$$

$$\frac{\partial \ln L}{\partial \sigma} = -\frac{T-1}{\sigma} + \frac{1}{\sigma^3}\sum_{t=2}^{t=T}(Y_t - \beta Y_{t-1})^2 = 0$$

The solutions are

$$\hat{\beta} = \frac{\displaystyle\sum_{t=2}^{t=T}(Y_t Y_{t-1})}{\displaystyle\sum_{t=2}^{t=T}(Y_{t-1}^2)} \qquad \text{and} \qquad \hat{\sigma}^2 = \frac{1}{T-1}\sum_{t=2}^{t=T}(Y_t - \hat{\beta} Y_{t-1})^2$$

8.10   INFORMATION MATRIX AND GENERALIZED CRAMER-RAO INEQUALITY

In this section we extend the concept of information and derive a generalized Cramer-Rao (CR) lower bound. First define the following expectation that depends on θ.

$$I_{jl} = -E\left[\frac{\partial^2 \ln L}{\partial\theta_j\,\partial\theta_l}\right] \qquad j, l = 1, 2, \ldots, k$$

The expression in the square brackets is the cross partial derivative of $\partial \ln L(\theta;x)/\partial\theta_j$ with respect to θ_l. Let $I(\theta)$ denote the matrix $I(\theta) = ((I_{jl}))$. This $k \times k$ matrix is called the **information matrix** and is the generalization of the information obtained for the single parameter case. We now derive a number of results for the information matrix.

Theorem 8.7

The score vector S has zero mean and covariance matrix $I(\theta)$, that is, $E(S) = 0$ and $E(SS') = E(-\partial S/\partial\theta) = I(\theta)$. [See Appendix A for matrix differentiation.]

Proof: As in the single parameter case, $\int L(\theta;x)\,dx = 1$. Differentiate both sides with respect to θ_j. We get the analogous result

$$\int \frac{1}{L}\frac{\partial L}{\partial\theta_j}L\,dx = \int \frac{\partial \ln L}{\partial\theta_j}L\,dx = 0$$

which gives $E(S) = 0$. Now differentiate with respect to θ_l.

$$\int \frac{\partial^2 \ln L}{\partial\theta_j\,\partial\theta_l}L\,dx + \int \frac{\partial \ln L}{\partial\theta_j}\frac{\partial \ln L}{\partial\theta_l}L\,dx = 0$$

or

$$E\left[\frac{\partial \ln L}{\partial\theta_j}\cdot\frac{\partial \ln L}{\partial\theta_l}\right] = -E\left[\frac{\partial^2 \ln L}{\partial\theta_j\,\partial\theta_l}\right] = I_{jl}$$

The relationship just derived gives k^2 equations, one for each pair (θ_j, θ_l). Note that $E[(\partial \ln L/\partial\theta_j)\cdot(\partial \ln L/\partial\theta_l)]$, when arranged in a matrix form, is nothing but $E(SS')$, the covariance matrix of the score vector S. Thus $E(SS') = I(\theta)$. The covariance matrix of the score vector is thus the Fisher's information matrix. Referring to the rule for differentiating a vector such as S with respect to another vector such as θ, we note that

$$\frac{\partial S}{\partial \theta} = E\left[\frac{\partial^2 \ln L}{\partial \theta_j \, \partial \theta_l}\right]$$

Therefore, $E(SS') = E(-\partial S / \partial \theta)$.

We are now ready to generalize the Cramer-Rao inequality, but we need to prove a lemma first.

Lemma 8.1

Given random variables $T_1, T_2, \ldots, T_l$ and $S_1, S_2, \ldots, S_k$ with $E(S_j) = 0$, let

$$A_{ij} = \text{Cov}(T_i, T_j), \quad D_{ij} = \text{Cov}(S_i, S_j), \quad C_{ij} = \text{Cov}(S_i, T_j)$$

so that the covariance matrix of the $l + k$ random variables is the partitioned matrix $\begin{bmatrix} A & C' \\ C & D \end{bmatrix}$. If D is positive definite, then $A - C' D^{-1} C$ is positive semi-definite.

Proof: For any vectors u and v, by Theorem 5.4,

$$\text{Cov}^2(u'T, v'S) \leq \text{Var}(u'T)\,\text{Var}(v'S)$$

But,

$$\text{Cov}(u'T, v'S) = E(u'TS'v) = u'\,E(TS')\,v = (u'C'v)$$

$$\text{Var}(u'T) = E(u'TT'u) = u'\,E(TT')\,u = u'Au$$

$$\text{Var}(v'S) = E(v'SS'v) = v'\,E(SS')\,v = v'Dv$$

Hence

$$(u'C'v)^2 \leq (u'Au)(v'Dv)$$

Since D is positive definite, D^{-1} exists and we can choose v to be equal to $D^{-1}Cu$. Then

$$(u'C'D^{-1}Cu)^2 \leq (u'Au)(u'C'D^{-1}Cu)$$

But D positive definite implies that $u'C'D^{-1}Cu > 0$. Dividing both sides by this scalar we get $u'(A - C'D^{-1}C)u \geq 0$, which implies that $A - C'D^{-1}C$ is positive semi-definite.

Theorem 8.8

Let $T_j(x_1, x_2, \ldots, x_n)$ be an unbiased estimator of θ_j, $j = 1$, $2, \ldots, k$, so that $E(T) = \theta$ in vector notation. Let V be the covariance matrix of T. Then $V - [I(\theta)]^{-1}$ is positive semi-definite. In other words, the "lower bound" of the covariance matrix of an unbiased vector estimate of θ is the inverse of the information matrix.

Proof: Let $S_j = (1/L)\,\partial L/\partial\theta_j$, and $C_{jl} = \text{Cov}(S_j, T_l)$. Differentiate $E(T_j) = \int T_j L\,dx = \theta_j$ with respect to θ_j. We have

$$\int T_j \frac{1}{L}\frac{\partial L}{\partial \theta_j} L\,dx = \int T_j S_j L\,dx = 1$$

Therefore $C_{jj} = 1$. Similarly, differentiating $E(T_j) = \theta_j$ with respect to θ_l we get $C_{jl} = 0$ for $j \neq l$. The C matrix is therefore a $k \times k$ identity matrix. Apply Lemma 8.1 with $A = V$, $D = I(\theta)$, and $C = I_k$, the identity matrix of order k. It follows that $A - C'D^{-1}C = V - [I(\theta)]^{-1}$ is positive semi-definite, which is the desired result. Note that the diagonal elements of $[I(\theta)]^{-1}$ are the lower bounds of the corresponding unbiased estimators.

Example 8.11

Let us apply this theorem to the normal distribution. The log-likelihood function is (setting $\sigma^2 = \theta$)

$$\ln L(\mu, \theta; x) = -\frac{n}{2}\ln(2\pi) - \frac{n}{2}\ln\theta - \frac{1}{2\theta}(x-\mu)'(x-\mu)$$

$$\frac{\partial \ln L}{\partial \mu} = \frac{\Sigma(x_i - \mu)}{\theta}$$

$$\frac{\partial \ln L}{\partial \theta} = -\frac{n}{2\theta} + \frac{1}{2\theta^2}(x-\mu)'(x-\mu)$$

$$\frac{\partial^2 \ln L}{\partial \mu^2} = -\frac{n}{\theta}$$

whose expectation is $-n/\theta$.

$$\frac{\partial^2 \ln L}{\partial \mu\,\partial\theta} = -\frac{\Sigma(x_i - \mu)}{\theta^2}$$

whose expectation is zero.

$$\frac{\partial^2 \ln L}{\partial \theta^2} = \frac{n}{2\theta^2} - \frac{(x-\mu)'(x-\mu)}{\theta^3}$$

whose expectation is

$$\frac{n}{2\theta^2} - \frac{n\theta}{\theta^3} = -\frac{n}{2\theta^2}$$

Therefore the information matrix and its inverse are

$$I(\mu,\theta) = \begin{bmatrix} n/\theta & 0 \\ 0 & n/(2\theta^2) \end{bmatrix} \qquad [I(\mu,\theta)]^{-1} = \begin{bmatrix} \theta/n & 0 \\ 0 & 2\theta^2/n \end{bmatrix}$$

The CR lower bound matrix is $[I(\mu,\theta)]^{-1}$. We know from Section 8.1 that $\bar{x}$ and $s^2 = [\Sigma(x_i - \bar{x})^2]/(n-1)$ are unbiased estimators of μ and θ respectively. Thus

$$\mathrm{Var}(\bar{x}) \geq \theta/n \qquad \text{and} \qquad \mathrm{Var}(s^2) \geq 2\theta^2/n$$

We know that $\mathrm{Var}(\bar{x}) = \theta/n$ and hence the CR bound is attained for it. We also know from Theorem 6.2, Corollary 1, that $(n-1)s^2/\theta \sim \chi^2_{n-1}$, whose variance is $2(n-1)$. Thus

$$\mathrm{Var}\left[\frac{(n-1)s^2}{\theta}\right] = 2(n-1)$$

$$\mathrm{Var}(s^2) = \frac{2\theta^2}{n-1} > \frac{2\theta^2}{n}$$

Therefore, the CR bound is *not* attained for θ but is attained for μ. The fault here is not in the estimator but the bound which is unattainable.

Example 8.12

We present another application of the principles using a model most frequently used in econometrics. It is known as the **multiple regression model**, which is discussed in considerable detail in Part III. Let $y_1, y_2, \ldots, y_n$ be observations on a random variable whose *conditional density* given observations on a set of other variables is multivariate normal. More specifically, let

$$y \mid X \sim N[X\beta, \sigma^2 I_k]$$

where y is an $n \times 1$ vector of observations on the *dependent* variable (say the earnings of an employee in a firm), X is an $n \times k$ matrix of observations on k ($k < n$) *independent* variables (for example, the individual's characteristics such as age, number of years of education, and number of years of experience), β is a $k \times 1$ vector of unknown parameters, σ^2 is also an unknown parameter, and I_k is the identity matrix of order k. X is given and hence can be treated as nonrandom. Conditional density of y given X is normally distributed with mean $X\beta$ and covariance matrix $\sigma^2 I_k$. We also assume that the $k \times k$ matrix $X'X$ is non-singular so that its inverse exists. The likelihood function is (see Example 8.5)

$$L(\beta, \sigma; y) = \left[\frac{1}{\sigma\sqrt{2\pi}} \right]^n \exp\left[-\frac{(y - X\beta)'(y - X\beta)}{2\sigma^2} \right]$$

The log-likelihood is

$$\ln L = -n \ln(\sqrt{2\pi}) - n \ln\sigma - \frac{1}{2\sigma^2} (y - X\beta)'(y - X\beta)$$

Let $\underset{(k+1)\times 1}{\theta}$ be $\begin{bmatrix} \beta \\ \sigma \end{bmatrix}$. The score function is (using matrix derivatives)

$$S(\theta; y) = \begin{bmatrix} \dfrac{\partial \ln L}{\partial \beta} \\[2mm] \dfrac{\partial \ln L}{\partial \sigma} \end{bmatrix} = \begin{bmatrix} \dfrac{X'(y - X\beta)}{\sigma^2} \\[2mm] -\dfrac{n}{\sigma} + \dfrac{1}{\sigma^3}(y - X\beta)'(y - X\beta) \end{bmatrix}$$

MLE is obtained when $S(\theta; y) = 0$. This implies that

$$X'(y - X\beta) = 0 \quad \text{and} \quad \frac{(y - X\beta)'(y - X\beta)}{\sigma^3} = \frac{n}{\sigma}$$

The solutions to these equations are given by

$$\hat{\beta} = (X'X)^{-1} X'y \quad \text{and} \quad \hat{\sigma}^2 = \frac{1}{n}(y - X\hat{\beta})'(y - X\hat{\beta})$$

$$\frac{\partial^2 \ln L}{\partial \beta^2} = -\frac{X'X}{\sigma^2} \quad \Rightarrow \quad E\left[-\frac{\partial^2 \ln L}{\partial \beta^2} \right] = \frac{X'X}{\sigma^2}$$

$$\frac{\partial^2 \ln L}{\partial \sigma^2} = \frac{n}{\sigma^2} - \frac{3}{\sigma^4}(y - X\beta)'(y - X\beta)$$

Hence

$$E\left[-\frac{\partial^2 \ln L}{\partial \sigma^2} \right] = -\frac{n}{\sigma^2} + \frac{3}{\sigma^4} E[(y - X\beta)'(y - X\beta)]$$

Let $u = y - X\beta$.

$$E[(y - X\beta)'(y - X\beta)] = E(u'u) = E(\Sigma\, u_i^2) = nE(u_i^2) = n\sigma^2$$

because $E(u_i) = 0$. Hence,

$$E\left[-\frac{\partial^2 \ln L}{\partial\sigma^2}\right] = -\frac{n}{\sigma^2} + \frac{3n}{\sigma^2} = \frac{2n}{\sigma^2}$$

$$\frac{\partial^2 \ln L}{\partial\sigma\partial\beta} = -\frac{2}{\sigma^3} X'(y - X\beta)$$

But $E(y|X) = X\beta$ and hence $E\left[-\partial^2 \ln L/\partial\sigma\partial\beta\right] = 0$. The information matrix and its inverse are therefore

$$I(\theta) = \begin{bmatrix} X'X/\sigma^2 & 0 \\ 0 & 2n/\sigma^2 \end{bmatrix} \qquad [I(\theta)]^{-1} = \begin{bmatrix} \sigma^2(X'X)^{-1} & 0 \\ 0 & \sigma^2/2n \end{bmatrix}$$

which are block diagonal. This means that we can treat questions that arise about β and σ^2 separately. Let us examine whether $\hat{\beta}$ is unbiased or not. Substituting for y from the model into $\hat{\beta}$ we get

$$E(\hat{\beta}) = E[(X'X)^{-1}X'(X\beta + u)] = \beta + E[(X'X)^{-1}X'u] = \beta$$

because $E(u) = 0$ by assumption. Therefore $\hat{\beta}$ is unbiased. A similar derivation for $\hat{\sigma}^2$ is tedious and is postponed to Chapter 10 (Section 10.2). The covariance matrix of $\hat{\beta}$ is obtained as

$$V(\hat{\beta}) = E[(\hat{\beta} - \beta)(\hat{\beta} - \beta)'] = E[(X'X)^{-1}X'uu'X(X'X)^{-1}]$$

$$= (X'X)^{-1}X'E(uu')X(X'X)^{-1} = \sigma^2(X'X)^{-1}$$

We note that this is the same as the CRLB and hence the lower bound is attained for $\hat{\beta}$.

8.11   EXISTENCE, UNIQUENESS, AND CONSISTENCY OF MAXIMUM LIKELIHOOD ESTIMATORS

We now prove that under the regularity conditions specified earlier, the MLE of $\theta = (\theta_1, \theta_2, \ldots, \theta_k)$ exists, is unique, and is consistent. The proof involves the use of the **inverse function theorem**. A form of it, proved in Rudin (1964, p. 193), is restated here.

Restatement of the Inverse Function Theorem (IFT)

Let

$$\|M\|_{k \times k} = Sup \left\{ |MX|_{k \times k}, k \times k : |X|_{k \times 1} \leq 1 \right\}$$

be the norm of the matrix M, where $|z|$ *is defined as* $[\Sigma_{i=1}^{k} z_i^2]^{1/2}$. *Suppose g is a mapping from an open set* Θ *in* E^k *into* E^k, *the partial derivatives of g exist and are continuous in* Θ, *the matrix of partial derivatives* $g'(\theta^*)$ *has inverse* $[g'(\theta^*)]^{-1}$ *for some* $\theta^* \in \Theta$. *Define* $\lambda = 1/[4\|g'(\theta^*)^{-1}\|]$. *Choose* δ *sufficiently small so that, by the continuity assumption of* $g'(\theta)$, $\|g'(\theta) - g'(\theta^*)\| < 2\lambda$ *whenever* $\theta \in U_\delta = [\theta : |\theta - \theta^*| < \delta]$, *a small neighborhood of* θ^*. *Then (a) for every* θ_1, θ_2 *in* U_δ, $|g(\theta_1) - g(\theta_2)| \geq 2\lambda|\theta_1 - \theta_2|$ *and (b) the image set* $g(U_\delta)$ *contains the open neighborhood with radius* $\lambda\delta$ *about* $g(\theta^*)$. *Conclusion (a) insures that g is one-to-one on* U_δ *and that* g^{-1}, *the inverse transformation, exists and is well defined on the image set of* $g(U_\delta)$.

Theorem 8.9

Under the regularity conditions stated earlier, there exists a unique solution vector to the likelihood equations that is consistent. More formally, let

$$S_n(\theta) = \frac{\partial \ln L}{\partial \theta} = \sum_{i=1}^{i=n} \frac{\partial \ln f}{\partial \theta}$$

There exists a sequence $\hat{\theta}_n$ *such that* $S_n(\hat{\theta}_n) = 0$ *with probability going to 1 as* $n \to \infty$ *and* $\hat{\theta}_n \xrightarrow{p} \theta_0$, *the true value of* θ. *If another estimator* $\tilde{\theta}_n$ *also satisfies these, then* $\tilde{\theta}_n = \hat{\theta}_n$ *with probability going to 1 as* $n \to \infty$.

Proof: The proof is carried out in several steps and is only sketched here. For more details see Foutz (1977).

Step 1:   Because

$$\frac{S_n(\theta_0)}{n} = \frac{1}{n} \Sigma \frac{\partial \ln f(x_i; \theta_0)}{\partial \theta}$$

and $E[S_n(\theta_0)] = 0$, by the *weak law of large numbers* $S_n(\theta_0)/n \xrightarrow{p} 0$. Define

$$\sigma(\theta) = \lim_{n \to \infty} \left[-I(\theta)/n \right] \quad \text{and} \quad S_n^{'}(\theta) = \frac{\partial S_n}{\partial \theta}$$

where $I(\theta)$ is the information matrix. $S_n^{'}$ is not to be confused with the transpose of S_n (which is not used in this proof). Because $\sigma(\theta)$ is a negative definite matrix by assumption, its inverse exists. By the WLLN,

$$\frac{1}{n} S_n^{'} = \frac{1}{n} \sum \frac{\partial^2 \ln f(x_i; \theta)}{\partial \theta \partial \theta^{'}} \xrightarrow{p} \sigma(\theta)$$

In what follows we will assume that the above convergence is uniform in an open neighborhood of θ_0.

Step 2:   Let

$$\lambda = \frac{1}{4 \| [\sigma(\theta_0)]^{-1} \|} \quad \text{and} \quad \lambda_n = \frac{n}{4 \| [S_n^{'}(\theta_0)]^{-1} \|}$$

whenever $S_n^{'}(\theta_0)$ is negative definite. Choose δ small enough so that $\| \sigma(\theta) - \sigma(\theta_0) \| < \lambda/2$ whenever $|\theta - \theta_0| < \delta$. The uniform convergence of $(1/n)S_n^{'}(\theta)$ to $\sigma(\theta)$ in $|\theta - \theta_0| < \delta$, assumed earlier, implies that $\lambda_n \xrightarrow{p} \lambda$. Also, if $|\theta - \theta_0| < \delta$, then for a large n,

$$\| (1/n)S_n^{'}(\theta) - \sigma(\theta) \| < \lambda/4$$

We have

$$\| (1/n) S_n^{'}(\theta) - (1/n) S_n^{'}(\theta_0) \| = \| (1/n) S_n^{'}(\theta) - \sigma(\theta)$$
$$+ \sigma(\theta) - \sigma(\theta_0) + \sigma(\theta_0) - (1/n) S_n^{'}(\theta_0) \|$$

Therefore we can find n large enough so that

$$\| (1/n) S_n^{'}(\theta) - (1/n) S_n^{'}(\theta_0) \| \leq \| (1/n) S_n^{'}(\theta) - \sigma(\theta) \|$$
$$+ \| \sigma(\theta) - \sigma(\theta_0) \| + \| \sigma(\theta_0) - (1/n) S_n^{'}(\theta_0) \|$$
$$< \lambda/4 + \lambda/2 + \lambda/4 = \lambda$$

Step 3:   The conditions of the IFT are now satisfied. From part (b) of IFT, the image set $(1/n) S_n(\theta_s)$ contains the open neighborhood of radius $\lambda_n \delta$ about $(1/n) S_n(\theta_0)$. But $\lambda_n \delta \xrightarrow{p} \lambda \delta$ and hence $(1/n)S_n(\theta_s)$ also contains a neighborhood of radius $\lambda \delta/2$ about $(1/n)S_n(\theta_0)$. Also $(1/n) S_n(\theta_0) \xrightarrow{p} 0$. Hence for large n,

$$| (1/n) S_n(\theta_0) | < \lambda\delta/2$$

Thus, $(1/n) S_n(\theta_\delta)$ contains zero also.

Step 4:   Consider the inverse function $S_n^{-1}: S_n(U_\delta) \to U_\delta$. This function is well defined when S_n is one-to-one, which by the earlier steps is true for sufficiently large n. Because $0 \in U_\delta$ and S_n is one-to-one in U_δ, $S_n^{-1}(0)$ exists and is unique. We therefore conclude that the likelihood equation $S_n(\theta) = 0$ has a unique solution. This establishes existence and uniqueness.

Step 5:   From part (a) of IFT,

$$\|\hat\theta_n - \theta_0\| \leq \frac{1}{2\lambda_n} \|(1/n) S_n(\hat\theta_n) - (1/n) S_n(\theta_0)\|$$

for large n and

$$\|(1/n) S_n(\hat\theta_n) - (1/n) S_n(\theta_0)\| < \lambda\delta/2$$

where $(1/n) S_n(\theta_0) = 0$ and $\lambda < 2\lambda_n$. Hence for large n, $|\hat\theta_n - \theta_0| < (\lambda\delta/2)/\lambda = \delta/2$ with probability going to 1. This, by the definition of convergence in probability, implies that $\hat\theta_n \xrightarrow{p} \theta_0$. Thus $\hat\theta_n$ is also consistent.

We have thus shown that MLE of θ exists, is unique, and is consistent provided the regularity conditions are met. As they are satisfied for most distributions, the above properties hold for most distributions.

8.12   ASYMPTOTIC NORMALITY OF MAXIMUM LIKELIHOOD ESTIMATORS

We now trace the steps for providing the asymptotic normality of MLE in the multiparameter case. For more details, see Dhrymes (1970, pp. 121–123).

Theorem 8.10

Let $\Sigma(\theta) = \lim_{n\to\infty} I(\theta)/n$. *Then under the assumptions made in Theorem 8.9,* $\sqrt{n}\, (\hat\theta_n - \theta) \xrightarrow{d} N\,[0, \Sigma(\theta)^{-1}]$. *Furthermore,*

$$- \frac{1}{n} \frac{\partial^2 \ln L}{\partial \theta_j \partial \theta_l} = - \frac{1}{n} \sum_{i=1}^{i=n} \left[\frac{\partial^2 \ln f(x_i; \hat{\theta}_n)}{\partial \theta_j \partial \theta_l} \right] = - \frac{1}{n} \left[\frac{\partial S}{\partial \theta} \right]_{\theta = \hat{\theta}}$$

is a consistent estimator of $\Sigma(\theta)$.

Step 1: By the *generalized mean value theorem* [see Jennrich (1969), pp. 633–643]

$$S(\theta; x) = S(\hat{\theta}_n; x) + \frac{\partial S(\theta_n^*; x)}{\partial \theta} (\theta - \hat{\theta}_n)$$

where $|\theta_n^* - \theta| \leq |\hat{\theta}_n - \theta|$ and $\partial S / \partial \theta$ is the $k \times k$ matrix of partial derivatives defined in Section 8.10. Because $\hat{\theta}_n$ is MLE, $S(\hat{\theta}_n; x) = 0$, and hence

$$S(\theta; x) = \frac{\partial S(\theta_n^*; x)}{\partial \theta} (\theta - \hat{\theta}_n)$$

Premultiplying by $\sqrt{n} \left[-\frac{\partial S(\theta_n^*; x)}{\partial \theta} \right]^{-1}$ both sides of the equation, we have

$$\sqrt{n} (\hat{\theta}_n - \theta) = \left[-\frac{\partial S(\theta_n^*; x)}{\partial \theta} \right]^{-1} \sqrt{n} \, S(\theta; x)$$

$$= \left[-\frac{1}{n} \frac{\partial S(\theta_n^*; x)}{\partial \theta} \right]^{-1} \frac{1}{\sqrt{n}} S(\theta; x)$$

Step 2: Because $\hat{\theta}_n$ is consistent (Theorem 8.9), $\hat{\theta}_n \xrightarrow{p} \theta$ and hence $\theta_n^* \xrightarrow{p} \theta$. Also,

$$S = \frac{\partial \ln L}{\partial \theta} = \Sigma \frac{\partial \ln f(x_i; \theta)}{\partial \theta}$$

We have shown in Theorem 8.7 that $E(S) = 0$ and $\text{Var}(S) = I(\theta)$. Hence by the *multivariate central limit theorem*

$$\frac{1}{\sqrt{n}} S(\theta; x) \xrightarrow{d} N[0, \Sigma(\theta)]$$

Step 3: Note from the proof of Theorem 8.7 that

$$E\left[-\frac{\partial S}{\partial \theta} \right] = E(SS') = I(\theta)$$

Therefore by WLLN,

$$-\frac{1}{n}\frac{\partial S}{\partial \theta} \xrightarrow{P} \lim_{n \to \infty} \left[\frac{I(\theta)}{n}\right] = \Sigma(\theta)$$

which implies that

$$\left[-\frac{1}{n}\frac{\partial S}{\partial \theta}\right]^{-1} \xrightarrow{P} [\Sigma(\theta)]^{-1}$$

Because $\theta_n^* \xrightarrow{P} \theta$, we also have

$$\left[-\frac{1}{n}\frac{\partial S(\theta_n^*;x)}{\partial \theta}\right]^{-1} \xrightarrow{P} [\Sigma(\theta)]^{-1}$$

From Property 5.5, if $Y \sim N[0, \Sigma]$ then, $AY \sim N(0, A\Sigma A')$. Define

$$A = \left[-\frac{1}{n}\frac{\partial S}{\partial \theta}\right]^{-1} \quad \text{and} \quad Y = \frac{1}{\sqrt{n}}S(\theta;x)$$

Using Property 5.5 and Theorem 7.2 in the last line of Step 1, we have

$$\sqrt{n}\,(\hat{\theta}_n - \theta) \xrightarrow{d} N[0, \{\Sigma(\theta)\}^{-1}\Sigma(\theta)\{\Sigma(\theta)\}^{-1}] = N[0, \{\Sigma(\theta)\}^{-1}]$$

Because

$$\left[-\frac{1}{n}\frac{\partial S(\hat{\theta}_n;x)}{\partial \theta}\right]^{-1} \xrightarrow{P} [\Sigma(\theta)]^{-1}$$

the former is a consistent estimator of $[\Sigma(\theta)]^{-1}$, the covariance matrix of $\hat{\theta}_n$.

8.13   NUMERICAL PROCEDURE FOR OBTAINING MAXIMUM LIKELIHOOD ESTIMATORS

The linear approximation used in Step 1 of Theorem 8.10 suggests a numerical procedure for actually computing the maximum likelihood estimators and their covariance matrix. The procedure is called the **method of scores** (also known as the **Newton-Raphson method**) and is useful in situations where analytical solutions are difficult or impossible to obtain.

The first step is to choose an initial approximation θ_1 to the solution. Expand $S(\theta;x) = \partial \ln L / \partial \theta$ around θ_1, linearly.

$$S(\theta; x) = S(\theta_1; x) + \frac{\partial S}{\partial \theta_1}(\theta - \theta_1)$$

Because $E(-\partial S/\partial\theta) = I(\theta)$, $-\partial S/\partial\theta_1$ is an estimate of $I(\theta)$. Also, $S(\theta; x) = 0$ if θ is the MLE solution. Using this we get

$$0 = S(\theta_1; x) - I(\theta_1)(\theta - \theta_1)$$

It follows that

$$\theta - \theta_1 = [I(\theta_1)]^{-1}S(\theta_1; x)$$

Thus, a second round approximation to θ (call it θ_2) is

$$\theta_2 = \theta_1 + [I(\theta_1)]^{-1}S(\theta_1; x)$$

If this procedure converges, ultimately you get $\hat{\theta}$, and the corresponding $[I(\hat{\theta})]^{-1}$ is the desired asymptotic covariance matrix of $\hat{\theta}$. If the procedure does not converge, one often starts with a different θ_1 and iterates as above. Prior knowledge of a range of values of $\hat{\theta}$ often helps. The specific steps involve evaluating the first and second partial derivatives of the likelihood function at each step of the iteration. The method is obviously computer intensive.

EXERCISES

8.1   Consider the discrete geometric distribution $f(x; \theta) = k\,\theta^x$, $0 < \theta < 1$ and $x = 0, 1, 2, \ldots$.

(a)   Evaluate k in terms of θ. Then show that the characteristic function is $(1-\theta)/(1-\theta e^{it})$. From this derive the mean of the distribution.

(b)   Let x_i, $i = 1, 2, \ldots, n$ be a random sample. Write down the likelihood function and show that the sample mean $\bar{x}$ is a sufficient statistic for θ.

(c)   Show that the Fisher's information is

$$I(\theta) = n/[\theta(1-\theta)^2]$$

(d)   Derive the MLE $(\hat{\theta}_n)$ of θ.

(e)   Deduce directly from appropriate theorems (not from the consistency property of MLE) that $\hat{\theta}_n$ is a consistent estimator.

(f) Write down the CRLB for $T(x) = \bar{x}$ and examine whether it is attained.

8.2 Consider the density function $f(x; \theta) = \theta x^{\theta-1}$ for $0 < x < 1$ and $\theta > 0$.

(a) Show that $\int_0^1 x^{\theta-1}\, dx = 1/\theta$. From this, derive the mean of the distribution.

(b) Differentiate the above integral partially with respect to θ under the integral sign, assuming all regularity conditions, and show that [using the fact that $y^\alpha = \exp(\alpha \ln y)$]

$$\int_0^1 x^{\theta-1} \ln x\, dx = -\frac{1}{\theta^2}$$

Then derive $E(\ln x)$.

(c) Differentiate under the integral sign once more and obtain $\int_0^1 x^{\theta-1} (\ln x)^2\, dx$. From all of this obtain Var $(\ln x)$.

(d) Let $x_1, x_2, \ldots, x_n$ be a random sample. Write down the likelihood function. Is Σx_i a sufficient statistic for θ? Is $\Pi_{i=1}^n x_i$ a sufficient statistic? Carefully prove your answers.

(e) Derive the MLE of θ (call it $\hat{\theta}$).

(f) Let $\beta \equiv 1/\theta$. Deduce, that is, do not rederive, the MLE of β from (e). Prove carefully whether or not the MLE of β (call it $\hat{\beta}$) is unbiased, using your results above.

(g) Derive the variance of $\hat{\beta}$ and the Fisher information $I(\beta)$. Deduce the CRLB. Is the lower bound attained?

(h) Use the results derived so far to construct a statistic Z_n such that $Z_n \xrightarrow{d} N(0, 1)$. Justify your choice with appropriate limit theorems.

8.3 Let X be a random variable with the density function

$$f(x; \theta) = \frac{1}{\theta}(1+x)^{-(\theta+1)/\theta} \qquad x > 0, \quad 0 < \theta < \tfrac{1}{2}$$

(a)   Show that

$$\int_0^\infty (1 + x)^{-(\theta+1)/\theta} dx = \theta$$

From this, derive the mean of the distribution. Next compute $E[(x + 1)^2]$ and then derive the variance of the distribution.

(b)   Differentiate the above integral partially with respect to θ under the integral sign, assuming all regularity conditions, and derive $E[\ln(1 + x)]$. [Use the fact that $y^\alpha = \exp(\alpha \ln y)$.]

(c)   Differentiate under the integral sign once more and obtain the expected value of $[\ln(1 + x)]^2$. Next obtain $\text{Var}[\ln(1 + x)]$.

(d)   Using the above results obtain Fisher's information $I(\theta)$.

(e)   Let $x_1, x_2, \ldots, x_n$ be a random sample. Write down the likelihood function and examine whether the statistic $T = (1/n) \Sigma_i \ln(1 + x_i)$ is a sufficient statistic for θ. Carefully prove your answers.

(f)   Is T an unbiased estimator of θ? Derive the CRLB for $\text{Var}(T)$ and examine whether it is attained.

(g)   Is T the MLE of θ? What can you say about the consistency of T?

(h)   Construct a statistic Z_n that depends on T, θ, and n, such that $Z_n \xrightarrow{d} N(0, 1)$.

8.4   Consider the density function $f(x; \theta) = 1/\theta \, x^{(1-\theta)/\theta}$ for $0 < x < 1$ and $\theta > 0$.

(a)   Show that $\int_0^1 x^{(1-\theta)/\theta} dx = \theta$. From this, derive the mean of the distribution.

(b)   Differentiate the above integral partially with respect to θ under the integral sign, assuming all regularity conditions, and show that (using the fact that $y^\alpha = e^{\alpha \ln y}$)

$$\int_0^1 x^{(1-\theta)/\theta} \ln x \, dx = -\theta^2$$

Then derive $E(\ln x)$.

(c) Differentiate the above integral once more under the integral sign and obtain $\int_0^1 x^{(1-\theta)/\theta} (\ln x)^2 \, dx$. From all of this obtain $\text{Var}(\ln x)$.

(d) Let $x_1, x_2, \ldots, x_n$ be a random sample. Write down the likelihood function. Is Σx_i a sufficient statistic for θ? Is $\Pi_{i=1}^n x_i$ a sufficient statistic? Carefully prove your answers.

(e) Derive the MLE of θ (call it $\hat{\theta}$).

8.5 The density function of a random variable is $f(x, \theta) = (\ln \theta)\theta^{-x}$, for $x > 0$ and $\theta > 1$. A random sample of size n is drawn.

(a) Show that the sample mean $\bar{x}$ is a sufficient statistic for the parameter θ.

(b) Derive the score function and the MLE of θ.

(c) Show that the mean and variance of x are $E(x) = 1/(\ln \theta)$ and $V(x) = 1/(\ln \theta)^2$. Use this to derive the information $I(\theta)$ and the CRLB for the variance of $\bar{x}$. Is the lower bound attained?

(d) Construct a random variable Z_n, which is a function of $\bar{x}$ and θ, such that its asymptotic distribution is $N(0, 1)$.

8.6 Consider the following density function:

$$f(x; \theta) = (x/\theta^2)\exp(-x/\theta) \qquad x > 0, \quad \theta > 0$$

(a) Use the substitution $\eta = \theta/(1 - i\theta t)$ and show that the characteristic function is $(1 - i\theta t)^{-2}$. From this, derive the mean and variance of the distribution.

(b) Obtain $\partial \ln L /\partial \theta$ and $\partial^2 \ln L /\partial \theta^2$. Show that the former has zero expectation and that the information is $I(\theta) = 2n/\theta^2$.

(c) Suppose a single observation x is drawn. Construct a statistic $T(x)$ that is unbiased. State the variance of T in terms of θ. Is the CRLB for $V(T)$ attained?

(d) Check whether $T(x)$ is a sufficient statistic.

(e) Obtain the MLE θ and compare it to $T(x)$.

8.7 The conditional density of y given x is $N(\theta x, 1)$. Here θ, x, and y are scalars and x can be treated as nonrandom and hence is fixed. A random sample of observations (x_i, y_i) is obtained for x and y.

   (a) Derive the log-likelihood, score function, and the second derivative of $\ln L$ with respect to θ.

   (b) From these derive the MLE ($\hat{\theta}$) of θ and the information $I(\theta)$.

   (c) Show that $\hat{\theta}$ is an unbiased estimate of θ. The variance of $\hat{\theta}$ is given by (you need not prove it) $1/S_{xx}$, where $S_{xx} = \Sigma x_i^2$. State the distribution of $\hat{\theta}$.

8.8 The relation $Y = \alpha + \beta X + u$ is known as a **simple linear regression model**. Observations X_i and Y_i, $i = 1, 2, \ldots, n$ are drawn. The conditional distribution of u *given* X is assumed to be $N(0, \sigma^2)$, which is equivalent to assuming that the conditional distribution of Y given X is $N(\alpha + \beta X, \sigma^2)$ and that X is nonrandom.

   (a) Write down the joint density function of the u_i's. From this write down the likelihood function of α, β, and σ^2 in terms of X_i and Y_i.

   (b) Maximize the log-likelihood with respect to α and β and derive the first order conditions. Solve them and show that the MLE are

$$\hat{\beta} = S_{xy}/S_{xx} \qquad \hat{\alpha} = \overline{Y} - \hat{\beta}\,\overline{X}$$

   where

$$S_{xy} = \left[\sum_{i=1}^{i=n} X_i Y_i \right] - n\overline{X}\,\overline{Y}, \qquad S_{xx} = \left[\sum_{i=1}^{i=n} X_i^2 \right] - n\overline{X}^2$$

   $\overline{X}$ and $\overline{Y}$ being the corresponding sample means.

   (c) In the expression for $\hat{\beta}$, substitute for Y_i from the model and express $\hat{\beta}$ in terms of X_i, u_i, α, and β. Show that when X_i is given and nonrandom, $\hat{\beta}$ is an unbiased estimator of β. Next show that $\hat{\alpha}$ is also unbiased.

   (d) Show that $\text{Var}(\hat{\beta}) = \sigma^2/S_{xx}$. Derive the statistical distribution of $\hat{\beta}$ (again for given X_i), noting that it is a linear function of u_i.

(e) Show that the MLE of σ^2 is given by $\hat{\sigma}^2 = [\Sigma \hat{u}_i^2]/n$, where $\hat{u}_i = Y_i - \hat{\alpha} - \hat{\beta} X_i$.

(f) Consider an alternative estimator of β, $\tilde{\beta} = \Sigma_i C_i Y_i$ known as a *linear estimator*. Assuming that C_i are non-random derive the two conditions needed for $\tilde{\beta}$ to be unbiased for all β.

(g) Compute the variance of $\tilde{\beta}$ in terms of σ^2 and C_i^2. Minimize that variance with respect to C_i, subject to the two conditions derived in (f), and show that the **best linear unbiased estimator** (or **BLUE**) is the same as $\hat{\beta}$ derived in (b).

8.9 Consider the two-parameter density function

$$f(x; \alpha, \beta) = \frac{1}{\beta} e^{-(x-\alpha)/\beta} \qquad x > \alpha, \quad \beta > 0$$

(a) Show that $\int_\alpha^\infty e^{-(x-\alpha)/\beta} \, dx = \beta$. Differentiate both sides with respect to β twice and obtain $E(x - \alpha)$ and $E[(x-\alpha)^2]$. From this determine the mean and variance of the distribution.

(b) Obtain $S_1 = \partial \ln L / \partial \alpha$ and $S_2 = \partial \ln L / \partial \beta$. From this and (a) obtain the information matrix $I(\alpha, \beta) = ((E[S_i S_j]))$, for $i, j = 1, 2$. [Do not use the second partials.]

(c) Let x_i, $i = 1, 2, \ldots, n$ $(n \geq 2)$ be a random sample. Derive the MLE of the two parameters (call them T_1 and T_2).

(d) Show that T_1 are T_2 and both unbiased. Invert the information matrix and derive the CRLB.

(e) Using the results in (a) compute $\text{Var}(T_1)$ and $\text{Var}(T_2)$. Are the CR bounds attained here?

8.10 The Pareto distribution is frequently used as a model to study income distribution. It has the following CDF (not PDF). $F(x) = 1 - (\theta_1/x)^{\theta_2}$ for $x \geq \theta_1$, and zero elsewhere. θ_1 and θ_2 are both positive.

(a) Derive the density function $f(x)$ and verify that it integrates to 1.

(b)   Given a random sample of n observations, find the MLE of θ_1 and θ_2.

(c)   Derive the information matrix for θ_1, θ_2.

(d)   Derive the CR lower bounds for variances of estimators of θ_1 and θ_2.

[Do not attempt to verify whether the lower bounds are attained, the procedure is too tedious.]

8.11   Let the conditional distribution of y given X be $N(X\beta, \sigma^2\Omega)$. X is an $n \times k$ matrix of observations, y is an $n \times 1$ vector of observations, β is a $k \times 1$ vector of unknown parameters and Ω is an $n \times n$ symmetric positive definite matrix. Let $u = y - X\beta$, so that $E(uu') = \sigma^2\Omega$. Assume that $X'\Omega^{-1}X$ is nonsingular.

(a)   Show that the log-likelihood for this is of the form

$$\ln L = b - n \ln \sigma - (\tfrac{1}{2}\sigma^2)(y - X\beta)'\Omega^{-1}(y - X\beta)$$

where b does not depend on β or σ.

(b)   Show that the score function $S(\beta, \sigma; y)$ for β is given by $X'\Omega^{-1}u / \sigma^2$.

(c)   Derive the information matrix for β and σ.

(d)   Derive the MLE $(\hat{\beta})$ of β.

(d)   Show that $\hat{\beta} = \beta + (X'\Omega^{-1}X)^{-1}X'\Omega^{-1}u$.

(f)   Derive the covariance matrix of $\hat{\beta}$ and show that the generalized CRLB is attained.

(g)   Write down the distribution of $\hat{\beta}$.

9

TESTS OF HYPOTHESES

The testing of statistical hypotheses on the unknown parameters of a probability model is one of the most important steps of any empirical study. The initial formulation of a model is based on economic theory, intuition, past studies, and past experience. Investigators often formulate a number of alternative models and then subject them to a variety of diagnostic tests with a view to obtaining robust conclusions, that is, conclusions that are not model sensitive. Tests of hypotheses are conducted also when one wants to examine the effect of policy changes. For instance, a state legislature might want to test whether laws strengthening the punishment for drunken driving significantly reduce deaths and injuries in car accidents. In addition to the two types of situations in which hypothesis testing would be appropriate, one might want to test empirically the validity of a body of theory. This chapter presents in considerable detail the essentials of hypothesis testing in single parameter as well as multiparameter cases.

9.1 BASIC CONCEPTS IN HYPOTHESIS TESTING

Consider a family of distributions represented by the density function $f(x;\theta)$, $\theta \in \Theta$. The term **hypothesis** stands for a statement or conjecture regarding the values that θ might take (for example, $\theta = 0.75$). The testing of a hypothesis consists of three basic steps: (1) formulate two opposing hypotheses, (2) derive a test statistic and identify its sampling distribution, and (3) derive a decision rule and choose one of the opposing hypotheses.

Null and Alternative Hypotheses

A hypothesis can be thought of as a binary partition of the parameter space Θ into two sets, Θ_0 and Θ_1 such that $\Theta_0 \cap \Theta_1 = \varnothing$ and $\Theta_0 \cup \Theta_1 = \Theta$. The set Θ_0, which corresponds to the statement of the hypothesis, is called the **null hypothesis** (usually denoted by

H_0) and Θ_1, which is the class of alternatives to the null hypothesis, is called the **alternative hypothesis** (usually denoted by H_1). For instance, $\theta = 0.75$ may be the null hypothesis and $\theta \neq 0.75$ the alternative. The null and alternative hypotheses are the two opposing hypotheses mentioned earlier.

Simple and Composite Hypotheses

If the null hypothesis is of the form H_0: $\theta = \theta_0$ and the alternative is H_1: $\theta = \theta_1$, then we have a **simple hypothesis** and a **simple alternative**. If either H_0 or H_1 specifies a range of values for θ (for example, H_1: $\theta \neq \theta_0$), then we have a **composite hypothesis**. The following table presents examples of null and alternative hypotheses.

	(a)	(b)	(c)	(d)
H_0	$\theta = \theta_0$	$\theta = \theta_0$	$\theta \leq \theta_0$	$\theta \geq \theta_0$
H_1	$\theta = \theta_1$	$\theta \neq \theta_0$	$\theta > \theta_0$	$\theta < \theta_0$

Initially we consider a simple hypothesis and a simple alternative only. In this case the problem reduces to one of choosing between the two density functions $f(x;\theta_0)$ and $f(x;\theta_1)$.

Statistical Test

A decision rule that selects one of the inferences "accept the null hypothesis" or "reject the null hypothesis" for each foreseeable outcome of an experiment is called a **statistical test** or simply a **test**. A test procedure is usually described by a sample statistic $T(x) = T(x_1, x_2, \ldots, x_n)$, which is called the **test statistic**. The range of values of T for which the test procedure recommends the rejection of a hypothesis is called the **critical region**, and the range for accepting the hypothesis is called the **acceptance region**.

Type I and Type II Errors

In performing a test one may arrive at the correct decision or commit one of two types of errors. The errors can be classified into two groups labeled **Type I** and **Type II** errors.

Type I error: Rejecting H_0 when it is true
Type II error: Accepting H_0 when it is false

Power of a Test

The probability of rejecting the null hypothesis H_0 based on a test procedure is called the **power of the test**. This probability would obviously depend on the value of the parameter θ about which the hypothesis is formulated. It is thus a function of θ. This **power function** is denoted by $\pi(\theta)$.

Operating Characteristic

The probability of accepting the null hypothesis is known as the **operating characteristic** and is represented by $1 - \pi(\theta)$. This concept is widely used in *statistical quality control theory*.

Example 9.1

Suppose the life of a light bulb X is a random variable that follows the exponential distribution with $f(x;\theta) = (1/\theta)e^{-x/\theta}$, $x \geq 0$, and $\theta > 0$. Let the null hypothesis be $\theta = \theta_0$ and the alternative be $\theta = \theta_1$ ($> \theta_0$). To test the null hypothesis, we draw a single observation x. Suppose the decision rule is to reject H_0 if $x > d$, where d is a fixed number to be determined later. Then the open interval (d, ∞) is the critical region and the closed interval $[0, d]$ is the acceptance region. The probabilities of the Type I and Type II errors as well as the power function are as follows.

$$P(I) = P[x > d \mid \theta = \theta_0] = \int_d^\infty \frac{1}{\theta_0} e^{-x/\theta_0} = e^{-d/\theta_0}$$

$$P(II) = P[0 \leq x \leq d \mid \theta = \theta_1] = \int_0^d \frac{1}{\theta_1} e^{-x/\theta_1} = 1 - e^{-d/\theta_1}$$

$$\pi(\theta, d) = P[x > d \mid \theta] = \int_d^\infty \frac{1}{\theta} e^{-x/\theta} = e^{-d/\theta}$$

Level of Significance and the Size of a Test

When θ is in Θ_0, $\pi(\theta)$ gives the probability of Type I error. This probability, denoted by $P(I)$, will also depend on θ. The maximum value of $P(I)$ when $\theta \in \Theta_0$ is called the **level of significance** of a test, denoted by α. It is also known as the **size of a test**. Thus,

$$\alpha = \max_{\theta \in \Theta_0} P(I) = \max_{\theta \in \Theta_0} \pi(\theta)$$

The level of significance is hence the largest probability of a Type I error. The common sizes used are 0.01, 0.05, and 0.10. The probability of a Type II error is often denoted by $\beta(\theta)$. It is readily seen to be $1 - \pi(\theta)$ when $\theta \in \Theta_1$. Thus,

$$\beta(\theta) = P(II)_{\theta \in \Theta_1} = 1 - \pi(\theta)$$

Ideally we would want to keep both $P(I)$ and $P(II)$ to the minimum no matter what the value of θ is. But this is impossible because an attempt to reduce $P(I)$ generally increases $P(II)$. For instance, the decision rule "reject H_0 always," regardless of x, has $P(II) = 0$ but $P(I) = 1$ if $\theta \in \Theta_0$. Similarly, the rule "always accept H_0" implies that $P(I) = 0$ but $P(II) = 1$ when $\theta \in \Theta_1$. Thus for some values of θ, one decision rule will be better than another.

As another example, in Example 9.1 if we want to reduce $P(I)$, then d must be larger. But in that case we see that $P(II)$ goes up. The classical testing procedure resolves this dilemma by choosing an acceptable value for α and then selecting a decision rule (that is, a test procedure) that minimizes $P(II)$. In other words, given α, among the class of decision rules for which $P(I) \leq \alpha$, choose the one for which $P(II)$ is minimized or, equivalently, for which $\pi(\theta)$ is maximized. Thus the test procedure selects the decision rule that maximizes $\pi(\theta)$ subject to $P(I) \leq \alpha$. Such a test is called a **most powerful (MP) test**. If the critical region obtained this way is independent of the alternative H_1, then we have a **uniformly most powerful (UMP) test**.

Example 9.2

Consider a pharmaceutical company that has developed a new drug to cure a certain disease. The company claims that the drug increases the probability of a cure (from say θ_0 to θ_1). The maintained null hypothesis is that the drug does not raise the probability of a cure (that is, that $\theta = \theta_0$). A Type I error would occur if the probability of a cure is really no higher than θ_0, but we reject H_0 and conclude that the probability of a cure is higher. A Type II error occurs if we accept H_0 and conclude that the drug is no more effective when it really is. The classical test procedure is to specify an upper limit (such as 0.01) for the probability that an ineffective drug is certified as effective and then minimize the probability that an effective drug is rejected.

9.2   THE NEYMAN-PEARSON FUNDAMENTAL LEMMA

In the case of a simple hypothesis and a simple alternative, the problem is one of choosing between θ_0 and θ_1. We thus have to find the best test of size α (that is, level of significance α) for H_0: $\theta = \theta_0$ against H_1: $\theta = \theta_1$. The **Neyman-Pearson (NP) fundamental lemma** provides a procedure for selecting the best test.

Theorem 9.1

A test with critical region C having

$$\frac{L(\theta_1; x)}{L(\theta_0; x)} \geq k > 0 \quad \textit{for x in C}$$

$$< k \qquad \textit{for x in } \overline{C} \textit{ (that is, not in C)}$$

is a most powerful test of H_0: $\theta = \theta_0$ against H_1: $\theta = \theta_1$ ($\theta_1 > \theta_0$) for some size α. If the value k and the set C can be chosen to satisfy $\int_C L(\theta_0; x)\,dx = \alpha$ exactly, then C is most powerful among tests of size α.

Proof   The proof is given only for the continuous case. The discrete case is identical with summations replacing integrals. Let A be another critical region of the same size α. Then

$$\int_C L(\theta_0; x)\,dx = \int_A L(\theta_0; x)\,dx$$

Denote the complementary sets by $\overline{A}$ and $\overline{C}$. The power function is

$$\pi(\theta) = P(x \in C \mid \theta) = \int_C L(\theta; x)\,dx$$

To demonstrate that the test is most powerful, we want to show that

$$\int_C L(\theta_1; x)\,dx \geq \int_A L(\theta_1; x)\,dx$$

We have,

$$\int_C L(\theta_1; x)\,dx - \int_A L(\theta_1; x)\,dx = \int_{C \cap A} L(\theta_1; x)\,dx + \int_{C \cap \overline{A}} L(\theta_1; x)\,dx$$

$$- \int_{A \cap C} L(\theta_1; x)\,dx - \int_{A \cap \overline{C}} L(\theta_1; x)\,dx$$

$$= \int_{C \cap \overline{A}} L(\theta_1; x)\,dx - \int_{A \cap \overline{C}} L(\theta_1; x)\,dx$$

In C, $L(\theta_1; x) \geq kL(\theta_0; x)$, and in $\overline{C}$ the inequality becomes $-L(\theta_1; x) > -kL(\theta_0; x)$. Therefore the above difference is greater than

$$\int_{C \cap \overline{A}} kL(\theta_0; x)\, dx \;-\; \int_{A \cap \overline{C}} kL(\theta_0; x)\, dx$$

Adding $\int_{A \cap C} kL(\theta_0; x)$ to both integrals, the difference is greater than

$$\int_{C} kL(\theta_0; x)\, dx \;-\; \int_{A} kL(\theta_0; x)\, dx \;=\; 0$$

Therefore the critical region C is more (no less) powerful than any other critical region A of the same size.

In the rest of the chapter we denote $L(\theta_m; x)$ by L_m for $m = 0$ and 1.

Example 9.3

Let $x_i \sim N(\mu, \sigma_0^2)$ with known σ_0 and let the hypotheses be H_0: $\mu = \mu_0$ and H_1: $\mu = \mu_1$ with $\mu_1 > \mu_0$. The critical region must satisfy the condition

$$\frac{L_1}{L_0} = \exp\left[-\frac{1}{2\sigma_0^2} \sum (x_i - \mu_1)^2 + \frac{1}{2\sigma_0^2} \sum (x_i - \mu_0)^2 \right] > k$$

or, equivalently,

$$\ln\left[\frac{L_1}{L_0}\right] = \frac{1}{2\sigma_0^2}(\mu_1 - \mu_0)[\sum(2x_i - \mu_0 - \mu_1)] > \ln k$$

It is easy to see that this reduces to the form $\bar{x} > d$, where d is to be chosen to give a size α test. Thus the critical region is $C = \{x : \bar{x} > d\}$, where d is obtained so that $P(\bar{x} > d \mid \mu = \mu_0) = \alpha$.

Under the null hypothesis, $\bar{x} \sim N(\mu_0, \sigma_0^2/n)$. Hence

$$\frac{\bar{x} - \mu_0}{\sigma_0/\sqrt{n}} = z \sim N(0,1)$$

$$P(\bar{x} > d \mid \mu = \mu_0) = P\left[z > \frac{d - \mu_0}{\sigma_0/\sqrt{n}}\right]$$

Let z_α be the point on the standard normal such that $\int_{z_\alpha}^{\infty} N(0,1)\, dz = \alpha$, that is, the point z_α would have an area equal to α to the right of it (see Figure 9.1). Then

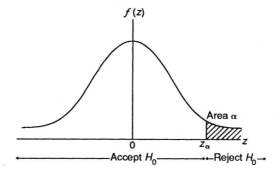

Figure 9.1 UMP test for the mean of a normal distribution

$$\frac{d - \mu_0}{\sigma_0 / \sqrt{n}} = z_\alpha \qquad \text{so that} \qquad d = \mu_0 + (\sigma_0 z_\alpha / \sqrt{n})$$

Therefore, the desired critical region is $z > z_\alpha$ or, equivalently, $C = \{x : \bar{x} > \mu_0 + (\sigma_0 z_\alpha / \sqrt{n})\}$. To use a numerical example, let the level of significance be $\alpha = 0.05$, $\bar{x} = 110$, $\mu_0 = 100$, $\sigma_0 = 20$, and $n = 25$. To compute the critical region, we need z_α. Appendix Table B.3 gives the areas under the normal curve to the right of specified values of z. We note that $z_\alpha = 1.645$ gives a 5 percent area in the right-hand tail. The critical value d is therefore given by $\mu_0 + (\sigma_0 z_\alpha / \sqrt{n}) = 100 + (20 \times 1.645/5) = 106.58$. Because $110 > 106.58$ we reject the null hypothesis that $\mu = 100$. Alternatively, compute

$$z_c = \frac{\bar{x} - \mu_0}{\sigma_0 / \sqrt{n}} = \frac{110 - 100}{4} = 2.5$$

Since $z_c > z_\alpha$ we reject H_0.

Note that the critical region C is independent of μ_1 and is the same for any $\mu_1 > \mu_0$. Thus the critical region C gives the *uniformly most powerful (UMP)* test for $H_0 : \mu = \mu_0$ against all alternatives of the form $H_1 : \mu > \mu_0$.

The *p*-value Approach to Hypothesis Testing

The test conducted in Example 9.3 can also be done in an equivalent way known as the *p*-value approach. To carry this out, first calculate the test statistic

$$z_c = \frac{\bar{x} - \mu_0}{\sigma_0/\sqrt{n}} = \frac{110 - 100}{4} = 2.5$$

Next compute

$$p\text{-value} = P(\bar{x} > d \mid \mu = \mu_0) = P[z > z_c] = P[z > 2.5]$$

This probability is the same as the probability of a Type I error, that is, the probability of rejecting a true hypothesis. A low value for this implies that $P(I)$ is small and hence we are "safe" in rejecting H_0. A high p-value implies that the consequences of making a Type I error are very high and hence we should not reject H_0. From Table B.3 we see that the area to the right of 2.5 is 0.0062. A level of significance of 0.0062 is quite small and hence we are safe in rejecting the null hypothesis that $\mu = \mu_0$.

To see the equivalence of the two test procedures, we note from Figure 9.1 that if $P[z > z_c]$ is *less* than the level α, then the point corresponding to z_c must necessarily be to the *right* of z_α, which puts it in the critical region of rejection. Similarly, if $P[z > z_c] > \alpha$, then z_c must be to the left of z_α, implying that we cannot reject H_0.

Practice Problems

9.1 For the normal case in Example 9.3, derive the MP test for $\mu = \mu_0$ against $\mu_1 < \mu_0$ and check whether it is also UMP.

9.2 For the exponential distribution presented in Example 9.1 (with a single observation x) derive the MP test of size α for $H_0: \theta = \theta_0$ against $H_1: \theta = \theta_1$ $(> \theta_0)$. Is it also a UMP test? Suppose $\theta_0 = 10$ and the observation is 15. Calculate the p-value given by $P(x > 15 \mid \theta = 10)$ and indicate whether or not you would accept the null hypothesis $\theta = 10$ against the alternative that $\theta > 10$.

Example 9.4

Let $x_i \sim$ Poisson (λ). Then

$$\frac{L_1}{L_0} = e^{(\lambda_0 - \lambda_1)n} \left[\frac{\lambda_1}{\lambda_0}\right]^{\Sigma x_i} \geq k$$

which reduces to $\Sigma x_i \geq d$. The distribution of Σx_i is Poisson $(n\lambda)$.

$$P(\Sigma x_i \geq d \mid \lambda_0) = \sum_{y=d}^{\infty} e^{-n\lambda_0} \frac{(n\lambda_0)^y}{y!}$$

Suppose $\lambda_0 = 1$ and $n = 2$. The sum of observations Σx_i is distributed as Poisson(2). From Table B.2, $P(\Sigma x_i \geq 6) = 0.0166$ and $P(\Sigma x_i \geq 5) = 0.0527$. Therefore, there is no value of d that will give exactly a 5% test (that is, $\alpha = 0.05$). In this case, a modified procedure known as a **randomized test** is applied. The test criterion is to reject H_0 if $L_1/L_0 > k$, accept H_0 if $L_1/L_0 < k$, and reject H_0 with a probability p if L_1/L_0 is exactly equal to k. The randomization probability p is obtained so as to make the overall probability of rejection equal to α. Thus

$$P\left[\frac{L_1}{L_0} > k\right] + pP\left[\frac{L_1}{L_0} = k\right] = \alpha$$

The value of p is therefore

$$p = \frac{\alpha - P\left[\dfrac{L_1}{L_0} > k\right]}{P\left[\dfrac{L_1}{L_0} = k\right]}$$

In the case of a continuous random variable, there will be no need to randomize because an exact k can be obtained.

In the Poisson example we get

$$p = \frac{0.05 - 0.0166}{0.0527 - 0.0166} = 0.925$$

The test is therefore: reject H_0 if $\Sigma x_i \geq 6$, accept H_0 if $\Sigma x_i \leq 4$, and reject H_0 92.5 percent of the time if Σx_i is exactly 5.

9.3   MONOTONE LIKELIHOOD RATIO

The simple hypothesis chooses between θ_0 and θ_1. This is unrealistic because we often have no basis to specify the alternative θ_1. We thus need a procedure to test composite hypotheses, preferably with a UMP test. For many distributions this is possible using the ratio $L(\theta_1; x)/L(\theta_0; x)$, called the **likelihood ratio**.

Uniformly Most Powerful

Definition 9.1

*The statistical model $f(x; \theta)$ is said to have the **monotone likelihood ratio (MLR)** form if there exists a real valued function $u(x_1, \ldots, x_n)$ such that the likelihood ratio $\lambda = L_1/L_0$ is a* *non-decreasing function of $u(x)$ for each choice of θ_0 and θ_1 with $\theta_1 > \theta_0$.*

Example 9.5

Consider the exponential family of distributions for which

$$L(\theta; x) = \exp\left[\Sigma\, U(x_i) + nB(\theta) + A(\theta)\Sigma\, T(x_i)\right]$$

The log of the likelihood ratio for this family is,

$$\ln \lambda = nB(\theta_1) - nB(\theta_0) + [\Sigma\, T(x_i)]\,[A(\theta_1) - A(\theta_0)]$$

Let $u = \Sigma\, T(x_i)$. Then note that $\partial \ln \lambda / \partial u = A(\theta_1) - A(\theta_0)$. If the function $A(\theta)$ is monotone, then λ is monotone in u. Furthermore, u is a sufficient statistic.

Theorem 9.2

For a model with MLR form, obtain u^* such that the level of significance $\alpha = P(u > u^* \mid \theta = \theta_0)$. Then the test: reject H_0 if $u(x) > u^*$, accept H_0 if $u(x) < u^*$, and randomize if $u(x) = u^*$ is a UMP test of size α for $H_0: \theta \le \theta_0$ against $H_1: \theta > \theta_0$.

The proof requires two lemmas that are proved first.

Lemma 9.1

For a model with MLR form, a most powerful test of θ_0 against $\theta_1\,(>\theta_0)$ has the form: reject the null hypothesis when $u(x) > u^*$ and accept it when $u(x) < u^*$ with possible randomization at the critical value u^*.

Proof   The lemma is quite obvious. The NP lemma says that the MP test has the form $\lambda > k$. If λ is monotonically increasing in u, then $\lambda > k$ is equivalent to the condition $u > u^*$.

Lemma 9.2

For a model with MLR form, the probability $P[u(x) > u^* \mid \theta]$ is a nondecreasing function of θ.

Proof   Although an algebraic proof is possible [see Fraser (1976), pp. 425–426], a geometric proof makes it more obvious intuitively (see Figure 9.2). Let $\theta_2 > \theta_1$. MLR property implies that as θ increases, the distribution of u shifts to the right, thus increasing $P(u > u^*)$, which is the area to the right of a given u^*.

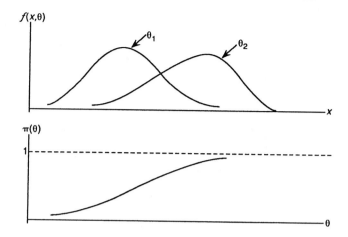

Figure 9.2   Power function for a model with MLR form

Proof of Theorem 9.2   By Lemma 9.1, the test is MP for $\theta = \theta_0$ with the level of significance α. Because of Lemma 9.2, the probability of rejecting H_0 when $\theta \leq \theta_0$ is less than or equal to α. Thus the test has size α for all $\theta \leq \theta_0$ and is hence MP for that null hypothesis. Because u^* is independent of θ_1, it is also UMP for all alternatives $\theta_1 > \theta_0$.

9.4   APPLICATIONS TO THE NORMAL DISTRIBUTION

Test on Mean with Known S.D.

We saw in Example 9.3 that, given $\sigma = \sigma_0$, the likelihood ratio for $N(\mu, \sigma_0^2)$ is monotonic in $\bar{x}$ for $\mu_1 > \mu_0$. Therefore the test: reject $H_0: \mu \leq \mu_0$ if $\bar{x} > \mu_0 + (\sigma_0 z_\alpha / \sqrt{n})$ is UMP against the alternative $\mu > \mu_0$. By a similar argument, the test: reject H_0 if $\bar{x} < \mu_0 - (\sigma_0 z_\alpha / \sqrt{n})$ is UMP for $H_0: \mu \geq \mu_0$ against $H_1: \mu < \mu_0$.

Test on S.D. with Known Mean

Let $\mu = \mu_0$ be known. The log-likelihood ratio is

$$\ln \lambda = \ln \left[\frac{L_1}{L_0} \right] = [\Sigma (x_i - \mu_0)^2] \left[\frac{\sigma_1^2 - \sigma_0^2}{2\sigma_0^2 \sigma_1^2} \right] + n \ln \left[\frac{\sigma_0}{\sigma_1} \right]$$

We note that the statistic $u(x) = \Sigma (x_i - \mu_0)^2$ makes this the MLR kind. The critical region for H_0: $\sigma = \sigma_0$, H_1: $\sigma = \sigma_1 > \sigma_0$, will therefore be $u > u^*$, where u^* is obtained so that $P(u > u^* \mid \sigma_0) = \alpha$. In order to evaluate u^*, we need the distribution of u under the null. From Theorem 5.21, $[\Sigma (x_i - \mu_0)^2]/\sigma_0^2 \sim \chi_n^2$ because it is the sum of squares of n independent normal variates. From this distribution find the point u^* such that $P(\chi_n^2 > u^*) = \alpha$ (Appendix Table B.4 has these values). The critical region is $[\Sigma (x_i - \mu_0)^2]/\sigma_0^2 > u^*$. Because u^* is independent of σ_1, this test is UMP against all $\sigma > \sigma_0$. The test for $\sigma \geq \sigma_0$ against $\sigma < \sigma_0$ is similar. Find u_* at the left tail of χ_n^2. The critical region is now $[\Sigma (x_i - \mu_0)^2]/\sigma_0^2 < u_*$.

Test on Mean with Unknown S.D.

We saw that the UMP test for H_0: $\mu \leq \mu_0$ against H_1: $\mu > \mu_0$ for a given σ_0 is $\bar{x} > \mu_0 + (\sigma_0 z_\alpha / \sqrt{n})$. But as the critical region depends on σ_0, a UMP for a general normal with unknown σ does not exist. In this case, however, an alternative test, called an **unbiased test**, is available. This is discussed in the next section.

Test on S.D. with Unknown Mean

As seen earlier, the likelihood ratio is monotone with respect to $u(x) = \Sigma (x_i - \mu)^2$, but because μ is unknown we cannot construct a UMP test. This and the previous case involve two unknown parameters and will be discussed later.

9.5   UNBIASED TESTS

So far, we have considered null hypotheses of the form $\theta \leq \theta_0$ or $\theta \geq \theta_0$. These are called **one-sided** hypotheses. Suppose we are interested in carrying out a **two-sided test**; H_0: $\theta = \theta_0$, H_1: $\theta \neq \theta_0$. We have seen that for a model with the MLR form the critical region $C_1 = \{ x: u(x) > u^* \}$ is UMP for $\theta > \theta_0$. By symmetry, the UMP test against the alternatives $\theta < \theta_0$ will be $C_2 = \{ x: u(x) < u^* \}$. The corresponding power functions are the solid lines in Figure 9.3. We note that when $\theta < \theta_0$, the power for the region C_1 is less than α, and similarly for C_2 when the alternative is $\theta > \theta_0$. It follows that if we use C_1 or C_2 for the two-sided alternative $\theta \neq \theta_0$, then we have the undesirable property that the probability of rejecting a true hypothesis is larger than the probability of rejecting a false one. Unfortunately, a UMP test does not exist in this case. We therefore

UMP = Uniformly Most Powerful

have to restrict our attention to a smaller class of tests. For instance, it is reasonable to require that the probability of rejection when a hypothesis is false be at least α, the level of significance. When the null hypothesis is $\theta = \theta_0$, this requires that the power function be a minimum at that point. The tests that satisfy this stipulation are called **unbiased tests,** and therefore we would search for the most powerful test in this restricted class.

Definition 9.2

A test of H_0: $\theta \in \Theta_0$ against the alternative H_1: $\theta \in \Theta_1$ is said to be **unbiased** *if*

$$P(I) \leq \alpha \ \ \text{for all} \ \ \theta \in \Theta_0 \quad \text{and} \quad \pi(\theta) \geq \alpha \ \ \text{for all} \ \ \theta \in \Theta_1$$

In other words, in an unbiased test the probability of rejecting the null hypothesis when it is false is at least as great as the probability of rejecting it when it is true. In the case of H_0: $\theta = \theta_0$, the power function will be the broken line illustrated in Figure 9.3.

Theorem 9.3

For a one-parameter exponential family of densities, the **UMP unbiased (UMPU) test** *of size α for H_0: $\theta = \theta_0$, H_1: $\theta \neq \theta_0$ is of the form: reject H_0 if $u(x) < u_1$ or $> u_2$, and accept H_0 if $u_1 < u(x) < u_2$, with possible randomization at u_1 and u_2 in the discrete case. The constants u_1 and u_2 are determined by the following conditions:*

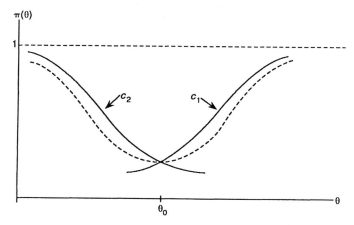

Figure 9.3   Illustration of an unbiased test

(1) the size of the test must be α, and (2) because of unbiasedness, $\pi(\theta)$ must have a minimum at $\theta = \theta_0$. If the distribution of $u(x)$ is symmetric about the origin, an equal tail test with $u_1 = -u_2$ is UMPU.

This theorem is proved in Lehmann (1991, pp. 135–137).

Example 9.6

Consider the case of a normal distribution (with known s.d.) for which H_0 is $\mu = \mu_0$ and H_1 is $\mu \neq \mu_0$. We have seen earlier that a one-sided test is of the form: reject H_0 if $z_c > z_\alpha$, where $z_c = (\bar{x} - \mu_0)/(\sigma_0/\sqrt{n})$, and z_α is the point on the standard normal such that the area to the right is α. For the two-sided test we use the fact that the normal distribution is symmetric around the mean and hence $z_{\alpha/2}$ is chosen so that the area to right of it is $\alpha/2$. The critical region now becomes $z_c < -z_{\alpha/2}$ or $z_c > z_{\alpha/2}$. This is equivalent to the critical region

$$ C = [\, x : |\bar{x} - \mu_0| \geq (\sigma_0\, z_{\alpha/2})/\sqrt{n}\,] $$

This test (known as a **two-tailed test**) is illustrated in Figure 9.4 in which the shaded area is the critical region. [See Casella and Berger (1990), pp. 374–375 for a rigorous derivation of this test.]

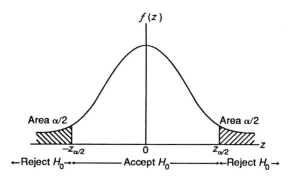

Figure 9.4   Two-tailed test for the mean of a normal distribution

Using the numerical values in Example 9.3, we have α = 0.05, $\bar{x}$ = 110, μ_0 = 100, σ_0 = 20, and n = 25. The value of z_c is the same as in Example 9.3 and is 2.5. From Appendix Table B.3 we see that $z_{\alpha/2}$ = 1.96. Because $z_c > z_{\alpha/2}$, we reject H_0 here also.

To use the *p*-value approach, we calculate $P(z \geq |z_c|)$ which is given in Table B.3 as 0.0124. Because this is below 2 percent, we reject the null hypothesis at less than 2 percent level.

Practice Problem

9.3   Show that (a) if a hypothesis is rejected at the 1 percent level, it will also be rejected at any level higher than that, and (b) if a hypothesis is accepted at the 10 percent level, it will be accepted at any level lower than that.

Uniform Most Powerful Unbias (UMPU)

9.6   UMPU TESTS FOR MULTIPARAMETER EXPONENTIAL FAMILIES

Although the test illustrated in Example 9.6 is a step in the right direction, σ_0 is generally unknown and is hence a **nuisance parameter**. UMPU tests do exist on θ in the presence of such nuisance parameters provided the density belongs to the exponential family. The procedure is described in Theorems 9.4 and 9.5 proved in Lehmann (1959), Section 4.4. The trick is to use the conditional distribution of the test statistic given sample estimators for the nuisance parameters.

Definition 9.3

A density function is said to belong to the **multiparameter exponential family** *if it can be expressed as*

$$f(x; \theta, \lambda) = A(\theta, \lambda) \exp\left[\theta u(x) + \sum_{j=1}^{k} \lambda_j S_j(x)\right] H(x)$$

where θ is the parameter of interest and $\lambda_1, \lambda_2, \ldots, \lambda_k$ are the remaining parameters of the distribution.

Theorem 9.4

For a model with the exponential form defined above, the UMPU test of size α for $\theta \leq \theta_0$ against $\theta > \theta_0$ has the form: reject H_0 if $u(x) > k(s)$, accept H_0 if $u(x) < k(s)$, with possible randomization *when $u(x) = k(s)$, where s is a set of estimators for λ. The critical value $k(s)$ is chosen so that $P(u > k(s) \mid s; \theta = \theta_0) = \alpha$ for each value of s.*

Example 9.7

Let $x_i \sim N(\mu, \sigma^2)$, $H_0: \mu \leq \mu_0$, $H_1: \mu > \mu_0$. The likelihood function is

$$L(\mu, \sigma; x) = \left[\frac{1}{\sigma\sqrt{2\pi}}\right]^n \exp\left[-\frac{\Sigma(x_i - \mu)^2}{2\sigma^2}\right]$$

$$= \left[\frac{1}{\sigma\sqrt{2\pi}}\right]^n \exp\left[-\frac{n\mu^2}{2\sigma^2}\right] \exp\left[\frac{\mu\Sigma x_i}{\sigma^2} - \frac{\Sigma x_i^2}{2\sigma^2}\right]$$

This has the exponential form with $u = \Sigma x_i$ and $S = \Sigma x_i^2$. Consider the sample variance $s^2 = [\Sigma(x_i - \bar{x})^2]/(n-1)$ and the test statistic

$$t_c = \frac{\sqrt{n}(\bar{x} - \mu_0)}{s}$$

Given s, there is a one-to-one correspondence between u and t_c. Also, the conditional distribution of $\bar{x}$ given s is the same as the distribution of t_c, which is the Student's t-distribution (Theorem 6.3). Therefore the test criterion is to reject H_0 if $t_c > t^*_{n-1}(\alpha)$, where t^* is the point on the Student's t-distribution with $n-1$ degrees of freedom such that the area to the right is α (see Appendix Table B.4 for specific values). By Theorem 9.4, this test (known as a t-test) is UMPU for $\mu \leq \mu_0$ against $\mu > \mu_0$.

Theorem 9.5

For the model with the exponential form in Definition 9.3, the UMPU test for $\theta = \theta_0$ against $\theta \neq \theta_0$ has the form: reject H_0 if $u(x) < k_1(s)$ or $u > k_2(s)$, accept H_0 if $k_1(s) < u < k_2(s)$, with possible randomization at $u = k_1$ or $u = k_2$. The critical values k_1 and k_2 are chosen so that the test has size α and that the power function is minimum at $\theta = \theta_0$. If the conditional distribution of u given s is symmetric around the origin, then an equal tail test (with $k_1 = -k_2$) is UMPU.

Example 9.8

To test the two-sided alternative $\theta \neq \theta_0$ in the normal case, the test is to reject H_0 if $|t_c| > t^*_{n-1}(\alpha/2)$, where t^* is the point on the t-distribution with $n-1$ d.f. such that the area to the right is $\alpha/2$, that is, one-half the level of significance. Equivalently, compute the p-value $P(t > |t_c|)$ and reject H_0 if it is less than α.

9.7   GENERALIZED LIKELIHOOD RATIO TESTS

A test procedure that gives reasonable tests for a variety of cases, including the two-sided case discussed above, is based on a generalized version of the likelihood ratio we encountered.

Definition 9.4

The **generalized likelihood ratio** *test statistic for testing* $\theta \in \Theta_0$ *against* $\theta \in \Theta_1$ *is given by*

$$\lambda(x) = \frac{\sup_{\theta \in \Theta_0} L(\theta; x)}{\sup_{\theta \in \Theta} L(\theta; x)}$$

where $H_0: \theta \in \Theta_0$ *is the null hypothesis and* Θ *is the entire parameter space.*

Essentially what we do when constructing this ratio is to maximize the likelihood over the restricted space Θ_0 and also without any restrictions. We readily note that the denominator of λ corresponds to the maximum likelihood estimator $\hat{\theta}$. It is obvious that $0 \le \lambda \le 1$. In the case of a discrete random variable, the numerator of λ is the maximum probability of the observed sample in the space restricted by the null hypothesis. The denominator is the maximum probability over all possible parameters. If λ is small, we can suspect that there are parameter values in the alternative hypothesis for which the observed sample is much more likely than for any parameter value in the null hypothesis. This would suggest that we should reject the null hypothesis in this case. This notion is formulated as the **likelihood ratio (LR) test**.

Definition 9.5

The critical region for testing $H_0: \theta \in \Theta_0$ *by a* **likelihood ratio test** *is* $0 < \lambda < k$, *where* k *is determined by the condition* $P(0 < \lambda < k \mid \theta \in \Theta_0) \le \alpha$.

Although the LR test is not based on any optimality procedure, it nevertheless gives remarkably reasonable tests that sometimes reduce to the tests that are UMP or UMPU. Even if the distribution of λ is unknown, that of some function of it can often be utilized or an asymptotic distribution can be derived. The test has also the advantage of being easily applied in the multiparameter case.

Theorem 9.6

> Let λ be the likelihood ratio just defined and let $u(\lambda)$ be a monotonic function of λ. Then the LR test based on λ is equivalent to an appropriate test on $u(\lambda)$. If $u(\lambda)$ is monotonic increasing, then the critical region is of the form: reject H_0 if $u_1 < u(\lambda) < u_2$, where the end points are determined to satisfy the size condition $\alpha = P(u_1 < u < u_2 \mid \theta \in \Theta)$. If $u(\lambda)$ is monotonic decreasing, then the critical region becomes $u < u_1$ and $u > u_2$.

The proof is trivial and is left as an exercise.

9.8   LR TESTS ON THE MEAN AND S.D. OF A NORMAL DISTRIBUTION

We now apply the likelihood ratio procedure to perform tests on the mean and standard deviation of a normal distribution when neither of them is known.

Test on the Mean with an Unknown S.D.

We have H_0: $\mu = \mu_0$ and H_1: $\mu \neq \mu_0$. The first step is to maximize the likelihood with no constraints. We note from Example 8.5 that the MLEs are $\hat{\mu} = \bar{x}$ and $\hat{\sigma}^2 = [\Sigma (x_i - \bar{x})^2]/n$. This gives

$$\hat{L} = L(\hat{\theta}; x) = \left[\frac{1}{\hat{\sigma}\sqrt{2\pi}} \right]^n e^{-n/2}$$

It is readily seen that, when $\mu = \mu_0$, the MLE of σ (call it $\tilde{\sigma}$) is given by $\tilde{\sigma}^2 = [\Sigma (x_i - \mu_0)^2]/n$. This gives

$$\tilde{L} = L(\tilde{\theta}; x) = \left[\frac{1}{\tilde{\sigma}\sqrt{2\pi}} \right]^n e^{-n/2}$$

The likelihood ratio is therefore

$$\lambda = \frac{\tilde{L}}{\hat{L}} = \left[\frac{\hat{\sigma}}{\tilde{\sigma}}\right]^n = \left[\frac{\sum(x_i - \bar{x})^2}{\sum(x_i - \mu_0)^2}\right]^{n/2} = \left[\frac{\sum(x_i - \bar{x})^2}{\sum(x_i - \bar{x} + \bar{x} - \mu_0)^2}\right]^{n/2}$$

$$= \left[\frac{\sum(x_i - \bar{x})^2}{\sum(x_i - \bar{x})^2 + n(\bar{x} - \mu_0)^2}\right]^{n/2} = \left[\frac{1}{1 + \dfrac{t_c^2}{n-1}}\right]^{n/2}$$

where $t_c^2 = \dfrac{(\bar{x} - \mu_0)^2}{s^2/n}$ and $s^2 = \dfrac{\sum(x_i - \bar{x})^2}{n-1}$.

We note that λ is a monotonic decreasing function of t_c^2 and hence the critical region $\lambda < k$ is equivalent to $t_c^2 > t^{*2}$, where the constant t^* is determined so that $P(t_c^2 > t^{*2}) = 2P(t_c > t^*) = \alpha$, the level of significance. The null hypothesis is thus rejected if $t_c < -t^*$ or $t_c > t^*$. From Theorem 6.3, t_c has the Student's t distribution (under the null hypothesis) with $n - 1$ degrees of freedom.

The test performed above is the t-test encountered before. The critical region is graphed in Figure 9.5. Suppose, $\alpha = 0.05$, $\mu_0 = 100$, $\bar{x} = 110$, $s = 16$, and $n = 25$. Then the computed t-statistic is

$$t_c = \frac{\sqrt{n}\,(\bar{x} - \mu_0)}{s} = \frac{50}{16} = 3.125$$

From Appendix Table B.4 we see that the critical t^* for a two-tailed test with 5 percent level of significance and 24 d.f. is given by 2.064. Because t_c is greater than this, we reject the null hypothesis.

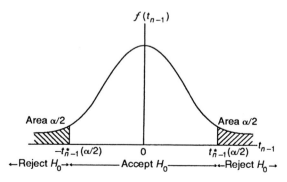

Figure 9.5   Two-tailed t-test for the mean of a normal distribution

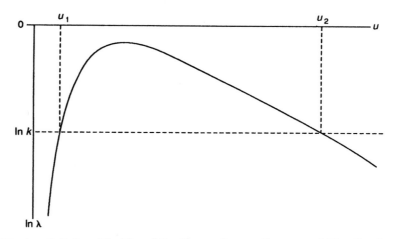

Figure 9.6 Log-likelihood for the variance of a normal distribution

Test on the S.D. with an Unknown Mean

Here H_0: $\sigma = \sigma_0$ and H_1: $\sigma \neq \sigma_0$. The likelihood function is

$$L = \left[\frac{1}{\sigma\sqrt{2\pi}}\right]^n e^{-\Sigma(x_i - \mu)^2/(2\sigma^2)}$$

As before, the MLEs are $\hat{\mu} = \bar{x}$ and $\hat{\sigma}^2 = [\Sigma(x_i - \bar{x})^2]/n$. This gives

$$\hat{L} = \left[\frac{1}{\hat{\sigma}\sqrt{2\pi}}\right]^n e^{-n/2}$$

Under the null hypothesis $\sigma = \sigma_0$, the MLE for μ is $\bar{x}$. The likelihood function is now

$$L_0 = \left[\frac{1}{\sigma_0\sqrt{2\pi}}\right]^n \exp\left[-\frac{\Sigma(x_i - \bar{x})^2}{2\sigma_0^2}\right] = \left[\frac{1}{\sigma_0\sqrt{2\pi}}\right]^n \exp\left[-\frac{n\hat{\sigma}^2}{2\sigma_0^2}\right]$$

The log-likelihood ratio becomes

$$\ln\lambda = \frac{n}{2}\ln u - \frac{nu}{2} + \frac{n}{2}$$

where $u = \hat{\sigma}^2/\sigma_0^2$. The above function of u graphs as in Figure 9.6 with the maximum at $u = 1$. It is readily seen that the critical region $\lambda < k$ translates into $u < u_1$ or $u > u_2$, where the end points are chosen appropriately. A simple way to do that is to assume equal tails so that $P(u < u_1) = P(u > u_2) = \alpha/2$. Because u has the chi-square distribution with $n - 1$ degrees of freedom, the critical values are obtained from χ^2_{n-1} and the test criterion is to reject the null hypothesis if $u < u_1$ or $u > u_2$. If $n = 25$, $\sigma_0 = 10$, and $\hat{\sigma} = 25$, then $u = 6.25$. From Appendix Table B.4, the points on χ^2_{n-1} with an area of 2.5 percent on each tail are given by $u_1 = 12.40$ and $u_2 = 39.36$. Because the test statistic is less than u_1, we reject the null hypothesis.

9.9   TESTING THE EQUALITY OF MEANS OF TWO NORMAL POPULATIONS

Let $x_1 \sim N(\mu_1, \sigma^2)$ and $x_2 \sim N(\mu_2, \sigma^2)$, with x_1 and x_2 independent. We draw random samples of sizes m and n respectively and use them to test the hypothesis $\mu_1 = \mu_2$. Note that the variances of the two populations are assumed to be equal. We will comment on this later.

The parameter space Θ is three dimensional (μ_1, μ_2, σ^2) and the space Θ_0 is two dimensional (μ, σ^2). The log-likelihood function is

$$\ln L = -\frac{m+n}{2} \ln(2\pi) - (m+n) \ln \sigma$$
$$- \frac{1}{2} \sum_{i=1}^{m} \frac{(x_{1i} - \mu_1)^2}{\sigma^2} - \frac{1}{2} \sum_{j=1}^{n} \frac{(x_{2j} - \mu_2)^2}{\sigma^2}$$

The MLE in Θ are:

$$\hat{\mu}_1 = \bar{x}_1, \qquad \hat{\mu}_2 = \bar{x}_2, \qquad \hat{\sigma}^2 = \frac{1}{m+n} \left[\sum_{i=1}^{m} (x_{1i} - \bar{x}_1)^2 + \sum_{j=1}^{n} (x_{2j} - \bar{x}_2)^2 \right]$$

so that

$$\hat{L} = \left[\frac{m+n}{2\pi(ms_1^2 + ns_2^2)} \right]^{(m+n)/2} e^{-(m+n)/2}$$

where $ms_1^2 = \sum_{i=1}^{m} (x_{1i} - \bar{x}_1)^2$ and $ns_2^2 = \sum_{j=1}^{n} (x_{2j} - \bar{x}_2)^2$. In Θ_0 we have

$$\tilde{\mu} = \frac{m\bar{x}_1 + n\bar{x}_2}{m+n}$$

$$\tilde{\sigma}^2 = \frac{1}{m+n} \left[\sum_{i=1}^{m} (x_{1i} - \tilde{\mu})^2 + \sum_{j=1}^{n} (x_{2j} - \tilde{\mu})^2 \right]$$

$$\tilde{\sigma}^2 = \frac{1}{m+n}\left[ms_1^2 + ns_2^2 + \frac{mn}{m+n}(\bar{x}_1 - \bar{x}_2)^2\right]$$

after substituting for $\tilde{\mu}$ and simplifying terms. This gives

$$L_0 = \left[\frac{m+n}{2\pi\left\{ms_1^2 + ns_2^2 + \frac{mn}{m+n}(\bar{x}_1 - \bar{x}_2)^2\right\}}\right]^{(m+n)/2} e^{-(m+n)/2}$$

Therefore

$$\lambda = \left[1 + \frac{\frac{mn}{m+n}(\bar{x}_1 - \bar{x}_2)^2}{ms_1^2 + ns_2^2}\right]^{-(m+n)/2}$$

Using Theorem 9.6 we can obtain a critical region based on a monotonic transformation of λ. First note the following results:

$$\bar{x}_1 \sim N\left[\mu_1, \frac{\sigma^2}{m}\right], \qquad \bar{x}_2 \sim N\left[\mu_2, \frac{\sigma^2}{n}\right]$$

and hence

$$\bar{x}_1 - \bar{x}_2 \sim N\left[\mu_1 - \mu_2, \sigma^2\left\{\frac{1}{m} + \frac{1}{n}\right\}\right]$$

Therefore under the null hypothesis $\mu_1 = \mu_2$

$$z = \frac{\bar{x}_1 - \bar{x}_2}{\sigma\left[\frac{1}{m} + \frac{1}{n}\right]^{1/2}} \sim N(0,1)$$

Furthermore, $ms_1^2/\sigma^2 \sim \chi_{m-1}^2$, $ns_2^2/\sigma^2 \sim \chi_{n-1}^2$, and hence

$$v = \frac{ms_1^2 + ns_2^2}{\sigma^2} \sim \chi_{m+n-2}^2$$

Also z and v are independent. Therefore by the definition of the Student's t–distribution,

$$t_c = \frac{z}{\sqrt{v/(m+n-2)}} \sim t_{m+n-2}$$

It is readily seen that the expression of λ in the square brackets is simply $1 + (z^2/v)$. Therefore the LR is

$$\lambda = \left[1 + \left[\frac{t_c^2}{m+n-2}\right]\right]^{-(m+n)/2}$$

which is monotonically decreasing in t_c^2. The critical region $\lambda < k$ is thus equivalent to $t_c^2 > t^{*2}$, where t^* is determined from the condition $P(t^2 > t^{*2} | \mu_1 = \mu_2) = \alpha$. The critical region is therefore given by $t > t^*$ or $t < -t^*$ where t^* is obtained as the $\alpha/2$ point of t_{m+n-2}. Thus, here also the appropriate test is a two-tailed t-test.

When the variances σ_1^2 and σ_2^2 of the two normal populations are unequal, the problem is much more difficult and does not have a universally accepted solution. This situation is called the **Behrens-Fisher problem** [see Scheffé (1970)]. If $L(\theta; x)$ is maximized under the null hypothesis $\mu_1 = \mu_2 = \mu$, the resultant equation in μ is a cubic. There is no way to determine the distribution of λ or any function of it because it involves the unknown variances σ_1^2 and σ_2^2. If the ratio of the variances is known, a solution is obtainable. It is possible to solve the cubic by numerical methods and obtain large sample distributions. Readers interested in exploring this problem further, should read Scheffé's paper and also relevant sections in Lehmann (1991).

9.10   TESTING THE EQUALITY OF VARIANCES OF TWO NORMAL POPULATIONS

Consider the null hypothesis H_0: $\sigma_1^2 = \sigma_2^2$ against the alternative $\sigma_1^2 \neq \sigma_2^2$. It can be verified by proceeding similarly that the LR is now

$$\lambda = \frac{\left[\frac{m+n}{ms_1^2 + ns_2^2}\right]^{(m+n)/2}}{\left[\frac{1}{s_1^2}\right]^{m/2}\left[\frac{1}{s_2^2}\right]^{n/2}} = k \frac{\left[\frac{m-1}{n-1}F_c\right]^{m/2}}{\left[1 + \frac{m-1}{n-1}F_c\right]^{(m+n)/2}}$$

where k is a constant depending only on m and n, and

$$F_c = \frac{(n-1)\Sigma(x_{1i} - \bar{x}_1)^2}{(m-1)\Sigma(x_{2j} - \bar{x}_2)^2}$$

is the ratio of the sample variances. We know from Theorem 6.5 that $F_c \sim F_{m-1,n-1}$. The graph of the log-likelihood ratio is similar to that in Figure 9.6 and the critical region consists of the two tails of the F-distribution. Because tabulations of F with equal tails are easier to obtain, the test usually involves such an equal tail test.

Practice Problem

9.4 Verify the expression derived above for the likelihood ratio.

9.11 THE WALD, LIKELIHOOD RATIO, AND LAGRANGE MULTIPLIER TESTS

In statistical theory, the classical procedure for testing hypotheses is the likelihood ratio (LR) approach used in Sections 9.7 through 9.10. Two other methods are also used frequently in econometrics, especially when the model has more than one unknown parameter. These are the **Wald test** and the **Lagrange multiplier (LM) test**. In all these methods, two models are formulated, a **restricted model** and an **unrestricted model**. The restricted model is obtained by imposing linear or nonlinear constraints on the parameters, and it corresponds to the null hypothesis. The unrestricted model is the alternative. In this section we discuss each of these in some detail and present a geometric comparison when there is only one parameter to be estimated.

Asymptotic Distribution of the Likelihood Ratio

Although the LR test gives determinate critical regions in many cases, in other situations the distribution of λ or a transformation of it is difficult to obtain. If the sample size is large, one can obtain the asymptotic distribution of λ and use it to obtain the critical region under the null hypothesis.

Theorem 9.7

Let $x_1, x_2, \ldots, x_n$ be an iid sample of observations from a random variable with density function $f(x, \theta_1, \theta_2, \ldots, \theta_k)$ satisfying the regularity conditions stated in Section 8.5. Suppose we wish to test the null hypothesis $\theta_1 = \theta_1^0, \theta_2 = \theta_2^0, \ldots, \theta_k = \theta_k^0$. Let λ be the likelihood ratio defined as $L_0/\hat{L}$. Then $-2 \ln \lambda$ is asymptotically distributed as χ_k^2. Thus for large n, χ_k^2 is an approximation to the distribution of $-2 \ln \lambda$.

Proof Let H_0: $\theta = \theta_0$ be true and let $\hat{\theta}_n$ be the MLE of θ. Using an argument similar to the one presented in Theorem 8.10, we can expand the log-likelihood in a small neighborhood of θ as follows.

$$\ln L(\theta) = \ln L(\hat{\theta}_n) + S'(\hat{\theta}_n)(\theta - \hat{\theta}_n) + \frac{1}{2}(\theta - \hat{\theta}_n)' \left[\frac{\partial S(\theta_n^*)}{\partial \theta} \right] (\theta - \hat{\theta}_n)$$

where $\ln L(\theta)$ is the log-likelihood denoted without the x for simplicity, $S(\theta) = \partial \ln L / \partial \theta$ is the score function used in Section 8.9, S' is its transpose, $\partial S / \partial \theta = \partial^2 \ln L / \partial \theta \partial \theta'$, and $|\theta - \theta_n^*| \leq |\theta - \hat{\theta}_n|$. Using the fact that $S(\hat{\theta}_n) = 0$ (because $\hat{\theta}_n$ is MLE) we have,

$$\ln L(\theta_0) - \ln L(\hat{\theta}_n) = \frac{1}{2}(\hat{\theta}_n - \theta_0)' \left[\frac{\partial S(\theta_n^*)}{\partial \theta} \right] (\hat{\theta}_n - \theta_0)$$

Hence

$$-2 \ln \lambda = \sqrt{n}(\hat{\theta}_n - \theta_0)' \left[-\frac{1}{n} \frac{\partial S(\theta_n^*)}{\partial \theta} \right] \sqrt{n}(\hat{\theta}_n - \theta_0)$$

The information matrix is $I(\theta) = E[-\partial S / \partial \theta]$. Let $\Sigma(\theta) = \lim_{n \to \infty} I(\theta)/n$. We showed in Theorem 8.10 that

$$\left[-\frac{1}{n} \frac{\partial S(\theta_n^*)}{\partial \theta} \right] \xrightarrow{p} \Sigma(\theta_0)$$

and that

$$\sqrt{n}(\hat{\theta}_n - \theta_0) \xrightarrow{d} N[0, \{\Sigma(\theta_0)\}^{-1}]$$

Therefore the quadratic form corresponding to the above distribution has the χ_k^2 distribution (see Theorem 5.23), for large n. It thus follows that $-2 \ln \lambda$ is asymptotically distributed as χ_k^2.

The LR Test Procedure

In stating Theorem 9.7, the null hypothesis H_0: $\theta = \theta_0$ was stated in terms of all the parameters. In practice, however, only a subset of θ may be in the null hypothesis, as for example, H_0: $\mu = \mu_0$ in a normal (μ, σ^2) population. In this case the test procedure should be modified slightly. Let $\theta' = (\theta_1', \theta_2')$ be a partition of θ into two sets of parameters, with k_1 parameters in the first set and k_2 parameters in the second set. For the null hypothesis $\theta_2 = \theta_2^0$, the test procedure is as follows. Let $\hat{L}(\hat{\theta}_1, \hat{\theta}_2)$ be the maximum likelihood with no restriction on θ, and let $\tilde{L}(\tilde{\theta}_1, \theta_2^0)$ be the maximum likelihood under the null. If $\lambda = \tilde{L}/\hat{L}$ is the likelihood ratio, then $\xi_{LR} = -2 \ln \lambda$ has the asymptotic $\chi_{k_2}^2$ distribution. We would reject the null hypothesis if $\xi_{LR} > \chi_{k_2}^2(\alpha)$, the point on the chi-square distribution with k_2 d.f. such that the area to the right is the level of significance given by α.

The Wald Test Procedure

The test procedure proposed by Wald (1943) is based directly on the fact that the maximum likelihood estimator of the parameter θ is asymptotically normally distributed. We saw in Theorem 8.10 that

$$\sqrt{n}(\hat{\theta} - \theta) \xrightarrow{d} N[0, \{\Sigma(\theta)\}^{-1}]$$

where $\Sigma(\theta) = \lim_{n \to \infty} I(\theta)/n$. A consistent estimator of $\Sigma(\theta)$ is $I(\hat{\theta})/n$. Therefore by Theorem 5.23, the corresponding quadratic form has a limiting chi-square distribution. Thus, under H_0,

$$\xi_W = (\hat{\theta} - \theta_0)' I(\hat{\theta})(\hat{\theta} - \theta_0) \xrightarrow{d} \chi_k^2$$

The test criterion is to reject the null hypothesis $\theta = \theta_0$ if the quadratic form ξ_W is greater than $\chi_k^2(\alpha)$, the point on the chi-square distribution with k d.f. and an area of α to the right of it. To carry out a test on a subset of θ, suppose that $\theta' = (\theta_1', \theta_2')$ and that the null hypothesis is $\theta_2 = \theta_2^0$. First partition the information matrix and its inverse as

$$I(\theta) = \begin{bmatrix} I_{11} & I_{12} \\ I_{21} & I_{22} \end{bmatrix} \quad \text{and} \quad [I(\theta)]^{-1} = \begin{bmatrix} I^{11} & I^{12} \\ I^{21} & I^{22} \end{bmatrix}$$

Applying the partitioned inverse property (see Appendix A, Property A.8), we have

$$I^{22} = [I_{22} - I_{21}I_{11}^{-1}I_{12}]^{-1}$$

It follows analogously that

$$\sqrt{n}(\hat{\theta}_2 - \theta_2^0) \xrightarrow{d} N\left[0, \lim_{n \to \infty} (n\, I^{22})\right]$$

The relevant quadratic form is now

$$\xi_W = (\hat{\theta}_2 - \theta_2^0)' [I^{22}(\hat{\theta})]^{-1} (\hat{\theta}_2 - \theta_2^0)$$

$$= (\hat{\theta}_2 - \theta_2^0)' [I_{22}(\hat{\theta}) - I_{21}(\hat{\theta})\{I_{11}(\hat{\theta})\}^{-1}I_{12}(\hat{\theta})] (\hat{\theta}_2 - \theta_2^0)$$

which has the large sample chi-square distribution with k_2 degrees of freedom. The null hypothesis $\theta_2 = \theta_2^0$ will be rejected if $\xi_W > \chi_{k_2}^2(\alpha)$.

The Lagrange Multiplier Test Procedure

In this method, the likelihood is maximized subject to the constraint $\theta_2 = \theta_2^0$ and a test statistic constructed from the Lagrange multiplier for the constrained maximization. Rao (1948) called this the **efficient score** test but it is now more commonly known as the **Lagrange multiplier** test [see Aitcheson and Silvey (1958), Silvey (1959), and Engle (1982, 1984)]. Here we carry out the test directly for the case of testing a subset of the parameters. Let

$$H = L(\theta_1, \theta_2; x) - l'(\theta_2 - \theta_2^0)$$

where l is a $k_2 \times 1$ vector of Lagrange multipliers. The first order condition for maximization is

$$\frac{\partial L}{\partial \theta_1} = 0 \quad \text{and} \quad \frac{\partial L}{\partial \theta_2} = l$$

which gives the transpose of the score vector as $S' = [0 \ \ l']$. Denote the solution to these equations by $\tilde{\theta}$. The test based on l is the same as that based on the score function $S(\theta; x)$. In Step 2 of Theorem 8.10 we showed that

$$\frac{1}{\sqrt{n}} S(\theta, x) \xrightarrow{d} N[0, \Sigma(\theta)]$$

where $\Sigma(\theta) = \lim_{n \to \infty} I(\theta)/n$. Therefore, a test statistic based on the Lagrange multiplier is given by the corresponding quadratic form

$$\xi_{LM} = S(\tilde{\theta})' [I(\tilde{\theta})]^{-1} S(\tilde{\theta}) = [0 \ \ l'] \begin{bmatrix} I^{11} & I^{12} \\ I^{21} & I^{22} \end{bmatrix} \begin{bmatrix} 0 \\ l \end{bmatrix} = l' I^{22}(\tilde{\theta}) l$$

The null hypothesis will be rejected if $\xi_{LM} > \chi_{k_2}^2(\alpha)$.

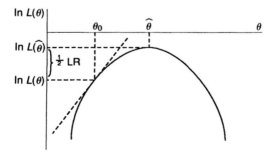

Figure 9.7   A comparison of the Wald, LR, and LM tests

A Geometric Comparison of the Tests

Figure 9.7 presents a geometric comparison of the Wald, LR, and LM tests when there is only one parameter to be estimated. The log of the likelihood is graphed against the parameter θ. The graph is below the axis because the likelihood, being a density function, is less than 1 and hence its logarithm is negative. The point θ_0 corresponds to the null hypothesis and $\hat{\theta}$ corresponds to the maximum likelihood. The LR test is based on the vertical difference, which is the same as one-half the log-likelihood ratio. The Wald test uses the horizontal distance instead of a vertical difference. Specifically, the squared horizontal distance $(\hat{\theta} - \theta_0)^2$, weighted by the information matrix, is used. The score function is the derivative of the log-likelihood and is hence the slope of the graph at the point θ_0. The alternative hypothesis corresponds to $S(\theta) = 0$, that is, the slope is zero. The LM test therefore measures how different the slope of the graph at θ_0 is from zero. Each of the test statistics is a reasonable measure of the distance between the null and alternative hypotheses. Engle (1984) and Buse (1982) have shown independently that when the log-likelihood function is a quadratic, all three test procedures give the same result. For a general linear model, there is an inequality among the three test criteria. This is given by

$$\xi_W \geq \xi_{LR} \geq \xi_{LM}$$

It follows that whenever the LM test rejects the null hypothesis, so will the others. Similarly, whenever the Wald test fails to reject the null, other tests will fail too. Computationally, the LR test is the most cumbersome, unless it can be converted to a t-, F-, or a χ^2 test as was done in Sections 9.8 through 9.10.

An Application to the Bernoulli Distribution

Consider a random sample of size n from the Bernoulli distribution for which $x_i = 1$ with probability θ and $x_i = 0$ with probability $1 - \theta$. For this random variable, the density function, mean, and variance are given by (see Table 4.1)

$$f(x) = \theta^x (1 - \theta)^{1-x} \; ; \qquad E(x) = \theta \; ; \qquad \text{Var}(x) = \theta(1 - \theta)$$

The log-likelihood function is given by

$$\ln L(\theta, x) = \sum [x_i \ln \theta + (1 - x_i) \ln (1 - \theta)]$$

The score function is

$$S(\theta, x) \;=\; \frac{\partial \ln L}{\partial \theta} \;=\; \frac{1}{\theta(1-\theta)} \sum (x_i - \theta)$$

which gives the maximum likelihood estimator as $\hat{\theta} = \bar{x}$. The information can be shown to be (verify it)

$$I(\theta) \;=\; \frac{n}{\theta(1-\theta)}$$

To test the null hypothesis $\theta = \theta_0$, the Wald, LR, and LM test statistics can be shown to be the following (verify them):

$$\xi_W \;=\; \frac{n\,(\bar{x} - \theta_0)^2}{\bar{x}(1-\bar{x})}$$

$$\xi_{LR} \;=\; 2\,n\bar{x}\,\ln\left[\frac{\bar{x}}{\theta_0}\right] + 2n(1-\bar{x})\ln\left[\frac{1-\bar{x}}{1-\theta_0}\right]$$

$$\xi_{LM} \;=\; \left[\frac{\sum (x_i - \theta_0)}{\theta_0(1-\theta_0)}\right]^2 \frac{\theta_0(1-\theta_0)}{n} \;=\; \frac{n\,(\bar{x} - \theta_0)^2}{\theta_0(1-\theta_0)}$$

Because each of these test statistics can be expressed in terms of the sufficient statistic $\bar{x}$, an exact test can be devised using that. In this case, the limiting chi-square approximation is not necessary.

An Application to the Multivariate Normal Distribution

We revisit the multivariate normal application presented in Example 8.12 and illustrate the Wald, LR, and LM tests in the context of the conditional probability model

$$y \mid X \;\sim\; \text{MVN}\,[X\beta, \sigma^2 I_k]$$

used most frequently in econometrics.

The $n \times 1$ vector y represents a random sample of size n drawn from the above conditional normal distribution. X is an $n \times k$ matrix of nonrandom and known values, θ is a $k \times 1$ vector of unknown parameters, σ^2 is an unknown scalar, and I_k is the identity matrix of order k. Defining $u = y - X\beta$ we obtain an $n \times 1$ vector that is multivariate normal with mean vector 0 and I_k as the covariance matrix. The likelihood function is (see Example 8.12)

$$L(\beta, \sigma) \;=\; \left[\frac{1}{\sigma\sqrt{2\pi}}\right]^n \exp\left[-\frac{(y - X\beta)'(y - X\beta)}{2\sigma^2}\right]$$

The log-likelihood function is

$$\ln L = \text{constant} - n \ln \sigma - \frac{1}{2\sigma^2}(y - X\beta)'(y - X\beta)$$

Let $\underset{(k+1)\times 1}{\theta}$ be $\begin{bmatrix} \beta \\ \sigma \end{bmatrix}$. Score function is

$$S(\theta; y) = \begin{bmatrix} \dfrac{\partial \ln L}{\partial \beta} \\[2mm] \dfrac{\partial \ln L}{\partial \sigma} \end{bmatrix} = \begin{bmatrix} \dfrac{X'(y - X\beta)}{\sigma^2} \\[2mm] -\dfrac{n}{\sigma} + \dfrac{1}{\sigma^3}(y - X\beta)'(y - X\beta) \end{bmatrix}$$

MLE is obtained when $S(\theta; y) = 0$. The solutions are given by

$$\hat{\beta} = (X'X)^{-1} X'y \qquad\qquad \hat{\sigma}^2 = \frac{(y - X\hat{\beta})'(y - X\hat{\beta})}{n}$$

The information matrix is

$$I(\theta) = \begin{bmatrix} X'X/\sigma^2 & 0 \\ 0 & 2n/\sigma^2 \end{bmatrix}$$

which is block diagonal. Let X be partitioned as (X_1, X_2) and θ' be partitioned as (θ_1', θ_2'), where X_i is $n \times k_i$ and θ_i is $k_i \times 1$, with $k_1 + k_2 = k$. Then

$$\underset{n \times k}{X} = [\underset{n \times k_1}{X_1}, \underset{n \times k_2}{X_2}]; \qquad \beta = \begin{bmatrix} \beta_1 \\ \beta_2 \end{bmatrix}; \qquad u = \begin{bmatrix} u_1 \\ u_2 \end{bmatrix}$$

The null hypothesis is $H_0 : \beta_2 = \beta_2^0$ and the alternative is $H_1 : \beta_2 \neq \beta_2^0$.

The Likelihood Ratio Test

Using the MLE $\hat{\beta}$ and $\hat{\sigma}^2$ in L we get

$$\hat{L} = \left[\frac{1}{\hat{\sigma}\sqrt{2\pi}}\right]^n \exp\left[-\frac{1}{2\hat{\sigma}^2}(y - X\hat{\beta})'(y - X\beta)\right] = \left[\frac{1}{\hat{\sigma}\sqrt{2\pi}}\right]^n e^{-n/2}$$

because $(y - X\hat{\beta})'(y - X\hat{\beta}) = n\hat{\sigma}^2$. MLE under H_0 gives (verify it)

$$\tilde{\sigma}^2 = \frac{1}{n}(y - X\beta_0)'(y - X\beta_0) \qquad \text{and} \qquad \tilde{L} = \left[\frac{1}{\tilde{\sigma}\sqrt{2\pi}}\right]^n e^{-n/2}$$

The likelihood ratio and its logarithm are

$$\lambda \; = \; \frac{\tilde{L}}{\hat{L}} \; = \; \left[\frac{\hat{\sigma}}{\tilde{\sigma}}\right]^n \qquad\qquad \ln\lambda \; = \; n\,\ln(\hat{\sigma}/\tilde{\sigma})$$

The critical region is given by

$$\ln\lambda \; < \; k \qquad \text{or} \qquad -2\ln\lambda \; > -2k \; = \; d$$

Also, $(-2\ln\lambda) \overset{d}{\to} \chi^2_{k_2}$. Hence reject H_0 if

$$\xi_{LR} \; = \; -2\ln\lambda \; = \; -2n\,\ln\left[\frac{\hat{\sigma}}{\tilde{\sigma}}\right] \; = \; n\,\ln\left[\frac{\tilde{\sigma}^2}{\hat{\sigma}^2}\right] \; > \; d$$

where $P(\chi^2_{k_2} > d) \; = \; \alpha$.

The Wald Test

In Example 8.12 of Section 8.10 we showed that $\hat{\beta}$ is a linear combination of normal variates and that $E(\hat{\beta}) = \beta$ and $\mathrm{Var}(\hat{\beta}) = \sigma^2\,(X'X)^{-1}$. Hence

$$\hat{\beta} \; \sim \; \mathrm{MVN}\,[\beta, \sigma^2(X'X)^{-1}]$$

The partition of $X'X$ is

$$(X'X) \; = \; \begin{bmatrix} X_1'X_1 & X_1'X_2 \\ X_2'X_1 & X_2'X_2 \end{bmatrix}$$

Let

$$(X'X)^{-1} \; = \; \begin{bmatrix} A & B \\ C & D \end{bmatrix}$$

By Property A.8 on partitioned inverse,

$$\underset{k_2 \times k_2}{D} \; = \; [X_2'X_2 - X_2'X_1(X_1'X_1)^{-1}X_1'X_2]^{-1}$$

Hence

$$\hat{\beta}_2 \; \sim \; \mathrm{MVN}\,[\beta_2, \sigma^2 D]$$

which is an exact distribution and is not asymptotic. Under H_0,

$$\xi_W \; = \; \frac{(\hat{\beta}_2 - \beta_2^0)'\, D^{-1}\, (\hat{\beta}_2 - \beta_2^0)}{\sigma^2} \; \sim \; \chi^2_{k_2}$$

The χ^2 test is inapplicable here because σ^2 is unknown. However, σ^2 can be consistently estimated by $\hat{\sigma}^2$. Using that, we have

$$\xi_W = \frac{(\hat{\beta}_2 - \beta_2^0)' D^{-1} (\hat{\beta}_2 - \beta_2^0)}{\hat{\sigma}^2} \xrightarrow{d} \chi_{k_2}^2$$

The test criterion is to reject H_0 if $\xi_W > d$, where d is the point on the χ^2 distribution with k_2 d.f. such that the area to the right is α, the level of significance.

The Lagrange Multiplier Test

From the earlier derivation on the LM test statistic, we have

$$\xi_{LM} = l' I^{22}(\tilde{\theta}) l = l' \sigma^2 D l$$

where l is the Lagrange multiplier. If $\tilde{\beta}_1, \hat{\sigma}$ are MLE under H_0,

$$l = \frac{\partial \ln L}{\partial \beta_1} = \frac{X_2'(y - X_1\tilde{\beta}_1 - X_2\beta_2^0)}{\tilde{\sigma}^2} = \frac{X_2'\tilde{u}}{\tilde{\sigma}^2}$$

Therefore

$$\xi_{LM} = \frac{\tilde{u}' X_2}{\tilde{\sigma}^2} \tilde{\sigma}^2 D \frac{X_2' \tilde{u}}{\tilde{\sigma}^2}$$

$$= \frac{1}{\tilde{\sigma}^2} \tilde{u}' X_2 [X_2' X_2 - X_2' X_1 (X_1' X_1) X_1' X_2]^{-1} X_2' \tilde{u}$$

The test criterion is, as before, to reject $H_0: \theta = \theta_0$ if $\xi_{LM} > \chi_{k_2}^2(\alpha)$.

9.12 TEST OF GOODNESS OF FIT

We often wish to determine whether a set of data may be considered to be observations from a given distribution. For example, suppose it is believed that scores in the Graduate Record Exams are distributed normally and we wish to test that belief. As another example, a textile manufacturer who inspects cloth for imperfections suspects that the distribution of defects is Poisson. It would be desirable to test whether that is indeed the case. What is done in these cases is to assume the distribution that is expected to have generated the data, compute the theoretical probabilities, and compare them with the observed frequencies. We now develop a formal test procedure for this. As a preliminary, we need to study in more detail the multinomial distribution introduced in Section 5.13.

The Multinomial Distribution

Suppose in a given trial, exactly one of k mutually exclusive events must occur, call them $A_1, A_2, \ldots, A_k$. For instance, we could classify an observation into one of k different classes. Let $P(A_i) = p_i$ and define the random variable x_i that takes the value 1 if A_i occurs and 0 otherwise. Thus $p(x_i = 1) = p_i$. Note that $\Sigma\, p_i = 1$. The joint density of the x's can be written as $f(x_1, x_2, \ldots, x_n; p) = p_1^{x_1} p_2^{x_2} \cdots p_k^{x_k}$. This is called the **multinomial distribution**. The relevance of this to the goodness of fit may be seen as follows. Divide the range of a random variable y into k intervals. Based on the theoretical distribution (for example, Poisson or normal in the above illustration), compute the theoretical probabilities of each interval that y can fall into. This corresponds to p_i. We can test hypotheses such as $p_i = p_i^0$ for all i and accept it or reject it based on observed data. The probability of testing goodness of fit thus reduces to one of testing hypotheses on multinomial probabilities. We therefore turn our attention first to the estimation and then to the hypothesis testing of the parameters of a multinomial distribution.

Estimation of Multinomial Parameters

Let a sample of size n be drawn from a multinomial. For the ith observation we have $x_{i1}, x_{i2}, \ldots, x_{ik}$, only one of which is 1 and the rest are zero. The likelihood function is

$$L(p) = \prod_{i=1}^{n} p_1^{x_{i1}} p_2^{x_{i2}} \cdots p_k^{x_{ik}} = p_1^{\Sigma x_{i1}} p_2^{\Sigma x_{i2}} \cdots p_k^{\Sigma x_{ik}}$$

$$= p_1^{n_1} p_2^{n_2} \cdots p_k^{n_k} \qquad \sum_{j=1}^{k} n_j = n$$

where $n_j = \sum_{i=1}^{n} x_{ij}$ is the number of times the jth class or event occurs, that is, the number of successes of the jth class. To estimate p_i, we maximize $\ln L$.

$$\ln L = \sum_{i=1}^{k} n_i \ln p_i \, ; \qquad \sum_{i=1}^{k} p_i = 1 \, ; \qquad \sum_{i=1}^{k} n_i = n$$

Because $p_k = 1 - p_1 - p_2 - \cdots - p_{k-1}$, we have the first order condition

$$\frac{\partial \ln L}{\partial p_i} = \frac{n_i}{p_i} - \frac{n_k}{p_k} = 0$$

Hence $n_i/p_i = n_k/p_k$ for all i, or $n_i = (p_i n_k)/p_k$. Because $\Sigma\, n_i = n$ and $\Sigma\, p_i = 1$, we have $n = n_k/p_k$. Hence $\hat{p}_i = n_i/n$ is the MLE of p_i. In other words, the sample frequency of the ith class is the MLE of p_i, which is not surprising. The next step is to get the information matrix for this distribution. Recall that the information matrix is defined as

$$I(\theta) \;=\; -E\left[\frac{\partial^2 \ln L}{\partial \theta_i \partial \theta_j}\right]$$

Differentiating $\ln L$ with respect to p_i twice, we get

$$\frac{\partial^2 \ln L}{\partial p_i^2} \;=\; -\frac{n_i}{p_i^2} - \frac{n_k}{p_k^2} \qquad \left(\text{using } \sum_{i=1}^{k-1} p_i = 1 - p_k\right)$$

Hence

$$-E\left[\frac{\partial^2 \ln L}{\partial p_i^2}\right] \;=\; E\left[\frac{n_i}{p_i^2} + \frac{n_k}{p_k^2}\right]$$

It is easy to verify that $E(\hat{p}_i) = p_i$ and hence $E(n_i) = np_i$. Therefore the above expectation, which gives the ith diagonal element of $I(p)$, is

$$-E\left[\frac{\partial^2 \ln L}{\partial p_i^2}\right] \;=\; \frac{n}{p_i} + \frac{n}{p_k} \qquad i = 1, 2, \ldots, k-1$$

For the off-diagonal element, differentiate $\partial \ln L / \partial p_i$ with respect to p_j, $i \neq j$. We have

$$\frac{\partial^2 \ln L}{\partial p_i \partial p_j} \;=\; -\frac{n_k}{p_k^2} \qquad \text{and} \qquad E\left[-\frac{\partial^2 \ln L}{\partial p_i \partial p_j}\right] = \frac{n}{p_k}, \qquad i \neq j$$

Thus the information matrix can be written as $I(p) = n\,(\delta_{ij}/p_i) + (n/p_k)$, where $\delta_{ij} = 1$ for $i = j$, and $\delta_{ij} = 0$ for $i \neq j$. In matrix form this can be written as follows.

$$I(p) \;=\; D + \frac{n}{p_k}\, ll'$$

where l' is the row vector of 1's $= (1, 1, \ldots, 1)$ and D is the $(k-1) \times (k-1)$ diagonal matrix

$$D \;=\; \begin{bmatrix} n/p_1 & \cdot & \cdot & 0 \\ \cdot & n/p_2 & \cdot & \cdot \\ \cdot & & \cdot & \cdot \\ 0 & \cdot & \cdot & n/p_{k-1} \end{bmatrix}$$

Test of Multinomial Probabilities

The next step is to test the hypothesis $H_0: p_i = p_i^0$ for all i against the alternative $p_i \neq p_i^0$. In vector notation this becomes $H_0: p = p^0$, where p is $(k-1) \times 1$. We know from the asymptotic normality of MLE estimators (Theorem 8.10) that

$$\sqrt{n}(\hat{p}_n - p^0) \overset{d}{\to} N\left[0, \lim_{n \to \infty} n \{I(p^0)\}^{-1}\right]$$

under the null hypothesis. Therefore

$$Q = (\hat{p} - p^0)' I(p^0) (\hat{p}_n - p^0) \overset{d}{\to} \chi^2_{k-1}$$

[note that the d.f. is $k-1$ because of the condition $\sum_i^k p_i = 1$ and hence $I(p)$ is $(k-1) \times (k-1)$ with full rank.]

$$Q = (\hat{p} - p^0)' D (\hat{p} - p^0) + \frac{n}{p_k} (\hat{p} - p^0)' ll' (\hat{p} - p^0) = Q_1 + Q_2 \quad \text{(say)}$$

Because D is diagonal, Q_1 becomes $\sum_{i=1}^{k-1} n (\hat{p}_i - p_i^0)^2 / p_i^0$. Also,

$$l'(\hat{p} - p^0) = \sum_{i=1}^{k-1} (\hat{p}_i - p_i^0) = p_k^0 - \hat{p}_k$$

Hence Q_2 becomes $n(\hat{p}_k - p_k^0)^2 / p_k^0$. Combining Q_1 and Q_2 we have

$$Q = \sum_{i=1}^{k} n \frac{(\hat{p}_i - p_i^0)^2}{p_i^0}$$

Using $\hat{p}_i = n_i / n$, we get

$$Q = \sum_{i=1}^{k} \frac{(n_i - np_i^0)^2}{np_i^0} = \sum_{i=1}^{k} \frac{(O_i - E_i)^2}{E_i}$$

where $E_i = np_i^0$ is the "expected" frequency of the ith observation under the null hypothesis and O_i is the "observed" frequency n_i. The asymptotic property of Q implies that

$$Q = \sum_{i=1}^{k} \frac{(O_i - E_i)^2}{E_i} \overset{d}{\to} \chi^2_{k-1}$$

The large sample Wald test of $p_i = p_i^0$ is to reject H_0 if $Q > \chi^{2*}_{k-1}(\alpha)$. For goodness of fit, assume the distribution (say Poisson). Prepare an empirical frequency table (say O_1 for 1 defect, O_2 for 2 defects, and so on). Based on the null hypothesis, compute p_i^0 (for Poisson). Then construct Q and use the above criterion for large samples.

9.13 CONFIDENCE INTERVALS

When we discussed estimation we concentrated on giving a single value that θ is estimated to have. This is called **point estimation**. It might be more useful to give a range of values that θ is very likely to take. Thus, for instance, rather than saying that the inflation rate is expected to be 5 percent, we might want to say that it will be between 4.9 percent and 5.1 percent with a certain probability. The latter is known as **interval estimation**. To illustrate, in Theorem 6.3 we showed that

$$t_c = \frac{\bar{x} - \mu}{s / \sqrt{n}} \sim t_{n-1}$$

Choose a level of significance α (say 0.05). Let $t_{n-1}^*(0.025)$ be the point on the t_{n-1} distribution to the right of which which there is an area of 2.5%. Then it is evident that

$$P\left[-t_{n-1}^*(0.025) \leq \frac{\bar{x} - \mu}{s / \sqrt{n}} \leq t_{n-1}^*(0.025) \right] = 0.95$$

or

$$P\left[\bar{x} - \frac{s}{\sqrt{n}} t_{n-1}^* \leq \mu \leq \bar{x} + \frac{s}{\sqrt{n}} t_{n-1}^* \right] = 0.95$$

This implies that the true parameter μ lies in the interval $\bar{x} \pm (s / \sqrt{n}) t_{n-1}^*$ with probability 0.95. Such an interval is called a 95% **confidence interval** for μ. Considering that μ is not a random variable but a fixed number, one may wonder what the above probability statement means. The interpretation is as follows. Draw a random sample $x_1, \ldots, x_n$ and obtain $\bar{x}$ and s. Then construct the above confidence interval. Now repeat this experiment indefinitely constructing the interval every time. These intervals will obviously vary from sample to sample and are hence random intervals. The interpretation is that 95% of these confidence intervals will contain the true value μ. In the general case, we construct a $1 - \alpha$ confidence interval where α is the level of significance.

Confidence Intervals and Hypothesis Testing

There exists a very close relationship between interval estimation and hypothesis testing. Consider the normal population (μ, σ^2) and the test $H_0: \mu = \mu_0$ against $\mu \neq \mu_0$. The acceptance region is

$$-t^*_{n-1}(\alpha/2) \le t_c = \frac{\bar{x} - \mu_0}{s\sqrt{n}} \le t^*_{n-1}(\alpha/2)$$

which can be rewritten as

$$\bar{x} - \frac{s}{\sqrt{n}} t^*_{n-1}(\alpha/2) \le \mu_0 \le \bar{x} + \frac{s}{\sqrt{n}} t^*_{n-1}(\alpha/2)$$

which is the confidence interval we had. This suggests the following interpretation. Consider the set of all μ's for which the hypothesis is *accepted*. It is nothing but the above confidence interval. Thus a confidence interval can be thought of as the set of all parameter values μ_0 for which the null hypothesis $\mu = \mu_0$ will be accepted against $\mu \ne \mu_0$. It is evident from this that we can also carry out a test of hypothesis using a confidence interval. For the mean of the above normal population, the confidence interval is first constructed. If μ_0 in the null hypothesis is within that interval, the hypothesis is accepted. If the interval does not contain μ_0 then $H_0: \mu = \mu_0$ is rejected.

Confidence Interval for the Variance

From Theorem 6.2, Corollary 1, for a normal population, $[(n-1)s^2/\sigma^2] \sim \chi^2_{n-1}$. We can thus look for points a and b such that

$$P\left[a \le \frac{(n-1)s^2}{\sigma^2} \le b\right] = 1 - \alpha$$

which gives the interval

$$\frac{(n-1)s^2}{b} \le \sigma^2 \le \frac{(n-1)s^2}{a}$$

as the $1-\alpha$ confidence interval for σ^2. The problem, however, is that a and b are not unique. We could minimize the length of the confidence interval with respect to a (that is, minimize $1/a - 1/b$), but the computation is quite tedious. What is often done is to choose a and b such that the tails are equal, that is, $P(\chi^2_{n-1} < a) = \alpha/2$ and $P(\chi^2_{n-1} > b) = \alpha/2$. Thus

$$a = \chi^{2*}_{n-1}(1-\alpha/2) \quad \text{and} \quad b = \chi^{2*}_{n-1}(\alpha/2)$$

Confidence Interval for a Binomial Probability

Let $x_1, x_2, \ldots, x_n$ be a random sample from the Bernoulli distribution with $x = 1$ or 0. The density is $f(x, p) = p^x(1-p)^{1-x}$. Let

k be Σx_i. We know from the derivation of the binomial distribution that the density of k is $\binom{n}{k}p^k q^{n-k}$. To get the 5% confidence interval of p, we need a p_1 such that

$$\sum_{y=0}^{k} \binom{n}{y} p_1^y (1-p_1)^{n-y} = \alpha/2 = 0.025$$

and a p_2 such that

$$\sum_{k}^{n} \binom{n}{y} p_2^y (1-p_2)^{n-y} = \alpha/2 = 0.025$$

The appropriate confidence interval is $p_1 < p < p_2$. In practice, however, the end points may not give an exact $\alpha/2$ probability and hence we may have to settle for an approximate $1-\alpha$ confidence interval in the case of discrete random variables.

Joint Confidence Intervals

In the normal case $N(\mu, \sigma^2)$ we constructed confidence intervals separately for μ and σ^2. To construct **joint confidence intervals**, we proceed as follows. Define

$$z = \frac{\bar{x} - \mu}{\sigma/\sqrt{n}} \quad \text{and} \quad u = \frac{\Sigma (x_i - \bar{x})^2}{\sigma^2}$$

We know from Theorem 6.2 that z and u are independent. Let values a, b_1, and b_2 be determined so that

$$P[-a < z < a \quad \text{and} \quad b_1 < u < b_2] = 0.95$$

The interval on μ and σ^2 implied by the above statement is called the joint confidence interval for μ and σ^2 (or σ). Because of independence, we have

$$0.95 = P[-a < z < a] \cdot P[b_1 < u < b_2]$$

Assuming equal probability we get

$$P[-a < z < a] = \sqrt{0.95} \approx 0.975$$
$$P[b_1 < u < b_2] = \sqrt{0.95} \approx 0.975$$

We get $|z| < a$ or $z^2 \le a^2$, which implies that $(\bar{x} - \mu)^2 \le a^2\sigma^2/n^2$. Also, $\sigma_1^2 < \sigma^2 < \sigma_2^2$ where $\sigma_1^2 = [\Sigma (x_i - \bar{x})^2]/b_2$ and $\sigma_2^2 = [\Sigma (x_i - \bar{x})^2]/b_1$. The joint confidence interval for μ and σ^2 is illustrated in Figure 9.8a. A 95% confidence area for μ and σ is illustrated in Figure 9.8b.

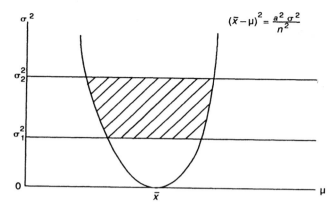

Figure 9.8a   Joint confidence interval for mean and variance

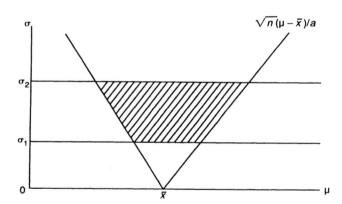

Figure 9.8b   Joint confidence interval for mean and s.d.

The confidence region we just constructed does not have minimum area but is easy to construct from standard normal and χ^2 tables.

Confidence Regions for Large Samples

Consider the multiparameter case with $\theta' = (\theta_1, \theta_2, \ldots, \theta_k)$. We have seen that under certain conditions the MLE $\hat{\theta}$ has the asymptotic normal distribution. Hence the quadratic form

$$Q = (\hat{\theta}_n - \theta)' I(\hat{\theta}) (\hat{\theta}_n - \theta) \overset{d}{\to} \chi_k^2$$

Let $\chi_k^{2*}(\alpha)$ be the point on χ_k^2 to the right of which is an area of α. Then, for large n,

$$P\,[Q < \chi_k^{2*}(\alpha)]\;=\;1-\alpha$$

Thus a $1-\alpha$ confidence region is given by the equation

$$Q\;=\;(\hat{\theta}_n - \theta)'\,I(\hat{\theta})\,(\hat{\theta}_n - \theta)\;=\;\chi_k^{2*}(\alpha)$$

which is an ellipsoid in the space $(\theta_1, \theta_2, \ldots, \theta_k)$ with center at $(\hat{\theta}_1, \hat{\theta}_2, \ldots, \hat{\theta}_k)$.

Example 9.9

Let us apply this to the normal case discussed in Example 8.11. From our earlier discussion we see that $I(\theta)$ is now

$$I(\mu, \sigma^2)\;=\;\begin{bmatrix} n/\sigma^2 & 0 \\ 0 & n/(2\sigma^4) \end{bmatrix}\quad\text{estimated by}\quad\begin{bmatrix} n/\hat{\sigma}^2 & 0 \\ 0 & n/(2\hat{\sigma}^4) \end{bmatrix}$$

where $\hat{\sigma}$ is MLE of σ.

$$\hat{Q}\;=\;\frac{n}{\hat{\sigma}^2}\,(\bar{x} - \mu)^2\;+\;\frac{n}{2\hat{\sigma}^4}\,(\hat{\sigma}^2 - \sigma^2)^2\;\sim\;\chi_2^2$$

$\chi_2^{2*}(0.05) = 5.99$. Therefore the confidence ellipsoid for μ and σ^2, for large n, is

$$\frac{n(\bar{x} - \mu)^2}{\hat{\sigma}^2}\;+\;\frac{n}{2\hat{\sigma}^4}\,(\hat{\sigma}^2 - \sigma^2)^2\;=\;5.99$$

which is an ellipse in the (μ, σ^2) space. It can be shown that the above confidence region based on MLE and $I(\theta)$ will be smaller, on average, than regions obtained by other methods.

EXERCISES

Because estimation and hypothesis testing go hand in hand, the questions here generally have estimation first and then the parts on tests of hypotheses follow.

9.1   First carry out Exercise 8.1 and then proceed with the following parts.

(a) Suppose you draw a single observation x from this. To test the null hypothesis $\theta = \theta_0$ against the alternative $\theta = \theta_1$ $(> \theta_0)$, you use the criterion: Reject H_0 if $x > d$, where d is a positive integer determined appropriately. Derive the power function in terms of θ and d.

(b) Explicitly derive the most powerful test of H_0 against H_1, again when the sample size is 1. [For simplicity, ignore the problem of randomization here.] State the null and alternative hypotheses for which this test will also be uniformly most powerful (explain your reasons and cite appropriate theorems).

(c) Derive the Wald, LR, and LM test statistics for $H_0: \theta = \theta_0$ against $H_1: \theta \neq \theta_0$. Use them to indicate the criterion for acceptance or rejection of the null hypothesis.

9.2 Assume that the random variable X has the same density as in Exercise 8.2 but that $n = 1$, with x as the single observation.

(a) Prove that $f(x; \theta)$ exhibits the monotone likelihood ratio property with $u(x) = x$.

(b) To test $H_0: \theta = \theta_0$ against $H_1: \theta = \theta_1$ $(\theta_1 > \theta_0)$, the critical region $\{x > d\}$ is used. Derive the power function in terms of θ and d and graph it for a fixed d $(0 < d < 1)$.

(c) Construct a UMP test of size α for $H_0: \theta \geq \theta_0$ against $H_1: \theta < \theta_0$. Explain why it is UMP. Express the critical region in terms of θ_0 and α explicitly. What is the rejection criterion when $\theta_0 = 2$ and $\alpha = 0.01$?

(d) Now assume that we draw a random sample $x_1, x_2, \ldots, x_n$. For $H_0: \theta = \theta_0$, derive the generalized likelihood ratio (λ). Explicitly derive the LR, Wald, and LM test statistics in terms of the MLE of θ, θ_0, and n.

9.3 This exercise follows that in Exercise 8.5.

(a) Use the random variable Z_n obtained in Exercise 8.5d to construct a large sample test for the hypothesis $H_0: \theta = \theta_0$ against $H_1: \theta \neq \theta_0$. Clearly describe the test procedure.

(b)   Consider H_0: $\theta \leq \theta_0$ and H_1: $\theta > \theta_0$. Suppose a single observation is drawn and the hypothesis is rejected if $x < d$ (d is fixed). Explicitly (that is, carry out actual integration) derive the power function $\pi(\theta, d)$. [Use the fact that $\theta^{-x} = e^{-x(\ln \theta)}$.]

(c)   Given the level of significance as α, derive the value of d, in terms of θ_0 and α, that will give the most powerful test. Explain why this is most powerful (cite relevant theorems). Is the test also uniformly most powerful? Why or why not?

9.4   First do Exercise 8.7 and then derive the Wald, LM, and LR test statistics for H_0: $\theta = \theta_0$ against H_1: $\theta \neq \theta_0$. For each test, describe step by step how you would go about testing the above hypothesis. [*Note: Do not simply reproduce the text material with matrix notation. Your test statistic must be specific to the model.*]

9.5   Let x_i be a random sample of size n from a population with density function $f(x;\theta) = \theta \exp(-x\theta)$, with $x, \theta > 0$. The mean and variance are respectively, $1/\theta$ and $1/(\theta^2)$.

(a)   Construct a random variable that depends on the sample mean and θ and converges in distribution to $N(0,1)$. Justify your claim with appropriate theorem(s).

(b)   Explain, step by step, how you will use your variable to perform a large sample test of H_0: $\theta = \theta_0$ against H_1: $\theta \neq \theta_0$.

9.6   Let x_i be a random sample of size n from a population with density function $f(x;\theta) = (1/\theta) \exp(-x/\theta)$, with $x, \theta > 0$. The mean and variance are respectively, θ and θ^2.

(a)   Construct a random variable that depends on the sample mean and θ and converges in distribution to $N(0,1)$. Justify your claim with appropriate theorem(s).

(b)   Explain, step by step, how you will use your variable to perform a large sample test of H_0: $\theta = \theta_0$ against H_1: $\theta \neq \theta_0$.

9.7 Let x_i be a random sample of size n from $N(\mu, \sigma^2)$. The maximum likelihood estimates are given by

$$\hat{\mu} = \bar{x} \qquad \text{and} \qquad \hat{\sigma}^2 = [\Sigma (x_i - \bar{x})^2]/n$$

You want to test the hypothesis $H_0: \sigma = \sigma_0$ against the alternative $H_1: \sigma \neq \sigma_0$.

(a) Derive the *generalized likelihood ratio*

$$\lambda = \frac{L(\hat{\mu}, \sigma_0)}{L(\hat{\mu}, \hat{\sigma})}$$

(b) Let $u(x) = n\hat{\sigma}^2/\sigma_0^2$. Express λ as a function of u, graph it, and show that it has a unique extremum.

(c) State the generalized likelihood ratio test criterion in terms of u.

(d) What is the distribution of $u(x)$ under H_0? How would you use this to test the above hypothesis? Describe the step by step procedure.

9.8 Do Exercise 8.6 first.

(a) To test the hypothesis $H_0: \theta \leq 1$ against $H_1: \theta > 1$, the following critical region was used; reject H_0 if $x > d (d > 0)$. Show that the power function is given by

$$\pi(\theta; d) = e^{-d/\theta} (d + \theta)/\theta$$

[*Note*: The formula for integration by parts is $\int u\, dv = uv - \int v\, du$.]

(b) Does a UMP test of size α exist? If yes, derive it (that is, determine how d can be obtained). If not, explain why.

(c) Suppose we want to test $H_0: \theta = 1$ against $H_1: \theta \neq 1$. Consider the test criterion, reject H_0 if $x < d_1$ or $x > d_2$, where d_1 and d_2 are suitably chosen. Derive the power function $\pi(\theta; d_1, d_2)$.

(d) Describe the procedure to obtain a UMPU test, that is, explain how d_1 and d_2 may be determined. [Do not try to solve for them explicitly.]

(e) Suppose a random sample x_i of size n is drawn. Construct a statistic Z_n that has the property $Z_n \xrightarrow{d} N(0,1)$.

(f) Derive the MLE of θ ($\hat{\theta}$) based on the random sample. From this derive the generalized likelihood ratio $\lambda = L(x;1)/L(x;\hat{\theta})$ for testing $H_0: \theta = 1$ against $H_1: \theta \neq 1$.

(g) Describe, step by step and with appropriate graphs if needed, how a large sample test may be constructed for the hypothesis in (f).

9.9 Let the conditional distribution of y given X be $N(X\beta, \sigma^2\Omega)$. X is an $n \times k$ matrix of observations, y is an $n \times 1$ vector of observations, β is a $k \times 1$ vector of unknown parameters, and Ω is an $n \times n$ symmetric positive definite matrix. Let $u = y - X\beta$, so that $E(uu') = \sigma^2\Omega$. Assume that $X'\Omega^{-1}X$ is nonsingular.

(a) Show that the log-likelihood for this is of the form

$$\ln L = b - n \ln\sigma - \frac{(y - X\beta)' \, \Omega^{-1} \, (y - X\beta)}{2\sigma^2}$$

where b does not depend on β or σ.

(b) Show that the score function $S(\beta, \sigma, y)$ for β is given by $X'\Omega^{-1}u/\sigma^2$.

(c) Derive the information matrix for β and σ.

(d) Show that the maximum likelihood estimate $(\hat{\beta})$ of β is $(X'\Omega^{-1}X)^{-1}X'\Omega^{-1}y$.

(e) Show that $\hat{\beta} = \beta + (X'\Omega^{-1}X)^{-1}X'\Omega^{-1}u$.

(f) Derive the covariance matrix of $\hat{\beta}$ and show that the generalized Cramer-Rao lower bound is attained.

(g) Write down the distribution of $\hat{\beta}$. From this, construct a test statistic for the Wald test for $\beta = \beta_0$. Next construct the LM and LR test statistics. Assume, for simplicity, that σ^2 is known for these.

Part III

ECONOMETRICS

The tools of probability and statistical analyses of data developed in the first two parts are applied in the rest of the book to the estimation of econometric models and testing hypotheses on the parameters. In addition, various issues that arise in connection with them are explored. Only what is considered the "core" of econometrics is discussed in this part. Chapter 10 presents the basics of the procedures of estimation and hypothesis testing. Chapter 11 extends the analysis to include nonlinear estimation, measuring the effects of qualitative independent variables, the consequences of erroneous model specification, and the problems created by explanatory variables that are highly correlated among each other. Special issues that arise when dealing with cross-section and time-series data are discussed in detail in Chapter 12.

10

MULTIPLE REGRESSION

The statistical foundations developed in the first two parts of this book have as their ultimate goals the estimation of economic relationships, testing hypotheses on economic theory and econometric models, and forecasting. When discussing Theorem 5.9 in Section 5.4, we introduced the *simple linear regression* which was nothing but the conditional expectation of a random variable Y, given another random variable X, that took the linear form if X and Y were bivariate normal. Thus, $E(Y \mid X) = \alpha + \beta X$. Defining $u = Y - E(Y \mid X)$, we have the following relationship between Y and X, which is the standard specification of a **simple linear regression model** that relates a variable Y to another variable X.

$$Y = \alpha + \beta X + u$$

The random variable Y is referred to as the **dependent variable** or as the **regressand**. Because we generally look only at the *conditional distribution* of Y given X, X is commonly treated as non-random. Initially we do the same but this assumption is later relaxed. X is referred to as the **independent variable** or as a **regressor**. The variable u is stochastic and is referred to as the **error term** or as the **disturbance term**.

Example 10.1

Suppose Y is the annual earnings of an employee in a firm and X is the number of years of education he or she has had. Assume that the relationship between Y and X is linear, in particular, let it be $\alpha + \beta X$. In spite of this, if we select two employees with the same number of years of schooling, their earnings may not be the same. This is because they might differ in other characteristics such as experience and age or because of pure randomness. To allow for these differences, an econometric model is specified with a random error term so that $Y = \alpha + \beta X + u$. Thus $\alpha + \beta X$ is a "statistical average" relationship. To illustrate the idea, suppose we fix the number of years of education, enumerate all the employees with that education, and measure their earnings. If the pairs of values X and Y are plotted, they would form a graph like the one in Figure 10.1. Next compute the average earnings of all employees with a given level of education. In the figure, these average points are denoted by ×. The assumption behind the simple linear regression model is that these average points lie on a straight line. The deviation of an actual observation (X, Y) from the average line is denoted by u.

In Example 8.12 of Section 8.10 we introduced the **multiple regression model** that related Y to several independent variables $(X_1, X_2, \ldots,$ and $X_k)$. The extension of the simple linear regression model is the **multiple linear regression model** given by

$$Y = \beta_1 X_1 + \beta_2 X_2 + \cdots + \beta_k X_k + u$$

Here also the X's are treated as given and nonrandom, at least for the time being. The unknown parameters β_i are referred to as the **regression coefficients**. The present chapter explores the multiple regression model in considerable detail and covers all aspects of estimation and testing. The assumption that the relationship between Y and the X's is linear is maintained throughout this chapter but is relaxed in the next chapter.

The first task is to obtain observations on Y and the X's and use them to obtain estimates of the unknown parameters. Let $Y_1, Y_2, \ldots, Y_t, \ldots, Y_T$ be the observations on Y. It should be noted that we are using t rather than i to denote a typical observation, and T rather

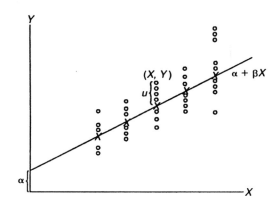

Figure 10.1   The simple linear regression model

than n to denote the sample size. This switch is to accommodate time series observations which are generally indexed this way. The observations on X_i will be denoted by X_{ti}. The relationship among the observations can therefore be specified as follows.

$$Y_t = \beta_1 X_{t1} + \beta_2 X_{t2} + \cdots + \beta_i X_{ti} + \cdots + \beta_k X_{tk} + u_t$$

We readily note that the simple linear regression model is a special case of this general specification with $k = 2$, $\beta_1 = \alpha$, and $\beta_2 = \beta$. To allow for an "intercept" such as α in the multiple regression case also, X_{t1} is generally set to 1 for all t.

10.1 ASSUMPTIONS OF THE MODEL

The notation can be simplified considerably if we use vectors and matrices as defined below.

$$
y = \begin{bmatrix} Y_1 \\ Y_2 \\ \cdot \\ \cdot \\ \cdot \\ Y_t \\ \cdot \\ Y_T \end{bmatrix}, \quad
u = \begin{bmatrix} u_1 \\ u_2 \\ \cdot \\ \cdot \\ \cdot \\ u_t \\ \cdot \\ u_T \end{bmatrix}, \quad
\beta = \begin{bmatrix} \beta_1 \\ \beta_2 \\ \cdot \\ \cdot \\ \beta_i \\ \cdot \\ \beta_k \end{bmatrix}, \quad
X = \begin{bmatrix} X_{11} & X_{12} & .. & X_{1i} & .. & X_{1k} \\ X_{21} & X_{22} & .. & X_{2i} & .. & X_{2k} \\ & & \cdots \cdots & & \\ X_{t1} & X_{t2} & .. & X_{ti} & .. & X_{tk} \\ & & \cdots \cdots & & \\ X_{T1} & X_{T2} & .. & X_{Ti} & .. & X_{Tk} \end{bmatrix}
$$

The model can now be expressed in the compact form specified below.

Assumption 10.1

The multiple linear regression model is given by

$$y = X\beta + u$$

where y is a $T \times 1$ vector of observations on the dependent variable Y, X is a $T \times k$ matrix of observations on a set of independent variables, β is a $k \times 1$ vector of unknown coefficients, and u is a $T \times 1$ vector of unknown stochastic disturbances whose statistical properties are specified later. $X\beta$ is known as the **deterministic part** *and u is known as the* **stochastic part**.

Example 10.2

Table 10.1 has the relevant data for the following econometric model that relates the number of subscribers to a cable TV system in 40 metropolitan areas to several independent variables.

$$\text{SUB} = \beta_1 + \beta_2\text{HOMES} + \beta_3\text{PCINCM} + \beta_4\text{INSTFEE}$$
$$+ \beta_5\text{SVC} + \beta_6\text{NCABLE} + \beta_7\text{NTV} + u$$

where

SUB	=	Number of subscribers
HOMES	=	Number of homes in the area
PCINCM	=	Per capita income for each television market with cable
INSTFEE	=	Installation fee
SVC	=	Monthly service charge
NCABLE	=	Number of television signals carried by each cable system
NTV	=	Number of television signals received with good quality without cable

Differentiating the above equation partially with respect to SVC, we get $\beta_5 = \partial SUB / \partial SVC$. This means that *holding all other variables constant*, if the monthly service charge is increased by one dollar, then the number of cable subscribers will change, on average, by β_5. Thus, β is the *marginal effect* of SVC on SUB. As we would expect the demand for cable TV to decrease as the monthly charge for its use increases, we would expect β_5 to be negative. The other regression coefficients have similar interpretations. The expected signs for the regression coefficients (excluding the constant term β_1 whose sign is unknown *a priori*) are as follows.

$$\beta_2 > 0, \quad \beta_3 > 0, \quad \beta_4 < 0, \quad \beta_5 < 0, \quad \beta_6 > 0, \quad \beta_7 < 0$$

Assumption 10.2   *X is given and is nonrandom.*

By treating the matrix X as *given*, we are focusing our attention only on the *conditional distribution of Y given X*. Although considering the matrix X as nonrandom appears unrealistic, since most economic data are *nonexperimental*, an investigator typically has no experimental control over the independent variables. Assumption 10.2 is not crucial, however, and will be relaxed later on.

Table 10.1   Data on Demand for Cable TV and Its Determinants

SUB	HOMES	PCINCM	INSTFEE	SVC	NCABLE	NTV
105000	350000	9839	14.95	10	16	13
90000	255631	10606	15	7.5	15	11
14000	31000	10455	15	7	11	9
11700	34840	8958	10	7	22	10
46000	153434	11741	25	10	20	12
11217	26621	9378	15	7.66	18	8
12000	18000	10433	15	7.5	12	8
6428	9324	10167	15	7	17	7
20100	32000	9218	10	5.6	10	8
8500	28000	10519	15	6.5	6	6
1600	8000	10025	17.5	7.5	8	6
1100	5000	9714	15	8.95	9	9
4355	15204	9294	10	7	7	7
78910	97889	9784	24.95	9.49	12	7
19600	93000	8173	20	7.5	9	7
1000	3000	8967	9.95	10	13	6
1650	2600	10133	25	7.55	6	5
13400	18284	9361	15.5	6.3	11	5
18708	55000	9085	15	7	16	6
1352	1700	10067	20	5.6	6	6
170000	270000	8908	15	8.75	15	5
15388	46540	9632	15	8.73	9	6
6555	20417	8995	5.95	5.95	10	6
40000	120000	7787	25	6.5	10	5
19900	46390	8890	15	7.5	9	7
2450	14500	8041	9.95	6.25	6	4
3762	9500	8605	20	6.5	6	5
24882	81980	8639	18	7.5	8	4
21187	39700	8781	20	6	9	4
3487	4113	8551	10	6.85	11	4
3000	8000	9306	10	7.95	9	6
42100	99750	8346	9.95	5.73	8	5
20350	33379	8803	15	7.5	8	4
23150	35500	8942	17.5	6.5	8	5
9866	34775	8591	15	8.25	11	4

(Continued)

Table 10.1 (Continued)

SUB	HOMES	PCINCM	INSTFEE	SVC	NCABLE	NTV
42608	64840	9163	10	6	11	6
10371	30556	7683	20	7.5	8	6
5164	16500	7924	14.95	6.95	8	5
31150	70515	8454	9.95	7	10	4
18350	42040	8429	20	7	6	4

Source: *Broadcasting Year Book* (1980) and *State and Metropolitan Area Data Book* (1982).

Assumption 10.3

X has rank k ($< n$) so that the matrix $X'X$ is nonsingular.

If this condition is not fulfilled, then there is a linear dependency among the X's implying that one of them can be expressed as a linear combination of the others. In this case, the parameters may not be estimable. This point will be clearer when we discuss the estimation of the regression coefficients.

Assumption 10.4

u is a random vector with $E(u \mid X) = 0$ or $E(y \mid X) = X\beta$.

In Figure 10.1 we note that some of the observed points lie above the straight line $\alpha + \beta X$ and some lie below it, implying that some errors are positive and others are negative. It seems reasonable to assume that these errors average to zero *in the population*, which is what Assumption 10.4 means.

Assumption 10.5

$E(uu' \mid X) = \sigma^2 I$, where I is the identity matrix of order k. Because $E(u \mid X) = 0$, this implies that $Var(u_t \mid X) = \sigma^2$ for all t and that $Cov(u_t, u_s \mid X) = 0$ for all $t \neq s$.

Assumption 10.6: $u \mid X \sim N(0, \sigma^2 I)$.

Figure 10.2 graphically presents the assumptions made here for the case of the simple two-variable regression model. The X and Y

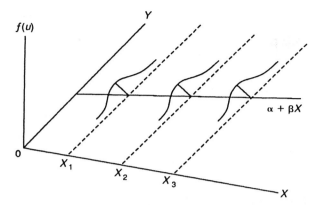

Figure 10.2 Graphic representation of the simple linear regression model

axes represent the values of the random variables X and Y. The Z-axis is the probability density function $f(u)$ of the random error term u. The straight line $\alpha + \beta X$ is the conditional mean of Y given X, assumed to be linear. The statistical distributions drawn around the mean line for the three values X_1, X_2, and X_3 are the corresponding conditional distributions. The assumption that $\text{Var}(u_t) = \sigma^2$ is called **homoscedasticity**, which means "equal scatter" (to simplify the notation we henceforth omit the conditionality with respect to X). Figure 10.2 depicts this constancy of the error variance for all observations. If these variances are not constant but vary with t [thus $Var(u_t) = \sigma_t^2$], we have **heteroscedasticity** (unequal scatter). Figure 10.3 illustrates the case of heteroscedasticity in which the variance increases as X increases. This case is examined in considerable detail in Chapter 12.

The assumption that $Cov(u_t, u_s) = 0$ for $t \neq s$ is called **serial independence** and its violation is known as **serial correlation** or **autocorrelation**. This topic is explored in depth in Chapter 12.

10.2 PROCEDURES FOR ESTIMATING THE PARAMETERS

The next step after the model specification is to estimate the unknown parameters β and σ^2, given the data values Y and X.

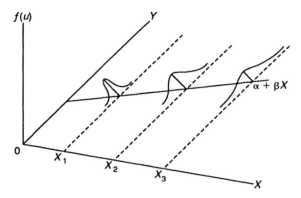

Figure 10.3 Illustration of heteroscedasticity

The Method of Maximum Likelihood

The first procedure we apply is the method of maximum likelihood described in Chapter 8. In fact, this has already been done in Example 8.12 of Section 8.10. We showed there that the MLE for the parameters are

$$\hat{\beta} = (X'X)^{-1}X'y \qquad\qquad \hat{\sigma}^2 = \frac{1}{T}(y - X\hat{\beta})'(y - X\hat{\beta})$$

We readily see that the Assumption 10.3 that $X'X$ is nonsingular is crucial in making β estimable.

The Method of Least Squares

This method is one of the most frequently used procedures in econometrics. Consider an estimate $\hat{\beta}$ of β. Then $\hat{y} = X\hat{\beta}$ is the **predicted** or **fitted value** of y. The vector $\hat{u} = y - \hat{y} = y - X\hat{\beta}$ is known as the **residual vector** and measures the deviation of the observed values of the dependent variable from the predicted values. Square each of the residuals $\hat{u}_t$ and sum them over the observations to obtain the **error sum of squares**

$$\text{ESS} = \sum_{t=1}^{t=T} \hat{u}_t^2 = \hat{u}'\hat{u}$$

The least squares procedure [referred to as the **ordinary least squares (OLS) procedure** to distinguish it from other types of least squares methods] chooses those parameter estimates $\hat{\beta}$ that minimize the error sum of squares ESS. The first order condition for minimization is given by (refer to vector and matrix differentiation in Appendix A)

$$0 = \frac{\partial\,(\hat{u}'\,\hat{u})}{\partial\hat{\beta}} = 2\left[\frac{\partial\hat{u}}{\partial\hat{\beta}}\right]'\hat{u}$$

This condition reduces to the equations (known as the **normal equations**)

$$X'(y - X\hat{\beta}) = 0 \qquad or \qquad X'X\hat{\beta} = X'y$$

The solution to the normal equations is identical to that of the maximum likelihood estimates of β. Thus, we have the following property.

Property 10.1

Under Assumptions 10.1 through 10.6, OLS estimates of β are also MLE and hence have all the properties that MLEs have (see Sections 8.7 and 8.8).

It will be noted, however, that the assumption of normality (Assumption 10.6) is not needed to obtain OLS estimates of β. In other words, OLS estimates of β can be obtained without any knowledge of the exact distribution of the error terms u. Because the least squares method does not provide an estimate of σ^2, we could use the MLE for this. However, as will be shown presently, it is biased and hence a modified estimator is used instead.

Estimation of the Residual Variance

$$\hat{u} = y - X\hat{\beta} = X\beta + u - X(X'X)^{-1}X'y$$
$$= X\beta + u - X(X'X)^{-1}X'(X\beta + u) = Mu$$

where $M = I - X(X'X)^{-1}X'$. It follows from this that the error sum of squares is given by

$$\text{ESS} = \hat{u}'\hat{u} = u'M'Mu = u'Mu$$

because M is symmetric and idempotent (verify it).

$$E(\text{ESS}) = E(u'Mu) = E\,[trace\,(u'Mu)]$$

because $u'Mu$ is a scalar. But by Property A.14 of Appendix A,

$$E \, [trace \, (u'Mu)] \; = \; E \, [trace \, (Muu')] \; = \; \sigma^2 \, trace \, (M)$$

$$= \; \sigma^2 \, trace \, (I) - \sigma^2 \, trace \, [X(X'X)^{-1}X']$$

$$= \; T\sigma^2 - \sigma^2 \, trace \, [(X'X)^{-1}X'X] \; = \; (T-k)\sigma^2$$

Therefore,

$$E\,(ESS) \; = \; E\,(\hat{u}'\hat{u}) \; = \; (T-k)\sigma^2$$

It follows that

$$s^2 \; = \; \frac{1}{T-k} \, \hat{u}'\hat{u}$$

is an unbiased estimator of σ^2. It is easily verified that because the maximum likelihood estimator of σ^2 is consistent, so is s^2. The term s is referred to as the **standard error of estimate** or as the **standard error of residuals**.

Practice Problem

10.1   Show that $X'\hat{u} = 0$ and that $\hat{y}'\hat{u} = 0$.

Best Linear Unbiased Estimation (BLUE) and the Gauss-Markov Theorem

In Practice Problem 8.3 we introduced an estimator called the *linear estimator* which was a linear combination of the observations, and in Section 8.1 we introduced the concept of a *uniformly minimum variance unbiased estimator*. Those notions can be extended to the multiple regression model also. Consider the parameter θ which is a linear combination of the β's, $\theta = c'\beta$, where c is a column vector of dimension k with known and fixed values. Next consider the linear combination of the y's

$$\hat{\theta} \; = \; \sum_{t=1}^{t=T} w_t Y_t \; = \; w'y$$

where w is a $T \times 1$ vector and the "weights" w_t are chosen appropriately. We can ask the question "what values of w_t will make $\hat{\theta}$ an unbiased estimator of θ with the smallest variance?" Such an estimator, if it exists, is known as the **best linear unbiased estimator (BLUE)**. Thus among the class of linear estimators we choose the one with the smallest variance as the "best" estimator.

$$E\,(\hat{\theta}) \; = \; E\,[w'(X\beta + u)] \; = \; w'X\beta + E(w'u) \; = \; w'X\beta$$

because X is given and nonrandom and $E(u) = 0$, by Assumptions 10.2 and 10.4. For unbiasedness, we need $E(\hat{\theta}) = \theta = c'\beta$ for all β, which implies that $w'X = c'$.

$$\text{Var}(\hat{\theta}) = E[(\hat{\theta} - c'\beta)^2] = E[(w'X\beta + w'u - c'\beta)(w'X\beta + w'u - c'\beta)']$$

$$= E(w'uu'w) = w'E(uu')w = \sigma^2 w'w$$

because of the condition $w'X = c'$ and because $E(uu') = \sigma^2 I$ by Assumption 10.5. The best linear unbiased estimator of θ is therefore obtained by minimizing $w'w$ subject to the restriction $w'X = c'$. Define

$$H = w'w + (c' - w'X)\lambda$$

where λ is a $k \times 1$ vector of Lagrange multipliers. The first order condition is given by

$$0 = \frac{\partial H}{\partial w} = 2w - X\lambda$$

Therefore, $w = \frac{1}{2}X\lambda$. $c' = w'X = \frac{1}{2}\lambda'X'X$, from which we can solve for the Lagrange multipliers as $\lambda' = 2c'(X'X)^{-1}$. Therefore, $w' = \frac{1}{2}\lambda'X' = c'(X'X)^{-1}X'$, which gives $\hat{\theta} = c'(X'X)^{-1}X'y = c'\hat{\beta}$, $\hat{\beta}$ being the OLS estimator of β. If we choose $c' = (1\ 0\ 0\ \ldots\ 0)$ which is the first row of the identity matrix of order k, then $\theta = \beta_1$ and it is easy to see that its best linear unbiased estimator is the same as the OLS estimator obtained earlier. By setting c' equal to the different rows of the identity matrix of order k we can establish that for each β_i, the corresponding OLS estimator is BLUE. This result is stated in the following property.

Property 10.2 (Gauss-Markov Theorem)

Under Assumptions 10.1 through 10.5, the OLS estimators of β are identical to the best linear unbiased estimators and hence are most efficient among linear estimators.

The OLS estimators of β therefore have a number of desirable properties. First, they are unbiased. Second, they are MLE and hence have all the properties of MLEs (in particular, consistency). Finally, they are minimum variance unbiased linear estimators and hence are most efficient among linear unbiased estimators.

Example 10.3

We used the data in Table 10.1 and estimated the model presented in Example 10.2 as follows (you are encouraged to use a regression program and do the same).

$$\text{S}\hat{\text{U}}\text{B} \; = \quad -28961 + 0.438 \text{ HOMES} + 5.462 \text{ PCINCM}$$
$$-159.568 \text{ INSTFEE} + 609.212 \text{ SVC}$$
$$+ 1129.869 \text{ NCABLE} - 5696.541 \text{ NTV}$$

The signs of all the regression coefficients agree with our prior intuition except for that of SVC. Its sign is positive, which is unexpected because our intuition says that if the monthly service charges increase, then the demand for cable TV would decrease. This result might be due to model misspecification or a poor reliability of the parameter estimates. In later sections we examine this more closely. The marginal effect of INSTFEE is −159.568. This means that, other things held constant, if the installation fee is increased by one dollar, then the number of cable subscribers is expected to decrease, on average, by 160. It is interesting to note that the marginal effect of the number of television signals available without cable (NTV) is larger in numerical value than that of the number of cable channels available (NCABLE). An increase of one TV signal is expected to decrease the number of cable subscribers, on average, by 5697 (again holding other variables constant). In contrast, an increase of one cable channel is expected to increase the average number of cable subscribers by only 1130.

Although we have used a number of procedures to estimate a multiple regression model, there are still numerous questions to be answered. Most importantly, we would like to know how reliable the estimates are. We would also like to know whether the linear regression model "adequately" captures the underlying data generating process. Finally, we would be interested in testing hypotheses on the unknown parameters of the model. These issues are examined one by one in the rest of this chapter.

10.3   PRECISION OF THE ESTIMATES

We have seen in Chapter 8 that a measure of the precision of an estimate is its variance. A smaller variance (or standard deviation) implies that, on average, the estimate is closer to the expected value of the statistic and hence a small variance or standard deviation is desirable.

In Example 8.12 of Section 8.10, we derived the covariance matrix of the estimated coefficients as

$$Var(\hat{\beta}) \; = \; \sigma^2 (X'X)^{-1}$$

The covariance matrix can be estimated by using s^2 for σ^2 so that

$$\hat{Var}(\hat{\beta}) \;=\; s^2 \, (X'X)^{-1}$$

The diagonal elements of the covariance matrix are the estimated variances of the regression coefficients and their square roots are known as the **standard errors of regression coefficients**. The off-diagonal elements are the covariances between the estimates of two different regression coefficients.

Example 10.4

The standard errors for the regression coefficients of the model presented in Examples 10.2 and 10.3 are given below.

$s_{\hat{\beta}_1} = 26671$ $s_{\hat{\beta}_2} = 0.034$ $s_{\hat{\beta}_3} = 3.286$

$s_{\hat{\beta}_4} = 480.911$ $s_{\hat{\beta}_5} = 2178.625$ $s_{\hat{\beta}_6} = 715.103$

$s_{\hat{\beta}_7} = 1600.140$

As a direct measure of precision, the usefulness of the above standard errors is limited because they are very sensitive to a change in units. This point can be understood better by doing the following practice problems. Standard errors are, however, extremely important in testing hypotheses.

Practice Problems

10.2   In Table 10.1 suppose the number of homes is measured in thousands rather than in actual numbers. Examine the consequences of this change in units on the parameter estimates and their standard errors.

10.3   Now examine the consequences of measuring the number of subscribers in thousands, but measuring the number of homes in actual numbers.

10.4   THE GOODNESS OF FIT

The first assumption we made about the multiple regression model is that the relationship between the dependent and independent variables is linear. It would be desirable to have a numerical measure that tells us whether the linear approximation adequately describes the DGP or not. Here we develop a measure of the **goodness of fit** of a model. In trying to make predictions about Y, if the only information we have is the set of sample observations Y_t, then perhaps the best we can do is to use the sample mean $\bar{Y}$ as the "average" forecast. The error in predicting observation t is $Y_t - \bar{Y}$. Squaring this and summing over all the observations gives us the **total sum of squares (TSS)**

$$\text{TSS} = \sum_{t=1}^{t=T} (Y_t - \bar{Y})^2$$

Suppose we are now told that Y is related to X according to Assumption 10.1. Then we can expect that this additional information will help us produce a "better" prediction of Y. In particular, given that the independent variables take the value x_t (which is the tth row of the X matrix), our prediction of the dependent variable would be $\hat{Y}_t = x_t \hat{\beta}$. The error in this is $\hat{u}_t = Y_t - \hat{Y}_t$. The **error sum of squares (ESS)** is given by

$$\text{ESS} = \sum_{t=1}^{t=T} (Y_t - x_t \hat{\beta})^2$$

We can also compute the sum of squared deviations of the predicted values Y_t from the mean $\bar{Y}$. This gives the **regression sum of squares (RSS)**

$$\text{RSS} = \sum_{t=1}^{t=T} (\hat{Y}_t - \bar{Y})^2$$

In the next section we show that the following decomposition of the three sums of squares holds *provided the model has a constant term in it.*

$$\text{TSS} = \text{RSS} + \text{ESS}$$

Linear Regression Model in Deviation Form

If the multiple linear regression model has a constant term, it can be eliminated by expressing the **model in deviation form**. The model is

$$Y_t = \beta_1 + \beta_2 X_{t2} + \beta_3 X_{t3} + \cdots + \beta_k X_{tk} + u_t$$

Using the bar notation to denote the corresponding means,

$$\bar{Y} = \beta_1 + \beta_2 \bar{X}_2 + \beta_3 \bar{X}_3 + \cdots + \beta_k \bar{X}_k + \bar{u}$$

$$Y_t - \bar{Y} = \beta_2 (X_{t2} - \bar{X}_2) + \beta_3 (X_{t3} - \bar{X}_3) + \cdots + \beta_k (X_{tk} - \bar{X}_k) + v_t$$

Let l be a $T \times 1$ column of ones. Define

$$M_0 = I_T - \frac{1}{n} l\, l' = \begin{bmatrix} 1 & 0 & 0 & \cdots & 0 \\ 0 & 1 & 0 & \cdots & 0 \\ & & \cdots\cdots & & \\ 0 & 0 & 0 & \cdots & 1 \end{bmatrix} - \frac{1}{n} \begin{bmatrix} 1 & 1 & 1 & \cdots & 1 \\ 1 & 1 & 1 & \cdots & 1 \\ & & \cdots\cdots & & \\ 1 & 1 & 1 & \cdots & 1 \end{bmatrix}$$

It is easy to show that the vector $((Y_t - \bar{Y}))$ can be written as $M_0 y$ and that the matrix M_0 is idempotent (verify them). Hence we have

$$\text{TSS} = \sum_{t=1}^{t=T} (Y_t - \bar{Y})^2 = (y - \bar{y})'(y - \bar{y}) = y'M_0 y$$

Because $y = \hat{y} + \hat{u}$, we have $M_0 y = M_0 \hat{y} + M_0 \hat{u}$. Transposing this and multiplying, we get

$$y'M_0 y = \hat{y}'M_0\hat{y} + \hat{u}'M_0\hat{u} + 2\hat{y}'M_0\hat{u}$$

The left-hand side is TSS and the first term on the right-hand side is RSS. The third term vanishes because, as is shown below, $M_0\hat{u} = \hat{u}$ and $\hat{y}'\hat{u} = 0$ (see Practice Problem 10.1). Minimizing $\sum \hat{u}_t^2$ with respect to the constant term β_1, we have

$$\sum Y_t = T\hat{\beta}_1 + \hat{\beta}_2 \sum X_{t2} + \cdots + \hat{\beta}_k \sum X_{tk} = \sum \hat{Y}_t$$

Therefore, $\sum \hat{u}_t = l'\hat{u} = l'(y - \hat{y}) = \sum Y_t - \sum \hat{Y}_t = 0$, and hence

$$M_0\hat{u} = [I - (1/n)\, l l'\,]\,\hat{u} = \hat{u}$$

It follows that

$$y'M_0 y = \hat{y}'M_0\hat{y} + \hat{u}'\hat{u} \qquad \text{or} \qquad \text{TSS} = \text{RSS} + \text{ESS}$$

Dividing both sides of the equation by TSS we get

$$1 = \frac{\text{RSS}}{\text{TSS}} + \frac{\text{ESS}}{\text{TSS}}$$

The error sum of squares ESS is due to the error in the model and has the interpretation that it is the **unexplained sum of squares**. The regression sum of squares RSS is referred to as the **explained sum of squares**. The following ratio is known as the **coefficient of multiple determination** and is used as a measure of the goodness of fit of the model.

$$R^2 = 1 - \frac{\text{ESS}}{\text{TSS}} = \frac{\text{RSS}}{\text{TSS}}$$

We readily note that R^2 is between 0 and 1. If the model approximates the DGP very well, we would expect the value of R^2 to be close to 1, and similarly if the "fit" is poor, we would expect R^2 to be low. This measure of the goodness of fit has a serious drawback, however. If we add more variables to the model and minimize the new error sum of squares, it cannot increase and hence ESS is likely to be lower. This implies that R^2 will generally increase as more and more variables are added to the model. To avoid the indiscriminate addition of variables to improve the fit, a modified measure is often used. To develop this measure, note that R^2 is the "explained" fraction of the total sum of squares which is proportional to an estimate of the variance of Y. A natural measure (known as the **adjusted R^2**) is therefore

$$\bar{R}^2 = 1 - \frac{\hat{Var}(u_t)}{\hat{Var}(Y_t)}$$

We know from the previous section that an unbiased estimator of $\sigma^2 = Var(u_t)$ is $\text{ESS}/(T-k)$, and (from Section 8.1) that an unbiased estimator of $Var(Y)$ is $\text{TSS}/(T-1)$. Therefore,

$$\bar{R}^2 = 1 - \frac{\text{ESS}/(T-k)}{\text{TSS}/(T-1)} = 1 - \frac{\text{ESS}(T-1)}{\text{TSS}(T-k)} = 1 - \frac{T-1}{T-k}(1-R^2)$$

The addition of a variable leads to a gain in the **unadjusted R^2** but also a loss of one d.f. because we are estimating another parameter. The adjusted R^2 is therefore a better measure of goodness of fit as it permits a trade-off between increased R^2 and decreased d.f. A model with a higher $\bar{R}^2$ is preferable to one with a lower value. It should be noted that $\bar{R}^2$ is inversely proportional to $\hat{\sigma}^2$ (verify it) and therefore a lower $\hat{\sigma}^2$ is equivalent to a higher $\bar{R}^2$. Also note that because $(T-1)/(T-k)$ is never less than 1, $\bar{R}^2$ can never be higher than R^2. However, while R^2 cannot be negative, $\bar{R}^2$ can be negative. For instance, suppose $T = 31$, $k = 11$, and $R^2 = 0.3$. Then $\bar{R}^2 = -0.05$. A negative $\bar{R}^2$ is an indication that the model may not be an adequate description of the DGP.

Example 10.5

In the cable TV example, $T - 1 = 39$, $T - k = 33$, TSS = 438.65e+8, and ESS = 57.91e+8, which gives the values of 0.868

for R^2 and 0.844 for $\bar{R}^2$. This means that 84.4 percent of the variation in the number of cable TV subscribers is explained by the variables included in the model. This is quite good for a cross section study for which R^2 values typically tend to be lower than those for time series studies.

10.5 TESTS OF HYPOTHESES

One of the important goals of empirical econometrics is to subject a body of theory to hypothesis testing to see if the data support the theory or not. In addition, one may want to use diagnostic testing to decide whether a model or methodology is appropriate. In this section we develop the methodology for conducting a variety of tests on the unknown parameters of the econometric model. Before doing that, however, we need to derive the statistical distributions of the estimators $\hat{\beta}$ and s^2.

Statistical Distributions of the Estimators

The estimator of β can be written as

$$\hat{\beta} = (X'X)^{-1}X'y = (X'X)^{-1}X'(X\beta + u) = \beta + (X'X)^{-1}X'u$$

Because this is a linear combination of the u's which are normally distributed, by Property 5.7 of Section 5.10, $\hat{\beta}$ is also multivariate normal. We have already shown that (Section 8.10, Example 8.12)

$$E(\hat{\beta}) = \beta \qquad Var(\hat{\beta}) = \sigma^2 (X'X)^{-1}$$

Therefore,

$$\hat{\beta} \sim N[\beta, \sigma^2 (X'X)^{-1}]$$

We showed in Section 10.2 that $\hat{u}'\hat{u} = u'Mu$, which is a quadratic form with a symmetric idempotent matrix. The rank of M is equal to its trace (see Property A.15 of Appendix A). Therefore, by Theorem 5.24 of Section 5.12,

$$\frac{\hat{u}'\hat{u}}{\sigma^2} \sim \chi^2_{T-k}$$

Also, $X'M = 0$, and hence, by Theorem 5.29, $\hat{\beta}$ and s^2 are statistically independent.

Testing Individual Coefficients

Let a_{ii} be the ith diagonal element of $(X'X)^{-1}$. Then $\hat{\beta}_i \sim N(\beta_i, \sigma^2 a_{ii})$. This implies that

$$\frac{\hat{\beta}_i - \beta_i}{\sigma \sqrt{a_{ii}}} \sim N(0, 1)$$

Also,

$$\frac{(T - k) s^2}{\sigma^2} = \frac{\Sigma \hat{u}_t^2}{\sigma^2} \sim \chi_{T-k}^2$$

By the definition of a Student's t-distribution (Section 6.3),

$$\frac{\hat{\beta}_i - \beta_i}{\sigma \sqrt{a_{ii}}} \div \left[\frac{\Sigma \hat{u}_t^2}{\sigma^2 (T - k)} \right]^{1/2} \sim t_{T-k}$$

Since $s^2 = (\Sigma \hat{u}_t^2)/(T - k)$, this reduces to

$$t_c = \frac{\hat{\beta}_i - \beta_i}{s \sqrt{a_{ii}}} \sim t_{T-k}$$

The denominator of t_c is nothing but the standard error of the regression coefficient. We can use a t-test similar to the one used in Section 9.8. If the null and alternative hypotheses are $H_0: \beta_i = \beta_i^0$ and $H_1: \beta_i > \beta_i^0$, we would use a *one-tailed test*. The procedure is to find the point $t_{T-k}^*(\alpha)$ on the t-distribution with $T - k$ d.f. such that the area to the right is the level of significance α. We would reject the null hypothesis if the computed t-value (t_c) is greater than t^*. Equivalently, we could use the p-value approach (see Section 9.2) and compute p-value $= P(t > t_c)$. Then reject H_0 if p-value $< \alpha$.

For a two-sided hypothesis with $H_1: \beta_i \neq \beta_i^0$, we use a *two-tailed test*. Find $t_{T-k}^*(\alpha/2)$ and reject H_0 if $|t_c| > t^*$. Alternatively, compute $2P(t > t_c)$ and reject if it is less than the level of significance.

The most common null hypothesis is to test whether a given regression coefficient is zero or not. Thus, β_i^0 is usually set to zero. This makes the t-statistic even simpler because it then reduces to the ratio of the estimated regression coefficient divided by the corresponding standard error.

Example 10.6

The following table presents the estimates, the standard errors, t-statistics, and p-values for the example on the demand for cable subscriptions that we have been using in this chapter.

Variable	$\hat{\beta}$	$s_{\hat{\beta}}$	t_c	$P(t > \mid t_c \mid)$
Constant	− 28961.162	26671.205	− 1.086	0.285
HOMES	0.438	0.034	12.764	< 0.0001
PCINCM	5.462	3.286	1.662	0.106
INSTFEE	− 159.568	480.911	− 0.332	0.742
SVC	609.212	2178.625	0.280	0.782
NCABLE	1129.869	715.103	1.580	0.124
NTV	− 5696.541	1600.140	− 3.560	0.001

Although the t-table can be used to perform this test, because the p-values are available, it is more convenient to use them. Recall that the p-value is the probability of a Type I error and hence a high value implies that the risk of this type of error is great. For instance, the p-value for SVC is 0.782, which is extremely high (the p-values in the table are for two-tailed tests). Therefore it is not safe to reject the null hypothesis that the corresponding regression coefficient is zero. We therefore conclude that the coefficient for SVC is *statistically not significantly different from zero*. The same holds for the coefficient of INSTFEE also. This result is somewhat surprising because we would have thought *a priori* that the installation fee and the monthly service charge would be important determinants of the demand for cable TV. Their insignificance might be due to model misspecification.

The p-values for the coefficients of PCINCM and NCABLE are 10.6 percent and 12.4 percent, respectively. If we use 10 percent as a strict upper bound for the level of significance, these p-values are still unacceptably high, although 10.6 percent is on the borderline. According to our criteria, only the coefficients for the number of homes in the area and the number of free TV channels have statistically significant coefficients. This is not a happy result and might be a further indication of possible model misspecification.

If we find variables with insignificant coefficients, should we eliminate all of them and reestimate the model? An argument in favor of doing that is that it will reduce the number of parameters to estimate and hence increase the d.f., the precision of the remaining parameter estimates, and the power of tests of hypotheses. Table 10.2 has the regression coefficients and a number of associated statistics for four different models. The numbers in the third box of the tables are explained Section 10.6.

Table 10.2   Estimated Models for Cable TV Subscriptions[†]

Variable	Model A	Model B	Model C	Model D
Constant	−28961	17636	−26451	−26930
	(−1.09)	(2.63)	(−1.07)	(−1.10)
	[0.29]	[0.01]	[0.29]	[0.28]
HOMES	0.438	0.436	0.440	0.438
	(12.76)	(13.49)	(13.28)	(13.76)
	[<0.001]	[<0.001]	[<0.001]	[<0.001]
PCINCM	5.462		5.532	5.315
	(1.66)		(1.71)	(1.72)
	[0.11]		[0.10]	[0.09]
INSTFEE	−159.568		−128.532	
	(−0.33)		(−0.28)	
	[0.74]		[0.78]	
SVC	609.212			
	(0.28)			
	[0.78]			
NCABLE	1129.869		1169.912	1200.412
	(1.58)		(1.69)	(1.78)
	[0.12]		[0.10]	[0.08]
NTV	−5696.541	−2900.060	−5651.155	−5606.246
	(−3.56)	(−2.62)	(−3.60)	(−3.64)
	[0.001]	[0.01]	[0.001]	[0.001]
ESS	57.91e+08	68.00e+08	58.05e+08	58.18e+08
R^2	0.868	0.845	0.868	0.867
$\bar{R}^2$	0.844	0.837	0.848	0.852
F	36.158	100.838	44.582	57.216
SGMASQ	1.75e+08	1.84e+08	1.71e+08	1.66e+08*
AIC	2.05e+08	1.98e+08	1.96e+08	1.87e+08*
FPE	2.06e+08	1.98e+08	1.96e+08	1.87e+08*
HQ	2.29e+08	2.07e+08	2.15e+08	2.02e+08*
Schwarz	2.76e+08	2.24e+08*	2.52e+08	2.31e+08
Shibata	1.95e+08	1.96e+08	1.89e+08	1.82e+08*
GCV	2.13e+08	1.99e+08	2.01e+08	1.90e+08*
Rice	2.23e+08	2.00e+08	2.07e+08	1.94e+08*

[†]   Entries in parentheses are corresponding *t*-statistics and entries in square brackets are *p*-values for two-tailed tests. An asterisk denotes the model that is "best" for that criterion (see Section 10.6).

The *p*-values for regression coefficients are in square brackets and small values are desirable because then the corresponding variables will significantly affect the dependent variable. Model A is the original specification with all the variables present. In Model B we excluded all the variables with insignificant regression coefficients (*p*-values greater than 10 percent) with the exception of the constant term, which is ignored. In Model C, only SVC (with the least significant coefficient) was omitted and in Model D, both SVC and INSTFEE were eliminated. It is interesting to note that although PCINCM and NCABLE had *p*-values above 10 percent in Model A, they are below 10 percent in Model D. This is because the elimination of INSTFEE and SVC has improved the precision of the remaining coefficients and made them significant. In terms of the significance of the regression coefficients (at 10 percent), Model D appears to be preferable.

Table 10.2 also has the adjusted measure $\bar{R}^2$. The model with the highest $\bar{R}^2$ can be considered the "best" in terms of goodness of fit. This criterion also will choose Model D over the others.

Testing a Linear Combination of Coefficients

In some situations we may want to test a linear combination of several regression coefficients. For instance, consider the following consumption function

$$C = \beta_1 + \beta_2 W + \beta_3 D + u$$

where C is consumption expenditure, W is total wage income, and D is income from dividends and interest. The coefficients β_2 and β_3 are the *marginal propensities to consume* out of wage income and dividend income, respectively. A natural hypothesis to test is $\beta_2 = \beta_3$. In other models we may want to test a hypothesis of the type $\beta_2 + \beta_3 = 1$. These are examples where the null hypothesis in question is a linear combination of the parameters. In this section we develop the test for a general set of restrictions on the parameters given by the relationship $R\beta = r$, where R is a $q \times k$ matrix of known numbers, β is the $k \times 1$ vector of unknown regression coefficients, and r is a $q \times 1$ vector of known numbers. Several special cases of this are listed below.

(1) $R = [\,0 \ 0 \ \dots \ 1 \ 0 \ \dots \ 0\,]$ $r = 0$ $\beta_i = 0$

(2) $R = [\,0 \ 1 \ -1 \ 0 \ \dots \ 0\,]$ $r = 0$ $\beta_2 - \beta_3 = 0$

(3) $R = [\,0 \ 1 \ 1 \ 0 \ \dots \ 0\,]$ $r = 1$ $\beta_2 + \beta_3 = 1$

(4) $R = [\, 0 \quad I_q \,]$ $\qquad\qquad r = 0 \qquad \beta_{k-q+1} = \beta_{k-q+2}$
$$= \cdots = \beta_k = 0$$

where I_q is the identity matrix of order q.

$$(5) \quad R = \begin{bmatrix} 0 & 1 & 0 & \cdots & 0 \\ 0 & 0 & 1 & \cdots & 0 \\ . & . & . & \cdots & . \\ 0 & 0 & 0 & \cdots & 1 \end{bmatrix} \qquad r = \begin{bmatrix} 0 \\ 0 \\ . \\ . \\ 0 \end{bmatrix} \qquad \beta_2 = \beta_3 = \cdots = \beta_k = 0$$

Consider $R\hat{\beta}$. If $R\hat{\beta}$ is very different from r, we would conclude that $R\beta \neq r$. The test statistic is therefore based on $R\hat{\beta} - r$. Because $\hat{\beta}$ is normally distributed, so is $R\hat{\beta}$. Also,

$$E(R\hat{\beta}) = R\beta \qquad Var(R\hat{\beta}) = R\, Var(\hat{\beta})\, R' = \sigma^2 R(X'X)^{-1}R'$$

Therefore,

$$R\hat{\beta} \sim N[\, R\beta, \sigma^2 R(X'X)^{-1}R' \,]$$

Under the null hypothesis $H_0: R\beta = r$,

$$R\hat{\beta} - r \sim N[0, \sigma^2 R(X'X)^{-1}R' \,]$$

Therefore, by Theorem 5.23, the quadratic form

$$\frac{(R\hat{\beta} - r)'\, [R\,(X'X)^{-1}\,R']^{-1}\, (R\hat{\beta} - r)}{\sigma^2} \sim \chi_q^2$$

where $q = \text{rank}(R)$. In the section on the statistical distributions of $\hat{\beta}$ and s^2 we showed that they are independent and hence the above quadratic form is also independent of s^2. Replacing σ^2 by $s^2 = (\hat{u}'\hat{u})/(T-k)$ we obtain the ratio of two independent quadratic forms each of which has a χ^2 distribution. Therefore, by the definition of the F-distribution (see Section 6.4), the test statistic

$$F_c = \frac{(1/q)\,(R\hat{\beta} - r)'\, [R\,(X'X)^{-1}\,R']^{-1}\, (R\hat{\beta} - r)}{(\hat{u}'\hat{u})/(T-k)} \sim F_{q,\,T-k}$$

The test procedure (known as an **F-test**) is to look up the F-table for q d.f. in the numerator and $T - k$ d.f. in the denominator and obtain the point F^* such that $P(F \geq F^*) = a$, the level of significance. Reject the null hypothesis if $F_c > F^*$. Alternatively, compute p-value $= P(F \geq F_c)$ and reject the null if it is less than the size of the test.

Practice Problems

10.4 Show that for the special case (1) given earlier, the F-statistic is $F_c = \hat{\beta}_i^2/(s^2 a_{ii})$, where a_{ii} is the ith diagonal element of $(X'X)^{-1}$. Next show that $\sqrt{F_c} = t_c$ derived earlier for a single regression coefficient, and hence the F-test is equivalent to the two-tailed t-test.

10.5 Derive the F-statistics for cases (2) and (3) stated above. Also show how those tests can be done with t-tests.

Special Case of a Subset of Coefficients (Wald test)

In Example 10.6 we tested each of the regression coefficients individually for significance. It is possible to construct a single **joint test** for several of them. For instance, we could have tested the joint null hypothesis $\beta_4 = \beta_5 = 0$ in the model in Example 10.2. In this section we develop a test statistic for a subset of the regression coefficients.

In the general multiple regression model consider the null hypothesis $\beta_{k-q+1} = \beta_{k-q+2} = \cdots = \beta_k = 0$, that is, the last q regression coefficients are zero. This situation will be recognized as special case (4) given above for R and r, with $r = 0$ and $R = [0 \ I_q]$. To test this, first partition the X matrix and the $\hat{\beta}$ vector as

$$X = [X_p \quad X_q] \qquad \hat{\beta} = \begin{bmatrix} \hat{\beta}_p \\ \hat{\beta}_q \end{bmatrix}$$

where p refers to the first set of terms and q refers to the last set of terms with $p + q = k$. We have

$$X'X = \begin{bmatrix} X_p'X_p & X_p'X_q \\ X_q'X_p & X_q'X_q \end{bmatrix}$$

The model (known as the **unrestricted model**) can be written as

$$\text{(U)} \qquad y = X_p\beta_p + X_q\beta_q + u$$

Because $r = 0$ and $R = [0 \ I_q]$, $R\hat{\beta} - r = \hat{\beta}_q$. Let $C_{qq} = R(X'X)^{-1}R'$ be the submatrix at the bottom right-hand corner of $(X'X)^{-1}$. By the partitioned inverse property (Property A.8 in Appendix A),

$$C_{qq} = [X_q'X_q - X_q'X_p(X_p'X_p)^{-1}X_p'X_q]^{-1}$$
$$= [X_q'\{I - X_p'X_p)^{-1}X_p'\}X_q]^{-1} = [X_q'M_pX_q]^{-1}$$

where

$$M_p = I - X_p(X_p'X_p)^{-1}X_p'$$

The numerator of F_c now becomes $\hat{\beta}_p'(X_q'M_pX_q)\hat{\beta}_q/q$. The F-statistic F_c can be derived in an equivalent way that is considerably easier to compute. Impose the condition $R\beta = r$ on the unrestricted model and obtain the following **restricted model**.

(R) $$y = X_p\beta_p + u_R$$

Let $\tilde{\beta}_p$ be the estimate of β_p using the restricted model. Then

$$\tilde{\beta}_p = (X_p'X_p)^{-1}X_p'y \qquad \tilde{u}_R = y - X_p\tilde{\beta}_p = M_py$$

Let

$$\text{ESS}_U = \hat{u}'\hat{u} \qquad \text{ESS}_R = \tilde{u}_R'\tilde{u}_R$$

be the error sums of squares of the unrestricted and restricted models. We will presently show that

$$\text{ESS}_R - \text{ESS}_U = \hat{\beta}_q'(X_q'M_pX_q)\hat{\beta}_q$$

We have

$$y = X_p\hat{\beta}_p + X_q\hat{\beta}_q + \hat{u}$$

Multiplying each term by M_p defined earlier, we get

$$M_py = M_pX_p\hat{\beta}_p + M_pX_q\hat{\beta}_q + M_p\hat{u} = M_pX_q\hat{\beta}_q + M_p\hat{u}$$

because the first term vanishes due to the condition $M_pX_p = 0$. The last term can be written as

$$M_p\hat{u} = \hat{u} - X_p(X_p'X_p)^{-1}X_p'\hat{u} = \hat{u}$$

because, from an earlier derivation (see Practice Problem 10.1), $X'\hat{u} = X'Mu = 0$, and

$$\begin{bmatrix} 0 \\ 0 \end{bmatrix} = X'\hat{u} = \begin{bmatrix} X_p'\hat{u} \\ X_q'\hat{u} \end{bmatrix}$$

Therefore,

$$M_py = M_pX_q\hat{\beta}_q + \hat{u}$$

Transposing this and post-multiplying by it, we obtain (recall that M_p is idempotent and $X'\hat{u} = 0$)

$$y'M_p y = \hat{\beta}_q' X_q' M_p X_q \hat{\beta}_q + \hat{u}'\hat{u}$$

Also, we showed earlier that

$$\tilde{u}_R = M_p y$$

Therefore,

$$\tilde{u}_R'\tilde{u}_R = y'M_p y = \hat{\beta}_q' X_q' M_p X_q \hat{\beta}_q + \hat{u}'\hat{u}$$

from which it follows that

$$\hat{\beta}_q' X_q' M_p X_q \hat{\beta}_q = \tilde{u}_R'\tilde{u}_R - \hat{u}'\hat{u}$$

The expression for the F-statistic now becomes

$$F_c = \frac{(\tilde{u}_R'\tilde{u}_R - \hat{u}'\hat{u})/q}{\hat{u}'\hat{u}/(T-k)} = \frac{(\text{ESS}_R - \text{ESS}_U)/q}{\text{ESS}_U/(T-k)}$$

Computationally this is much easier because we simply run the two regressions and obtain the corresponding error sums of squares from which the above F-statistic can be calculated.

Example 10.7

We illustrate the F-test with the example of cable TV subscriptions we have been using in this chapter. In Table 10.2 the unrestricted model is Model A and we found that the coefficients for INSTFEE and SVC were statistically insignificant. To test them jointly the null hypothesis is $\beta_4 = \beta_5 = 0$. If we omit INSTFEE and SVC, we obtain Model D as the restricted model. The error sums of squares are, respectively, $\text{ESS}_R = 58.18\text{e}+08$ and $\text{ESS}_U = 57.91\text{e}+08$. Other values are, $T = 40$, $k = 7$, and $q = 2$. Calculated F-statistic is therefore

$$F_c = \frac{(\text{ESS}_R - \text{ESS}_U)/q}{\text{ESS}_U/(T-k)} = \frac{(58.18 - 57.91)/2}{57.91/(40-7)} = 0.077$$

Under the null hypothesis that $\beta_4 = \beta_5 = 0$, F_c has the F-distribution with 2 d.f. for the numerator and 33 d.f. for the denominator. From Appendix Table B.6c, for a 10 percent level of significance, the critical F^* for $F_{2,33}$ is between 2.44 and 2.49, which are higher than F_c. Therefore we cannot reject the null which implies that β_4 and β_5 are *jointly insignificant*. Thus INSTFEE and SVC are candidates for elimination from the model.

Practice Problem

10.6   Carry out a test for the joint hypothesis $\beta_3 = \beta_4 = \beta_5 = \beta_6 = 0$.
Table 10.2 has the required statistics for computing the test
statistic.

Lagrange Multiplier (LM) Test for a Subset of Regression Coefficients

In the previous chapter we introduced the Lagrange multiplier
test as an alternative to the Wald and likelihood ratio tests (see Section 9.11). That approach can also be applied to the case of testing a
subset of the regression coefficients (the model must include the constant term for this). Here we derive the test statistic and its distribution under the null. To conduct the LM test, the log-likelihood function $(\ln L)$ is maximized subject to the restriction $R\beta = 0$ (where $R =
[0 \; I_q])$ and the distribution of the score function $S(\beta) = \partial \ln L / \partial \beta$ is
used to derive the test statistic. The log-likelihood is

$$\ln L(\beta; y) = -\frac{n}{2} \ln(2\pi) - \frac{n}{2} \ln(\sigma^2) - \frac{1}{2\sigma^2} (y - X\beta)'(y - X\beta)$$

In Section 9.11 we showed that for multivariate normal errors
the score function (ignoring the part for σ^2) is given by

$$S = \frac{\partial \ln L}{\partial \beta} = \frac{X'u}{\sigma^2}$$

The covariance matrix of S is $\Sigma = X'X/\sigma^2$ and therefore

$$S \sim N(0, \Sigma) \qquad \text{and} \qquad S'\Sigma^{-1}S \sim \chi_q^2$$

$$S'\Sigma^{-1}S = \frac{u'X(X'X)^{-1}X'u}{\sigma^2}$$

To compute the test statistic ξ_c under H_0, we use the estimates
from the restricted model.

$$\xi_c = \tilde{S}'\tilde{\Sigma}^{-1}\tilde{S} = \frac{\tilde{u}'X(X'X)^{-1}X'\tilde{u}}{\tilde{\sigma}^2} \xrightarrow{d} \chi_q^2$$

where

$$\tilde{u} = y - X\tilde{\beta} \qquad \text{and} \qquad \tilde{\sigma}^2 = \frac{\tilde{u}'\tilde{u}}{T}$$

Computationally this can be done in an equivalent way. Consider the **auxiliary regression**

$$\tilde{u} = X\alpha + v$$

Regress the residuals of the restricted model ($\tilde{u}$) against *all the variables in the unrestricted model* and obtain

$$\hat{\alpha} = (X'X)^{-1}X'\tilde{u}$$

Unadjusted R^2 of the auxiliary regression is given by the regression sum of squares (that is, the sum of squares of the predicted values $X\hat{\alpha}$) divided by the total sum of squares (that is, the sum of squares of the values of the dependent variable $\tilde{u}$).

$$R^2 = \frac{(X\hat{\alpha})'(X\hat{\alpha})}{\tilde{u}'\tilde{u}} = \frac{\tilde{u}'X(X'X)^{-1}X'X(X'X)^{-1}X'\tilde{u}}{\tilde{u}'\tilde{u}}$$

$$= \frac{\tilde{u}'X(X'X)^{-1}X'\tilde{u}}{\tilde{u}'\tilde{u}} = \frac{\tilde{S}'\tilde{\Sigma}^{-1}\tilde{S}}{T}$$

We need not subtract the mean of the dependent variable in the sums of squares because, as we saw in the discussion of the deviation from the mean model, the average of $\tilde{u}$ is zero. It follows from this that

$$\xi_c = \tilde{S}'\tilde{\Sigma}^{-1}\tilde{S} = TR^2$$

The test statistic ξ_c therefore reduces to the product of the number of observations used in the auxiliary regression and the unadjusted R^2 of that regression. Note that if we had used only X_q in the auxiliary regression, the regression sum of squares would have had X_q in the expression and hence ξ_c will not have the simple structure.

Unlike the Wald test statistic which has the F-distribution, the LM test statistic has a large sample χ^2 distribution. The procedure for obtaining the test statistic is to estimate the restricted model first, compute the residual vector for this model, regress these residuals against all the variables in the unrestricted model, and then compute TR^2. The test criterion is to reject the null hypothesis if TR^2 exceeds the appropriate critical value for χ_q^2 or if the p-value $P(\chi_q^2 > TR^2)$ is less than the level of significance.

Example 10.8

The LM test is illustrated here for the cable TV example used throughout this chapter. The basic unrestricted model will contain only a constant term and the variable HOMES. The auxiliary

regression will relate the residuals from the basic model to
HOMES, PCINCM, INSTFEE, SVC, NCABLE, and NTV. The following table presents the relevant statistics for the auxiliary
regression.

Variable	$\hat{\beta}$	$s_{\hat{\beta}}$	t_c	$P(t > | t_c |)$
Constant	− 30481.102	26671.205	− 1.143	0.261
HOMES	0.043	0.034	1.261	0.216
PCINCM	5.462	3.286	1.662	0.106
INSTFEE	− 159.568	480.911	− 0.332	0.742
SVC	609.212	2178.625	0.280	0.782
NCABLE	1129.869	715.103	1.580	0.124
NTV	− 5696.541	1600.140	− 3.560	0.001

The unadjusted R^2 for the auxiliary regression is 0.282,
which gives the test statistic $\xi_c = TR^2 = 11.28$. Under H_0, this
has a large sample χ^2 distribution with 5 d.f. From Appendix
Table B.5 the critical value for a 5 percent test is 11.07, which is
smaller than ξ_c and hence we reject the null hypothesis that all
the added variables have insignificant regression coefficients. This
means that at least one of them belongs in the model. But which
one(s)? The auxiliary regression can give us a clue as to which of
these added variables might have significant effects. High
p–values in the above table suggest that the corresponding variables are probably not important. This would rule out INSTFEE
and SVC. If we exclude these variables, we obtain Model D in
Table 10.2, which was chosen earlier as the "best" model.

Special Wald Test for Overall Goodness of Fit

The Wald test for a subset of regression coefficients can be used
to test the overall goodness of fit of a regression model. To do this,
consider the null hypothesis, $\beta_2 = \beta_3 = \cdots = \beta_k = 0$. If this
hypothesis is true, then none of the variables in the model would have
any significant effect on the dependent variable. This corresponds to
case (5) on page 272 with $R = [0 \ I_{k-1}]$. If the above conditions are
imposed, the restricted model becomes $Y_t = \beta_1 + u_t$ and the OLS
estimate of β_1 is $\bar{Y}$, which means that $\hat{Y}_t = \bar{Y}$. Hence,

$$\text{ESS}_R = \Sigma (Y_t - \hat{Y}_t)^2 = \Sigma (Y_t - \bar{Y})^2 = \text{TSS}_U$$

Therefore the Wald F-statistic is

$$F_c = \frac{(\text{ESS}_R - \text{ESS}_U)/(k-1)}{\text{ESS}_U/(T-k)} = \frac{(\text{TSS}_U - \text{ESS}_U)/(k-1)}{\text{ESS}_U/(T-k)}$$

$$= \frac{\text{RSS}_U/(k-1)}{\text{ESS}_U/(T-k)} = \frac{R^2(T-k)}{(1-R^2)(k-1)}$$

Example 10.9

The F-statistic for the null hypothesis that all the regression coefficients except the constant term are zero is presented in Table 10.2 for each of the models. The number of observations (T) is 40 and the d.f. for each of the four models are (6, 33), (2, 37), (5, 34), and (4, 35). It is readily seen from Table B.6a that all the critical F^* values for the 1 percent level are well below the observed F_c. This implies that we reject the null hypothesis in each case. This result is not surprising because we have found from the t-statistics that several regression coefficients are indeed significant. The usefulness of the special F-test is in identifying poorly specified models in which none of the variables have significant effects on the dependent variable.

The various sums of squares and the F-statistic for overall goodness of fit are often presented in a tabular form called the **Analysis of variance ANOVA)** table which is presented below.

SOURCE	SUM OF SQUARES	D.F.	F-STATISTIC
Regression (RSS)	$\Sigma(\hat{Y}_t-\bar{Y})^2 = \hat{\beta}'X'y - T\bar{Y}^2$	$k-1$	$\dfrac{\text{RSS}/(k-1)}{\text{ESS}/(T-k)}$
Error (ESS)	$\Sigma\hat{u}_t^2 = \hat{u}'\hat{u}$	$T-k$	
Total (TSS)	$\Sigma(Y_t-\bar{Y})^2 = y'y - T\bar{Y}^2$	$T-1$	

A Model Without a Constant Term

Some modifications are needed if the model does not have a constant "intercept" term. First, the sum of squares decomposition TSS = RSS + ESS is no longer valid. The decomposition is instead (verify it),

$$y'y = \hat{y}'\hat{y} + \hat{u}'\hat{u}$$

The formula for computing the F-statistic will now be different. To see why this is so, consider the modified restricted and unrestricted models.

$$(U) \qquad Y = \beta_2 X_2 + \beta_3 X_3 + \cdots + \beta_k X_k + u$$

$$(R) \qquad Y = w$$

where the constant term $(X_1 = 1)$ has been excluded. The unrestricted model has only $k - 1$ parameters with d.f. $T - k + 1$ and the restricted model has none (with d.f. T). The Wald F is therefore given by

$$F_c = \frac{(\text{ESS}_R - \text{ESS}_U) / (k - 1)}{\text{ESS}_U / (T - k + 1)} = \frac{(y'y - \hat{u}'\hat{u}) / (k - 1)}{\text{ESS}_U / (T - k + 1)}$$

$$= \frac{\hat{y}'\hat{y} / (k - 1)}{\text{ESS}_U / (T - k + 1)}$$

Because of the modified decomposition, the formula for F_c is different. Under the null hypothesis $\beta_i = 0$ $(i = 2, 3, \ldots, k)$, this has an F-distribution with d.f. $k - 1$ and $T - k + 1$ and the test criterion is computed appropriately.

Should the goodness of fit measures R^2 and $\bar{R}^2$ also be modified when there is no constant term? Some computer programs compute R^2 as $1 - (\text{ESS} / y'y)$. The value computed this way is, however, not comparable to one calculated with TSS because the denominators are different. It is recommended that R^2 be computed as the square of the correlation coefficient between the observed value Y_t and the predicted value $\hat{Y}_t$ (see Exercise 10.1). The formula for the adjusted R^2 should be the same as before because it is computed as $\bar{R}^2 = 1 - [\hat{Var}(\hat{u}) / \hat{Var}(Y)]$, which is model independent.

10.6 MODEL SELECTION CRITERIA

When discussing the goodness of fit measure, we stated that one criterion that is used to choose a particular model over another is $\bar{R}^2$ or equivalently $\hat{\sigma}^2$. This measure takes into account the trade-off between gain in explanatory power and loss in d.f. There are, however, other ways to measure this trade-off between model complexity and goodness of fit, and this section gives a brief introduction to them. For a thorough analysis of many of these alternative criteria for model selection, refer to the book by Judge *et al.* (1985, Chapter 21, Section 7.5 and Section 16.6.1a). See also the paper by Engle and Brown (1985).

The criteria proposed by various researchers are based on some optimality principle as, for example, minimizing prediction error sum of squares or minimizing observed likelihood values. All these criteria are based on the mean squared error (ESS/T) multiplied by some penalty factor that depends on the model complexity as measured by the number of regression coefficients to be estimated (k). The measures are referred to as **Akaike information criterion (AIC)** (Akaike, 1974), **finite prediction error (FPE)** (Akaike, 1970), **HQ** (Hannan and Quinn, 1979), **Schwarz** (1978), **Shibata** (1981), **Rice** (1984), and **generalized cross validation (GCV)** (Craven and Wahba, 1979). The following table presents a summary of the criteria statistics. SGMASQ is the estimated residual variance $\hat{\sigma}^2$.

SGMASQ	$\dfrac{\text{ESS}}{T}\left[1-\dfrac{k}{T}\right]^{-1}$	HQ	$\dfrac{\text{ESS}}{T}(\ln T)^{2k/T}$
AIC	$\dfrac{\text{ESS}}{T}\,e^{2k/T}$	Rice	$\dfrac{\text{ESS}}{T}\left[1-\dfrac{2k}{T}\right]^{-1}$
FPE	$\dfrac{\text{ESS}}{T}\dfrac{T+k}{T-k}$	Schwarz	$\dfrac{\text{ESS}}{T}\,T^{k/T}$
GCV	$\dfrac{\text{ESS}}{T}\left[1-\dfrac{k}{T}\right]^{-2}$	Shibata	$\dfrac{\text{ESS}}{T}\dfrac{T+k}{T}$

A model with a lower value of a criterion statistic is judged to be preferable. Ideally we would like our model to have the lowest value for all the criteria, but that may not always happen in practice. In that case, a model that outperforms another one in more of these criteria would be preferred. Ramanathan (1992) has shown that in special cases some of the criteria might be redundant, that is, a model superior under one criterion might automatically be superior under another criterion.

Table 10.2 presents the above model selection statistics for the four models studied here. We note that Model D is superior to all the others in seven out of the eight criteria. The Schwarz criterion penalizes model complexity much more heavily than others and tends to pick the model with a parsimonious parameter specification (as is the case for Model B).

10.7 NONNORMALITY OF ERRORS

The maximum likelihood estimators as well as the test statistics were derived using Assumption 10.6 that the error terms u_t were normally distributed. However, several important properties hold even if the errors are nonnormal. First, the OLS estimators

$$\hat{\beta} = (X'X)^{-1}X'y \qquad\qquad s^2 = \hat{u}'\hat{u}/(T-k)$$

do not make use of any assumptions about the distribution of the u's. All that is required is that the cross-product matrix $X'X$ be non-singular and that $T > k$.

Unbiasedness and Consistency

The properties of unbiasedness and consistency, however, require additional assumptions on the stochastic properties of u.

$$\hat{\beta} = (X'X)^{-1}X'y = (X'X)^{-1}X'(X\beta + u) = \beta + (X'X)^{-1}X'u$$

If X is given, then

$$E(\hat{\beta}) = \beta + (X'X)^{-1}E(X'u) = \beta$$

provided $E(X'u) = 0$. Therefore unbiasedness requires that all the X's be uncorrelated with u. Alternatively, for a given X, $E(X'u) = X'E(u) = 0$, provided u has zero mean. Thus, unbiasedness requires Assumptions 10.2, 10.3, and 10.4, but not 10.5 or 10.6. These assumptions are also sufficient to show that $\hat{\beta}$ is a consistent estimator of β provided

$$\lim_{T \to \infty} [X'X / T] = \Sigma_x$$

which is positive definite. The probability limit of $\hat{\beta}$ is then obtained as

$$plim\ \hat{\beta} = \beta + plim \left[\frac{1}{T} X'X\right]^{-1} plim \left[\frac{1}{T} X'u\right]$$

$$= \beta + \Sigma_x^{-1}\ plim \left[\frac{1}{T} X'u\right]$$

By the law of large numbers (see Section 7.3),

$$plim\ (X'u/T) = E(X'u/T) = 0$$

and hence $\hat{\beta}$ is consistent. In Section 10.2 we showed that

$$E(s^2) = E\left[\frac{\hat{u}'\hat{u}}{T-k}\right] = \sigma^2$$

and hence s^2 is unbiased. By the law of large numbers, s^2 converges in probability to its expectation σ^2 and hence it is also consistent.

Efficiency

In Section 10.2 we proved the Gauss-Markov theorem, which states that $\hat{\beta}$ is best linear unbiased (BLUE), that is, among all linear combinations of y's that are unbiased, $\hat{\beta}$ has the smallest variance. Thus, $\hat{\beta}$ is most efficient. In order to prove this theorem, however, we needed also Assumption 10.5 that the u's are homoscedastic and serially uncorrelated. In other words, all the u's have the same variance σ^2 and every element of u is uncorrelated with every other element, but normality of errors is not required.

Asymptotic Distributions and Hypothesis Testing

It is easy to show that, under Assumptions 10.2 through 10.5 (again without assuming normality of errors) $\hat{\beta}$ has an asymptotically normal distribution. We have,

$$\hat{\beta} - \beta = \left[\frac{1}{T}X'X\right]^{-1}\left[\frac{1}{T}X'u\right]$$

and hence

$$\sqrt{T}(\hat{\beta} - \beta) = \left[\frac{1}{T}X'X\right]^{-1}\left[\frac{1}{\sqrt{T}}X'u\right]$$

It is readily verified that $X'u/\sqrt{T}$ has zero mean and covariance matrix $\sigma^2 X'X/T$ which converges to $\sigma^2 \Sigma_x$. By applying the multivariate central limit theorem (Theorem 7.19) we have the result that

$$X'u/\sqrt{T} \xrightarrow{d} \text{MVN}[0, \sigma^2 \Sigma_x]$$

from which it follows that (using Property 5.7, Section 5.10)

$$\sqrt{T}(\hat{\beta} - \beta) \xrightarrow{d} \text{MVN}[0, \sigma^2 \Sigma_x^{-1}]$$

For a large sample, the distribution of $\hat{\beta} - \beta$ can be approximated by $\text{MVN}[0, s^2(X'X)^{-1}]$. If a_{ii} is the ith diagonal element of $(X'X)^{-1}$, then for a typical β_i, the standardized version

$$z = \frac{\hat{\beta}_i - \beta_i}{s\sqrt{a_{ii}}}$$

has the asymptotic standard normal distribution. Therefore we can use that distribution instead of the t-distribution to test individual coefficients for significance.

The procedure for testing a set of linear restrictions of the type $R\beta = r$ also needs modification. From the large sample distribution of $\hat{\beta}$ we have (using Property 5.7 of Section 5.10)

$$R(\hat{\beta} - \beta) \approx \text{MVN}\,[0,\,s^2 R(X'X)^{-1}R']$$

By Theorem 5.23, the corresponding quadratic form has a chi-square distribution with d.f. equal to the number of rows (q) of R. Therefore, the following Wald test statistic can be used even when the errors are not normally distributed, provided the sample size is large enough.

$$W = (R\hat{\beta} - r)'\,[\,s^2 R(X'X)^{-1}R'\,]^{-1}(R\hat{\beta} - r) \approx \chi_q^2$$

The F-test is not applicable here because that test statistic does not have an F-distribution. The alternative is a chi-square test using the above Wald statistic.

Testing for Normality

The residuals $\hat{u}_t$ from a regression model can be used to carry out a test for the normality of the errors u_t. If the errors were normal, then the skewness will be zero and the kurtosis will be 3 (see Section 3.5). For a general random variable u with mean zero and variance σ^2, skewness and kurtosis are measured as follows.

$$\eta_1 = E\,[(u/\sigma)^3] \qquad\qquad \eta_2 = E\,[(u/\sigma)^4]$$

which can be estimated as

$$\hat{\eta}_1 = \frac{1}{T}\sum\left[\frac{\hat{u}_t}{s}\right]^3 \qquad\qquad \hat{\eta}_2 = \frac{1}{T}\sum\left[\frac{\hat{u}_t}{s}\right]^4$$

Under the null hypothesis of normality, each of the above statistics has a large sample normal distribution. The next step is to construct the following Wald test statistic and reject the null if it is larger than the corresponding critical value.

$$W = T\left[\frac{\hat{\eta}_1^2}{6} + \frac{(\hat{\eta}_2 - 3)^2}{24}\right] \approx \chi_2^2$$

For more on testing the residuals for normality, see Greene (1990, page 328) and Judge *et al.* (1985, page 826).

The Poisson Regression Model

Because the standard normal distribution is symmetric about the origin, the random variable can theoretically assume large negative values. When dealing with variables that always take positive values, the assumption of normality of the errors is really questionable. This objection is particularly strong when working with **count data**, that is, data that are numerical counts. Examples of count data are numbers of patents issued in a period, riders in a bus or train, union members, visits to the doctor, strikes, and the number of times a person has been laid off. In these cases, a popular alternative to the normality assumption is to postulate that the errors follow a Poisson distribution. The model is

$$P(Y = y_t) = \frac{e^{-\lambda_t} \lambda_t^{y_t}}{y_t!}$$

$$\ln(\lambda_t) = \beta' x_t \qquad \text{or} \qquad \lambda_t = e^{\beta' x_t}$$

where β is the $k \times 1$ vector of coefficients and x_t is the $k \times 1$ vector of observations on the X's. Note that λ_t is always positive because the exponential function gives only positive values. Also note that the mean and variance of a Poisson random variable is its parameter (see Table 4.1). Therefore,

$$E(y_t \mid x_t) = \text{Var}(y_t \mid x_t) = \lambda_t = \exp(\beta' x_t)$$

Although the Poisson model fits the nonlinear regression framework discussed in this section, maximum likelihood estimation using Newton's method of scores (see Section 8.13) is easier because the log-likelihood and score functions are readily obtained.

$$\ln L = \sum_{t=1}^{t=T} [-\lambda_t + y_t \beta' x_t - \ln(y_t!)]$$

$$\text{Score } S(y, \beta) = \frac{\partial \ln L}{\partial \beta} = \sum_{t=1}^{t=T} (-\lambda_t + y_t) x_t$$

For more discussion of the Poisson regression model and examples, see Gourieroux *et al* (1984), Carson and Grogger (1991) and the references given in those papers.

10.8   INTRODUCTION TO BAYESIAN ESTIMATION

In Chapter 2 we discussed the *Bayes theorem* which related the probability of an event A given another event B in terms of the probability of B given A. In the context of statistical inference, we can think of A as the parameter θ and B as the observations y. Rather than treat the parameter θ as an unknown fixed number, the Bayesian approach would start with some prior notions about it. In particular, it would be thought of as random with a **prior probability density function**; call it $h(\theta)$. $P(B \mid A)$ is simply $P(y \mid \theta)$ which is nothing but the familiar likelihood function $L(y; \theta)$. Applying Bayes theorem, we have the following density function known as the **posterior probability function**.

$$f_{\theta|y}(\theta \mid y) \;=\; \frac{f_{y|\theta}(y \mid \theta) f_\theta(\theta)}{f_y(y)} \;=\; \frac{L(y;\theta) h(\theta)}{\int L(y;\theta) h(\theta) d\theta}$$

The above equation enables us to revise our prior notions about θ based on the observations y. This idea of *Bayesian updating* of probabilities of parameters has spawned a whole new approach to statistical inference. In this section we provide an introduction to the **Bayesian estimation** of the multiple regression model. If you are interested in learning more about the approach, refer to Zellner (1971), Cyert and DeGroot (1987), Maddala (1977), Greene (1990), Raiffa and Schlaifer (1961), and Judge *et al* (1985).

In many situations, the investigator does not have any prior notion of the exact form of the distribution $h(\theta)$. What is done in such cases is to use a **diffuse prior** (also known as a **noninformative prior** or as a **Jeffrey's prior**) for which one assumes that the distribution is uniform. It is evident that in this case the posterior distribution will just be proportional to the likelihood function. The results will be the same as before except that the interpretation is now that the estmated distribution is for β centered around $\hat{\beta}$ rather than for β to be a single point.

A very common alternative to the diffuse prior is a **conjugate prior** that has the property that the functional form of the posterior distribution $f_{\theta|y}(\theta \mid y)$ is the same as that of the prior $h(\theta)$. This is analytically convenient and also facilitates the use of the posterior as the new prior when additional sample information becomes available. Raiffa and Schlaifer (1961) have several examples of conjugate priors.

In the context of a multiple regression model, a popular conjugate prior distribution for the parameters β and σ is the **gamma-normal prior** that has the density

$$f(\beta, \sigma) \propto \frac{1}{\sigma^m} \exp\left[-\frac{1}{2\sigma^2} [v + (\beta - \beta_0)' V (\beta - \beta_0)] \right]$$

where $v > 0$, $m > 0$, and V is a positive definite matrix. For a given σ, the vector β has a k-variate normal distribution with mean vector β_0 and covariance matrix $\sigma^2 V^{-1}$. For a given β, the distribution of σ is known as the **inverted gamma**. Integrating the joint density function with respect to each of the parameters in turn, the corresponding marginal densities can be shown to be the following.

$$f(\beta) \propto [v + (\beta - \beta_0)' V (\beta - \beta_0)]^{-(m-1)/2}$$

$$f(\sigma) \propto \frac{1}{\sigma^{m-k}} e^{-v/(2\sigma^2)}$$

The marginal distribution of σ is also an inverted gamma distribution and the marginal distribution of β is known as the **multivariate t-distribution**. Raiffa and Schlaifer (1961) have computed the first two moments and have shown that the marginal density of a typical element of β has the t-distribution. Zellner (1971) has the detailed derivation of the posterior distribution and other distributions such as F and t for hypothesis testing. Interested readers should also refer to Maddala (1977, Chapter 18) for more details.

EXERCISES

10.1   Consider the simple linear regression relation $Y_t = \alpha + \beta X_t + u_t$ where u_t is *iid* as $N(0, \sigma^2)$ with α, β and σ^2 unknown. You have T observations on Y_t and X_t but none on u_t. X_t can be treated as known and non-random.

    (a)   Write down the joint density function of u_t's. From this write down the likelihood function of α, β, and σ^2 in terms of X_t and Y_t. (You may wish to refer to a basic econometrics text for answering these questions.)

(b)   Maximize the log-likelihood with respect to α and β and derive the first order conditions. Solve them and show that the MLE of α and β are

$$\hat{\beta} = m_{XY}/m_{XX} \qquad \hat{\alpha} = \overline{Y} - \hat{\beta}\overline{X}$$

where $m_{XY} = \Sigma X_t Y_t - T\overline{X}\overline{Y}$ and $m_{XX} = \Sigma X_t^2 - T\overline{X}^2$. $\overline{X}$ and $\overline{Y}$ are the sample means.

(c)   In the expression for $\hat{\beta}$, substitute for Y_t from the model and express $\hat{\beta}$ in terms of X_t, u_t, and the unknown parameters. Show that when X_t is given and non-random, $\hat{\beta}$ is unbiased for β. Under that conditon is $\hat{\alpha}$ also unbiased? Justify your answers.

(d)   Show that $\text{Var}(\hat{\beta}) = \sigma^2/m_{XX}$. Derive the statistical distribution of $\hat{\beta}$, noting that it is a linear function of u_t and treating X_t as given and non-random.

(e)   Show that the maximum likelihood estimator of σ^2 is given by $\hat{\sigma}^2 = (\Sigma \hat{u}_t^2)/T$ where $\hat{u}_t = Y_t - \hat{\alpha} - \hat{\beta}X_t$.

(f)   Consider an alternative estimator of β, $\tilde{\beta} = C_1 Y_1 + C_2 Y_2 + \ldots + C_T Y_T$, that is, consider a linear combination of the Y_t's. Derive the two conditions needed for $\tilde{\beta}$ to be unbiased *for all* β.

(g)   Compute the variance of $\tilde{\beta}$ in terms of σ^2 and C_t^2 ($t = 1, 2, \ldots, T$). Minimize the variance of $\tilde{\beta}$ with respect to $C_1, C_2, \ldots, C_T$ subject to the two conditions in (f), and show that the best linear unbiased estimator (that is, BLUE or UMVU) of β is the same as $\hat{\beta}$. In other words, MLE of β is also BLUE.

(h)   It was proved in Section 10.2 that $s^2 = (\Sigma \hat{u}_t^2)/(T-2)$ is an unbiased estimator of σ^2 and that $(T-2)s^2/\sigma^2$ is distributed as χ_{T-2}^2. Use this and the result in (d) above to show that the quantity $t = (\hat{\beta} - \beta)/(s/\sqrt{m_{XX}}) \sim t_{T-2}$. From this derive the critical region for testing the hypothesis $H_0: \beta = \beta_0$ against $H_1: \beta \neq \beta_0$. Also derive the 95% confidence interval for β.

(i)   Suppose you wish to derive the likelihood ratio (LR) test for $H_0: \beta = \beta_0$ against $\beta \neq \beta_0$. Write down the generalized ln LR for the specific model given here. It can be shown that $\ln(\lambda)$ is monotonic in $F = \text{RSS}(T-2)/\text{ESS}$ where $\text{ESS} = \Sigma \hat{u}_t^2$, $\text{RSS} = \Sigma (\hat{Y}_t - \overline{Y})^2$, and $\hat{Y}_t = \hat{\alpha} + \hat{\beta}X_t$.

Also $F \sim F_{1,n-2}$. Using this describe the procedure for carrying out the above test. Can you conjecture the relationship between this test and the t-test of (h)?

10.2   In the regression model $y = X\beta + u$ estimated by OLS, show that the square of the correlation coefficient between the observed dependent variable Y_t and its predicted value $\hat{Y}_t$ is identical to the unadjusted R^2. Thus, R^2 has another interpretation, namely, it measures how strongly correlated the observed and predicted values of the dependent variable are. For simplicity, assume that all variables are expressed as deviations from the corresponding means so that the model is already is in deviation form.

10.3   Consider the restricted model (R) and the unrestricted model (U), which has more variables. Show that
$$\bar{R}_U^2 > \bar{R}_R^2$$
if and only if $F_c > 1$, where F_c is the value of the Wald F-statistic to test the joint hypothesis that the added variables have zero regression coefficients.

10.4   Suppose the true model is $y = X_p \beta_p + u$, but the estimated one had additional variables represented by the $T \times q$ matrix X_q, so that the estimated model is $y = X_p \beta_p + X_q \beta_q + v$.

   (a)   Derive an expression for the OLS estimator of β_p based on the *estimated* model.

   (b)   Using the *true* model compute the expected value of the above estimator and examine whether the estimator is unbiased or not.

   (c)   Derive the covariance matrix of the estimator in (a) and compare the variances of the estimators with those that you would have obtained if you had used the true model.

   (d)   From the above analysis what do you conclude about the consequences of adding redundant variables to a model?

10.5   Suppose the true model is $y = X_p \beta_p + X_q \beta_q + u$, but we estimated a model by omitting the variables in X_q, so that the estimated model is $y = X_p \beta_p + v$.

   (a)   Derive an expression for the OLS estimator of β_p based on the *estimated* model.

(b)   Using the *true* model compute the expected value of the above estimator and examine whether the estimator is unbiased or not.

(c)   Derive the covariance matrix of the estimator in (a) and compare the variances of the estimators with those that you would have obtained if you had used the true model.

(d)   From the above analysis what do you conclude about the consequences of excluding important variables from a model?

10.6   Suppose the true model is $y = X_p \beta_p + X_q \beta_q + u$. You obtain estimates of β_p and β_q in two steps. First you estimate β_p by regressing y against just X_p. Then obtain $y^* = y - X_p \hat{\beta}_p$ and regress y^* against X_q to get $\hat{\beta}_q$. Show that the estimators obtained by this procedure are biased. Under what conditions will they be unbiased?

10.7   Consider an alternative two-step procedure. As in 10.6 you regress y against X_p and obtain y^* the same way. Next you estimate several "auxiliary regressions" by regressing each of the X_q's on the set of X_p's and obtain $B = (X_p'X_p)^{-1} X_p'X_q$. Then obtain $\tilde{X}_q = X_q - X_p B = M_p X_q$, where $M_p = I - X_p (X_p'X_p)^{-1} X_p'$. $\tilde{X}_q$ will be recognized as the residuals of the auxiliary regressions in which the effect of X_p has been "removed" from X_q. Finally, regress y or y^* against $\tilde{X}_q$ and obtain $\tilde{\beta}_q$. Show that $\tilde{\beta}_q$ is identical to the OLS estimator of β_q when the full regression of y on X_p and X_q is run.

10.8   Assumption 10.3 on the multiple regression model was that the matrix $X'X$ is nonsingular. In some cases (common in the design of agricultural experiments) the number of parameters (k) is larger than the number of observations (T) that make the rank of $X'X$ less than full. In this case the normal equations $X'X\hat{\beta} = X'y$ will have an infinite number of solutions, making the individual parameters not estimable. However, a linear combination $t'\beta$ (t is $k \times 1$ with known elements) may be estimable. Thus there may exist a linear combination of the y's (say, $w'y$ where w is $T \times 1$) such that $E(w'y) = t'\beta$ for all β. In this exercise you are to obtain the best linear unbiased estimator of $t'\beta$ even though $X'X$ has no inverse. This can be done by carrying out the following steps.

(a) Derive the condition for $w'y$ to be a linear unbiased estimator of $t'\beta$ for all β.

(b) Compute the variance of $w'y$.

(c) Choose the w's to minimize $\text{Var}(w'y)$ subject to the condition in (a) and derive the relevant conditions.

(d) Show that $t'\hat{\beta}$, using *any* of the infinite solutions to the normal equations, will be the same as the estimator obtained in (c).

(e) Show that $t'\hat{\beta}$ is unique, that is, if two different solutions $\hat{\beta}_1$ and $\hat{\beta}_2$ are chosen, $t'\hat{\beta}_1$ will be identical to $t'\hat{\beta}_2$. To show this, prove that the variance of their difference $(t'\hat{\beta}_1 - t'\hat{\beta}_2)$ is zero.

10.9 Suppose we are interested in predicting y for a given X (say x_0'). Show that the best linear unbiased prediction of y_0 is given by $\hat{y} = x_0'\hat{\beta}$, where $\hat{\beta}$ is the OLS estimator of β. In other words, among all linear estimators of the type $a'y$ which have expected value $x_0'\beta$, $x_0'\hat{\beta}$ has the lowest variance.

10.10 In the standard linear model, suppose that it is known that the β's satisfy the restriction $t'\beta = c_0$, where t is a $k \times 1$ vector and c_0 is a scalar, all known. By minimizing $\hat{u}'\hat{u}$ with respect to $\hat{\beta}$, subject to the above restriction, show the following.

$$\tilde{\beta} = \hat{\beta} + Vt\,[t'Vt]^{-1}(c_0 - t'\hat{\beta})$$
$$\text{Var}(\tilde{\beta}) = \sigma^2\,[V - Vt\,(t'Vt)^{-1}\,t'V]$$

where $\hat{\beta}$ is the unrestricted OLS estimator and $V = (X'X)^{-1}$. What can you say about the variances of the restricted least squares estimators as compared to those of the unrestricted least squares estimators? Which are smaller?

10.11 Table 10.3 has data on the determinants of private housing building permits for 40 metropolitan areas. The variables are defined below and are based on the 1980 census data.

Y = Number of new private housing units authorized
X2 = Population density per square mile
X3 = Median value (in hundreds of dollars) of owner-occupied home
X4 = Median household income in hundreds of dollars
X5 = Average local taxes per capita
X6 = Average state taxes per capita

X7 = Percentage change in population between 1980 and 1982

X8 = Unemployment rate for the civilian labor force

(a) Use a regression program and regress Y against a constant term, X4, X5, X6, X7, and X8. Carry out an appropriate test for the overall goodness of fit. Test each of the regression coefficients (ignore the constant term) for significance at the 10 percent level. Based on this do you recommend that any of the variables be omitted? If yes, omit them and reestimate the new model. Are the results improved (define what you mean by "improved")?

(b) Using the new model obtained in (a) as the basic restricted model, perform an LM test for the addition of the remaining explanatory variables in the above list. What do you conclude? Does the auxiliary regression suggest any new variables to be added to the basic model?

(c) Next estimate a "kitchen sink" model in which all the X variables listed above are included. Perform an appropriate test for individual coefficients, omit the least significant one, and reestimate the new model. Compare the model selection statistics and comment on whether the omission was worthwhile in terms of these selection criteria. Continue this process until you have a model in which all the coefficients are significant at the 10 percent level. Which of the several models you estimated is the "best" (explain the criteria you use to choose the best)? In this final model do the regression coefficients have signs that correspond to intuition? Explain.

Table 10.3 Determinants of Private Housing Building Permits

Y	X2	X3	X4	X5	X6	X7	X8
3292	2887	313	120	247	650	−1.2	8.7
9878	2437	563	174	158	1007	4.3	5.6
7390	6384	961	157	227	1172	1.8	6.8
4972	2736	907	164	134	1172	4.6	7
1126	14633	1046	159	508	1172	1.9	6.1
1816	4452	637	155	378	990	2.6	5

(Continued)

Table 10.3 (Continued)

Y	X2	X3	X4	X5	X6	X7	X8
3143	10113	475	111	244	758	10.3	6.1
668	3244	318	113	251	770	0.7	8.1
2328	4196	472	153	245	1084	−0.3	9.8
4231	1991	351	173	180	744	1	7
564	4974	272	123	236	926	−1.7	9.9
252	3562	528	152	215	841	4.5	5.8
1685	2796	506	118	250	841	1.2	7
746	9798	289	128	424	1104	−1.6	10.8
473	11928	360	125	781	1243	−0.4	6.1
1171	8874	210	140	300	1075	−5.3	18.5
492	67322	545	144	203	1125	−0.5	4.8
846	5157	529	160	166	1125	0.1	4.7
465	1417	346	159	371	759	−0.6	6.5
302	7379	264	115	433	759	−3.4	11.1
805	3457	344	164	218	963	0.3	5.3
467	16934	306	128	350	1137	−0.3	9.8
384	13662	315	101	320	1137	−2.7	13.4
8436	23455	538	139	949	1494	0.2	7.7
2392	2251	460	169	177	748	2.7	4.4
215	7264	304	123	243	810	−2.6	11
2306	3123	414	148	164	810	1	6.4
3224	668	401	159	218	827	5.9	3.4
2334	1945	457	169	180	827	4	3.3
686	3547	548	148	217	979	−0.2	6.9
1906	12413	237	132	464	978	−1.4	11.4
247	7652	315	134	273	978	−2.1	9.2
564	2447	352	140	158	656	−0.1	8.5
1333	950	439	161	382	656	−0.1	5.1
8611	2978	474	147	158	806	6.4	3.8
7612	2715	442	162	239	806	4.3	3.4
13143	2867	491	185	221	806	8.2	3.6
5975	2992	278	138	99	806	4.2	5.5
1424	3415	659	163	247	989	−0.8	5.9
894	6641	455	160	142	1061	−0.8	6.9

Source: *County and City Data Book*, 1983.

11

FUNCTIONAL FORMS AND DUMMY VARIABLES

All the explanatory variables encountered so far have been **quantitative** in nature; that is, the attributes of variables are numerically measurable. However, variables that are **qualitative**, such as the gender or race of a person, the season that a variable might correspond to, the political party that a voter might be affiliated with, and so on, might also influence the dependent variable. Section 11.1 of this chapter examines how such qualitative variables can be incorporated in a regression model.

The specification of the regression model introduced in Assumption 10.1 is linear in the sense that the relationship between the dependent variable Y and the independent variables (X's) was assumed to be linear. This implies that the marginal effect of every explanatory variable is constant. This is obviously a serious restriction to the model specification and is often not justified. In particular, the marginal effect of an explanatory variable might depend on its value as well as those of other explanatory variables, thus making the regression relationship have a nonlinear functional form. Sections 11.2 and 11.3 address the issues that arise in modeling these nonlinear effects.

In Section 11.4 we explore the consequences of two types of errors, that of omitting important variables and that of including redundant variables in the model specification. The last section of the chapter deals with **multicollinearity**, the problems created by the high correlation among explanatory variables.

11.1 DUMMY (OR BINARY) INDEPENDENT VARIABLES

The effect of qualitative variables on the dependent variable is measured by creating **binary** (or **dummy**) variables that are numerical and correspond to qualitative characteristics. To illustrate the idea behind this, consider the two-variable simple regression model $Y = \alpha + \beta X + u$, where Y is the annual earnings of an employee in a particular firm and X is the number of years of experience the employee has. For simplicity of exposition we ignore, for the present, the effects of other variables such as the age of the employee or the number of years of education. In the simple model, β is the marginal effect of an extra year of experience and α is the intercept term. Suppose we believe that the above relationship is different between male and female employees. How can one estimate the earnings differential attributable to gender? A simple way to handle this is to define a new variable D (called a **dummy variable**) that takes the value 1 if the employee is female and 0 if the employee is male. The group for which D takes the value 0 (male in our example) is known as the **control group**. Next assume the following.

$$\alpha = \alpha_0 + \alpha_1 D \qquad \text{and} \qquad \beta = \beta_0 + \beta_1 D$$

Substituting these in the original model, we obtain the following revised model.

$$\begin{aligned} Y &= \alpha_0 + \alpha_1 D + (\beta_0 + \beta_1 D) X + u \\ &= \alpha_0 + \alpha_1 D + \beta_0 X + \beta_1 D X + u \end{aligned}$$

Note that the data for D will be a column of 1's and 0's (and hence the variable is binary) with the 1's corresponding to females. To estimate this model simply generate a new variable $Z = D \times X$ and regress Y against a constant, D, X, and Z. The estimated relationships for the two groups will be

$$\text{Male:} \qquad \hat{Y} = \hat{\alpha}_0 + \hat{\beta}_0 X$$

$$\text{Female:} \qquad \hat{Y} = (\hat{\alpha}_0 + \hat{\alpha}_1) + (\hat{\beta}_0 + \hat{\beta}_1) X$$

Suppose we find that $\hat{\alpha}_0$ and $\hat{\beta}_0$ are both positive and that $\hat{\alpha}_1$ and $\hat{\beta}_1$ are both negative. Then the estimated relationship will be as in Figure 11.1. We note that $\hat{\alpha}_1$ "shifts" the intercept and $\hat{\beta}_1$ shifts the slope of the regression. They measure the differentials with respect to the control group.

It is easy to test whether a significant gender differential exists. To test whether α_1 is zero or β_1 is zero, we simply carry out the standard t-test of Section 10.5. The test for the null hypothesis $\alpha_1 = \beta_1 = 0$ is the standard Wald test or a Lagrange multiplier test for the addition of the variables D and DX.

The choice of the control group for defining the dummy variable is arbitrary. It is easy to derive the implications with a different control group and examine what changes and what does not (see Practice Problem 11.1).

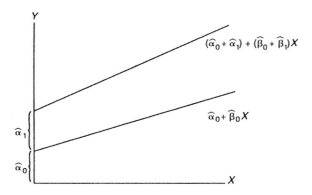

Figure 11.1   An example of a shift in the intercept and slope

Practice Problem

11.1   Suppose we had used a different dummy variable D^* (instead of D) that takes the value 1 for male and 0 for female. Show that, in terms of the F-test and R^2 measures, the two models are equivalent. How are the coefficients, standard errors, and t–statistics affected by this change? Examine the consequences of using a dummy variable that takes the value 10 for males and 20 for females.

Extension to the Case of Many Explanatory Variables

The above procedure is easily extended to the multiple regression case with many explanatory variables and dummy variables. For instance, suppose the earnings of an employee depend on the number of years of experience (now denoted by X_2), the age (X_3), the number of years of education (X_4), and whether or not the employee is nonwhite (denoted by W which takes the value 1 for "white" employees and 0 for all others). Then the basic restricted model will be

(R) $Y = \alpha + \beta X_2 + \gamma X_3 + \delta X_4 + u$

and the effect of the qualitative variables gender and race can be captured by assuming the following relationships.

$$\alpha = \alpha_0 + \alpha_1 D + \alpha_2 W$$
$$\beta = \beta_0 + \beta_1 D + \beta_2 W$$

$$\gamma = \gamma_0 + \gamma_1 D + \gamma_2 W$$

$$\delta = \delta_0 + \delta_1 D + \delta_2 W$$

When these are substituted into Model R, we get the unrestricted model

(U) $Y = \alpha_0 + \alpha_1 D + \alpha_2 W + (\beta_0 + \beta_1 D + \beta_2 W) X_2 +$

$(\gamma_0 + \gamma_1 D + \gamma_2 W) X_3 + (\delta_0 + \delta_1 D + \delta_2 W) X_4 + u$

This reduces to

(U) $Y = \alpha_0 + \alpha_1 D + \alpha_2 W + \beta_0 X_2 + \beta_1 D X_2 + \beta_2 W X_2 + \gamma_0 X_3$

$+ \gamma_1 D X_3 + \gamma_2 W X_3 + \delta_0 X_4 + \delta_1 D X_4 + \delta_2 W X_4 + u$

The procedure for estimating Model U is quite straightforward. Simply generate the new variables (known as **interaction terms**)

$$Z_1 = D X_2 \qquad\qquad Z_2 = W X_2 \qquad\qquad Z_3 = D X_3$$

$$Z_4 = W X_3 \qquad\qquad Z_5 = D X_4 \qquad\qquad Z_6 = W X_4$$

and regress Y against a constant, D, W, X_2, Z_1, Z_2, X_3, Z_3, Z_4, X_4, Z_5, and Z_6.

The null hypothesis $\alpha_1 = \alpha_2 = \beta_1 = \beta_2 = \gamma_1 = \gamma_2 = \delta_1 = \delta_2 = 0$ implies that there are no gender or racial differences in the relationship between earnings and age, experience, and education. The test is the general Wald F-test for which the test statistic is calculated as in Section 10.5. The d.f. is 8 for the numerator (which is the number of restrictions in the null hypothesis) and $T - 12$ for the denominator. To carry out this test using the LM test, we would first estimate Model R and save its residuals $\hat{u}_t$. Next obtain the auxiliary regression by regressing $\hat{u}_t$ against all the variables in Model U and compute the unadjusted R^2. Under the null hypotheses, TR^2 has the large sample χ^2 distribution with 8 d.f. and this can be used to perform a χ^2 test.

The Wald or LM test can be used to test other null hypotheses also. For example, $\alpha_2 = \beta_2 = \gamma_2 = \delta_2 = 0$ will test the hypothesis that there are no racial differences in the earnings profile.

In the above illustration we let the intercept and all the slope terms be different for males/females and white/nonwhite. This approach obviously adds numerous new terms to the model, thus reducing the d.f. and worsening the precision of estimates and the powers of tests. To avoid the proliferation of additional variables, the common procedure among researchers is to shift only the intercept terms. In the above example this would mean that the Z's will not be present in the unrestricted model. Whether or not it is justified to omit the Z's from the prior specification depends on the problem at hand. A better understanding of the underlying relationship would be helpful in the initial specification. Alternatively, one could specify the full model with all the shifts and perform diagnostic tests (LM and Wald tests) to examine which shifts are statistically significant. In the case of the wage profile example we have been using, it does make sense to shift all the slope coefficients. The interpretation for the significance of the coefficient for Z_1 is that the *additional* earnings for one *additional* year of experience is different, on average, for males and females. The interpretation for the other interaction terms is similar.

Models of the type considered here that mix qualitative and quantitative variables (especially where the slope shifts are due to qualitative variables) are known as **analysis of covariance** models. In agricultural experiments one frequently encounters models in which all the explanatory variables are binary. Such models are known as **analysis of variance** models.

Example 11.1

Table 11.1 has data on the sale price and characteristics of 57 single family homes sold in 1990 in a semirural area of San Diego County. It will be noted that the data include several dummy variables. The variables are as defined below.

Y   =   Sale price of the home in thousands of dollars
X2   =   Lot size in acres
X3   =   Size of the master bedroom in square feet
X4   =   Age of the house in years
X5   =   Number of bedrooms
X6   =   Number of bathrooms
X7   =   1 if the house has a nice view, 0 otherwise
X8   =   1 if the house has a swimming pool
X9   =   1 if horses are permitted to be kept on the property
X10   =   1 if the house has a spa
X11   =   1 if the property uses septic system for waste
X12   =   1 if farm animals are permitted on the property

Table 11.1   Data on Sale Price and Characteristics of Single Family Homes in San Diego County

Y	X2	X3	X4	X5	X6	X7	X8	X9	X10	X11	X12
179	0.25	143	50	2	2	0	0	0	0	0	0
200	0.25	158.75	12	3	2	0	0	0	1	0	0
200	0.25	168	4	3	2	1	0	0	0	0	0
205	0.25	162.84	40	2	1	0	0	0	0	0	0
212	0.25	190.4	2.5	3	2	0	0	0	0	0	0
215	0.25	140	5	4	2	1	0	0	0	0	0
210	0.25	175	5	3	2	1	0	0	0	0	0
222	0.25	225	7	3	2.5	1	0	0	0	0	0
225	0.25	149.16	14	2	2	1	0	0	0	1	0
239	0.25	189	16	3	2	1	0	0	0	0	0
240	0.25	226.8	3	2	1.5	1	0	0	0	0	0
259	0.25	204	4	3	2	1	0	0	0	0	0
255	0.25	216	0	3	2.5	1	0	0	0	0	0
250	0.25	180.18	20	3	2.5	1	0	0	0	0	0
279	0.25	182	0	3	2	1	0	0	0	0	0
279	0.25	182	0	3	2	1	0	0	0	0	0
286.5	0.25	204.4	0	3	2.5	0	0	0	0	0	0
290.5	0.25	269.136	0	3	2.5	0	0	0	0	0	0
270	2	144.16	73	3	2	1	0	1	0	1	0
285	0.25	182	0	3	2	1	0	0	0	0	0
304.9	0.45	192	0	4	2.5	1	0	0	0	0	0
300	1	227.8	0	3	2	0	0	1	0	1	1
306.5	0.45	192	0	4	2.5	1	0	0	0	0	0
335	1	146.16	21	3	2	1	0	1	0	1	1
324.9	0.45	216	0	4	3	1	0	0	0	0	0
305	0.45	210	15	3	2.5	1	1	1	0	0	0
329	0.45	216	0	3	2.5	1	0	0	0	0	0
305	1.19	282.75	11	4	2.5	1	0	1	1	1	1
330	0.45	357	16	3	2.5	1	1	1	1	1	1
320	0.5	360	9	3	2	1	0	0	0	0	0
325	1.1	151.8	26	3	2.5	0	0	1	0	1	1
335	0.5	225	5	3	2.5	0	0	0	0	0	0
340	0.45	216	1	4	3	1	0	0	0	0	0
360	1	196	2	3	2.5	0	0	0	0	1	0
372	0.45	233.6	2	3	2	1	0	1	0	0	0
377.5	1	234	0	3	2.5	1	0	1	0	1	1
399.5	1	420	0	4	2.5	1	0	1	0	1	1
425	4.3	240	0	4	3	1	0	1	0	1	0
450	1	185.6	15	4	2.5	1	1	1	1	1	1
465	5	336	9	3	2	1	1	1	0	1	0
450	15.5	168	40	3	3	1	0	1	0	1	0
500	5.3	228.8	5	1	2	1	0	1	0	1	0
540	0.75	266	0	4	3	1	0	0	0	0	0
529.5	1.7	252	0	4	2.5	1	0	1	0	1	1

(Continued)

Table 11.1 (Continued)

Y	X2	X3	X4	X5	X6	X7	X8	X9	X10	X11	X12
550	2.92	327.04	13	4	2	1	0	1	0	1	0
542	0.5	285	13	3	2	1	0	0	1	0	1
565	5.48	224	13	3	2	0	0	1	0	1	0
545	4.75	195	20	4	2	1	1	1	0	1	0
605	7.63	227.84	11	5	3	1	1	1	1	1	0
670	16.09	315	11	4	3.5	1	0	1	0	1	0
749	5	205.2	10	4	3	1	1	1	1	1	0
725	19	204	12	3	2.5	1	0	1	0	1	0
740	16.6	251.086	10	4	3	1	0	1	1	1	0
895	5	379	2	4	3	1	0	1	0	1	0
1320	11.09	506	2	3	3	1	1	1	1	1	0
1400	19.67	420	8	5	5.5	1	1	1	1	1	0
2285	79	306	3	4	4.5	1	1	1	0	1	0

Table 11.2 presents the estimated coefficients and associated statistics for a number of models starting from the "kitchen sink" specification that includes all the explanatory variables. Because there is no reason to believe that any of the dummy variables should affect the *marginal effect* of the lot size, number of baths, and so on, it is not meaningful to shift the slope terms here. We started with Model A, which has all the explanatory variables, and eliminated variables with insignificant coefficients a few at a time until we obtained a model in which all the coefficients (ignoring the constant term) are significant at 10 percent or lower levels. We note that AIC and other model selection criteria would choose Model E over all the others.

The models explain 90 percent of the variation in the price of homes. The price is significantly affected by the size of the lot the house is situated on, the size of the master bedroom, the number of bathrooms, and whether or not the property has a swimming pool. Other variables such as age, number of bedrooms, and so on do not significantly explain the variation in prices.

Practice Problem

11.2 Use a regression program to verify the entries in Table 11.1. Also carry out relevant Wald and LM tests.

Table 11.2 Estimated Models for Home Price Data[†]

Variable	Model A	Model B	Model C	Model D	Model E
X1 (Constant)	−64.455 (0.52)	−72.129 (0.38)	−108.310 (0.13)	−131.300 (0.07)	−129.528 (0.07)
X2 (LOTSIZE)	20.381 (< 0.0001)	20.399 (< 0.0001)	20.299 (< 0.0001)	20.175 (< 0.0001)	20.514 (< 0.0001)
X3 (MASTER)	1.070 (< 0.0001)	1.074 (< 0.0001)	1.149 (< 0.0001)	1.155 (< 0.0001)	1.201 (< 0.0001)
X4 (AGE)	−1.192 (0.36)	−1.104 (0.37)			
X5 (BEDRMS)	0.839 (0.98)				
X6 (BATHS)	59.679 (0.13)	59.644 (0.06)	65.426 (0.04)	75.857 (0.02)	76.735 (0.01)
X7 (VIEW)	−13.11 (0.75)				
X8 (POOL)	71.031 (0.19)	73.496 (0.14)	71.359 (0.15)	101.629 (0.03)	111.884 (0.01)
X9 (HORSE)	18.021 (0.78)				
X10 (SPA)	67.231 (0.19)	65.824 (0.18)	60.521 (0.21)		
X11 (SEPTIC)	49.902 (0.41)	62.259 (0.11)	51.322 (0.16)	36.509 (0.27)	
X12 (FARM)	−61.122 (0.20)	−57.853 (0.20)	−53.043 (0.24)		
$\bar{R}^2$	0.898	0.904	0.904	0.903	0.903
ESS	555,785	557,759	567,376	593,389	609,952
SGMASQ	12,351	11,620	11,579*	11,674	11,730
AIC	14,856	13,419	13,180	12,893	12,753*
FPE	14,951	13,455	13,204	12,903	12,759*
HQ	17,559	15,211	14,733	14,017	13,673*
SCHWARZ	22,840	18,527	17,556	15,987	15,256*
SHIBATA	13,856	12,875	12,748	12,644	12,578*
GCV	15,644	13,799	13,470	13,048	12,858*
RICE	16,842	14,302	13,838	13,231	12,978*

[†] Values in parentheses are *p*-values for two-tailed tests. An asterisk indicates the model that is best for that criterion (see Section 10.6).

Extension to the Case of Several Categories

The dummy variables D and W used in the previous section both involved only two categories, namely, male or female and white or nonwhite. In many situations a qualitative variable might involve several categories. For example, it is reasonable to expect the relationship between earnings and their determinants to be different between a clerical worker and a professional employee. Thus, the qualitative variable *occupational status* may be important in explaining differences in earnings patterns. Because employees can be classified into many occupational categories, the previous approach is not readily applicable here and needs to be modified. To incorporate the effect of occupational status on earnings, first we have to identify the various employment categories. Suppose they are, *unskilled workers, clerical, technical,* and *managerial.* The procedure here is to choose one of the groups as control (say unskilled workers) and define three dummy variables (in general, one less than the number of categories); $D_1 = 1$ if the employee is a clerical worker and 0 otherwise, $D_2 = 1$ if the employee is a technician and 0 otherwise, and $D_3 = 1$ if the employee is a manager. If we had defined a fourth dummy variable $D_4 = 1$ for unskilled workers and 0 for others, then $D_1 + D_2 + D_3 + D_4 = 1$, which is the constant term and hence there is an exact linear relationship among the X's. This would have made the rank of X less than k and hence $X'X$ would have been singular, thus violating Assumption 10.3 which was important for solving the normal equations. The common procedure to avoid the above **dummy variable trap** is to choose one of the categories as control (unskilled workers in our example) and to define a dummy variable for each of the other categories. If we specify that only the intercept is shifted by occupational status, we would assume that (ignoring race and gender for simplicity)

$$\alpha = \alpha_0 + \alpha_1 D_1 + \alpha_2 D_2 + \alpha_3 D_3$$

The α's can be estimated by regressing Y against a constant, D_1, D_2, D_3, and all the other variables in the model. To shift the slopes, we would make a similar assumption about β, γ, and δ and generate interaction terms such as $D_1 X_2$ and so on.

Practice Problem

11.3   Show that the above model, in which D_4 is excluded from the specification, is equivalent to another model in which D_4 is included but the constant term is excluded.

Modeling Seasonal Effects Using Dummy Variables

The dummy variable approach can be used to incorporate differences in relationships due to seasonality. To illustrate, consider the two-variable regression model $Y_t = \alpha + \beta X_t + u_t$ in which Y_t is a company's total profits at time t and X_t is its aggregate sales. If we estimate the model using quarterly data, we might suspect that the relationship will be different between summer and winter quarters. Thus, the qualitative variable *season* would be an important determinant. Using the principle suggested earlier, the modeling here is quite straightforward. Simply choose one of the quarters as control (say the fall) and define three dummy variables; WINTER = 1 for the winter quarter and 0 otherwise, SPRING = 1 for spring only, and SUMMER = 1 for summer only. Next we let

$$\alpha = \alpha_0 + \alpha_1 \text{WINTER} + \alpha_2 \text{SPRING} + \alpha_3 \text{SUMMER}$$

$$\beta = \beta_0 + \beta_1 \text{WINTER} + \beta_2 \text{SPRING} + \beta_3 \text{SUMMER}$$

The unrestricted model is given below.

$$Y = \alpha_0 + \alpha_1 \text{WINTER} + \alpha_2 \text{SPRING} + \alpha_3 \text{SUMMER} + \beta_0 X_t$$
$$+ \beta_1 \text{WINTER} X_t + \beta_2 \text{SPRING} X_t + \beta_3 \text{SUMMER} X_t + u_t$$

The rest of the analysis is identical in principle to that used earlier for the earnings profile example. The extension to the case of several X's is also straightforward. If the data were monthly instead of quarterly, we would choose one of the months as control and define 11 monthly dummy variables. Note that α_i and β_i $(i = 1, 2, 3)$ are interpreted as differences from the control group (fall quarter here).

Tests for Structural Change

When a regression model is specified, the assumption is that it represents a "statistical average" relationship between the dependent variable and the explanatory variables. With time series data, however, this relationship might not be constant over the sample period but might have undergone a **structural change**, that is, the relationship might have changed from one period to another. For instance, consider the demand for cigarettes and its determinants. Since the original U.S. Surgeon General's report linking cigarette smoking to lung cancer and other respiratory diseases was issued, the pattern of demand might have changed with a decreased demand at all income and price levels. The test to see whether there has been a significant shift in the demand can be carried out in two ways. One method proposed by Chow (1960) splits the sample into two or more parts and computes a Wald test statistic. The second method makes use of dummy variables. In this section we describe both approaches.

Test Based on Dividing the Sample (Chow test)

Suppose the null hypothesis is that there has been no change in the structure, and the alternative is that there has been a change after the time period T_1. The unrestricted model will assume that the relationship is different in the two periods $1 - T_1$ and $T_1 + 1$ through T, and the restricted model will assume that the coefficients are the same in the two periods. The first step is to estimate the model separately for each of the periods and compute the error sums of squares ESS_1 and ESS_2. The restricted error sum of squares ESS_R is then obtained by pooling the samples from both periods and estimating a single common model. The test statistic is the familiar Wald F-statistic

$$F_c = \frac{(ESS_R - ESS_1 - ESS_2) \div k}{(ESS_1 + ESS_2) \div (T - 2k)}$$

The procedure is to reject the null hypothesis if F_c exceeds F^*, the point on the F-distribution with k and $T - 2k$ d.f. such that the area to the right is equal to the level of significance.

Test Based on Dummy Variables

The test can also be carried out using the dummy variable approach. First define a dummy variable D that takes the value 1 in the second period and 0 in the first period. Using D generate all the interaction terms and estimate a complete unrestricted model in which the intercept and all the slope terms depend on D. We can then impose the restriction that the regression coefficients involving D are all zero and compute the standard Wald F-statistic. The extension of this method for the case in which the structure is different for three or more periods is straightforward.

11.2   ALTERNATIVE FUNCTIONAL FORMS

As mentioned earlier, the assumption that the relationship between the dependent and independent variables is linear is often unrealistic. Nonlinearities in relations usually occur in one of two forms. Some nonlinear models can be transformed into a form that can be estimated by OLS using the linear regression framework. These intrinsically linear formulations are discussed in the present section. The next section deals with models that are not amenable to conversion to the linear regression specification.

The Linear-log Model

In the simple regression model $Y = \alpha + \beta X + u$, suppose Y represents the harvest of wheat and X is the number of acres of wheat cultivated. The slope coefficient β is the marginal product of one acre cultivated and is a constant. In reality, however, one might expect a **diminishing marginal product** so that as X increases, the *additional* output might become smaller and smaller as less fertile land is brought under cultivation. In other words, β might actually decrease as X increases. The following model explicitly incorporates this idea of a diminishing marginal product.

$$Y = \beta_1 + \beta_2 \ln X + u$$

In this specification, $\partial Y / \partial X = \beta_2 / X$, which is not constant but decreases as X increases, provided $\beta_2 > 0$. The model is easily estimated by simply generating a new variable $Z = \ln X$ and regressing Y against a constant and Z. Because the model involves Y linearly and X in logarithmic form, it is known as a **linear-log model**. In the home price example, one might expect that the marginal effects of additional bedrooms, baths, lot size, and so on, are also diminishing. Thus a linear-log specification would be appropriate in that case also. Because the *model is linear in the parameters*, the specification fits nicely in the multiple regression framework.

The Log-linear Model

In some situations, the dependent variable might be in logarithmic form while the independent variables are in linear form. This is the **log-linear model**

$$\ln Y = \beta_1 + \beta_2 X_2 + \beta_3 X_3 + \cdots + u$$

In the earnings profile example, suppose Y_t is the earnings at time t and r is the rate of return to one additional year of schooling. Then $Y_2 = Y_1(1 + r)$ from which it follows by repeated substitution that $Y_S = Y_0(1 + r)^S$, where S is the number of years of schooling. Taking logarithms of both sides, we get the following model which has the structure of a log-linear model.

$$\ln Y_S \;=\; \ln Y_0 \,+\, S \ln (1{+}r) \;=\; \alpha \,+\, \beta S$$

It should be pointed out that it is not proper to compare the goodness of fit measures $\bar{R}^2$ between linear and log-linear models because their dependent variables are different. In the former, the measure explains the variation in Y, whereas in the latter model it explains the variation in $\ln(Y)$. A heuristic way of comparing the models would be to use the log-linear model to obtain the predicted value ($\hat{\ln Y}$), convert it to $\hat{Y}$, compute the new error sum of squares, and then proceed to calculate the model selection statistics based on this transformed prediction. Since the two predicted values measure the same thing, the selection statistics are comparable.

Practice Problem

11.4   Describe how you would use the log-linear model to fit an exponential time path for the population of a region. More specifically, let $P_t = P_0 e^{gt}$, where t is time in years and g is the rate of growth. Explain how g can be estimated from time series data on population.

The Double-log Model

This model combines the features of both the linear-log and log-linear models and is the preferred specification to estimate production functions as well as demand relations. To illustrate, consider the **Cobb-Douglas production function**

$$Q \;=\; Q_0 \, K^\alpha L^\beta$$

in which Q is the output of a production process, K represents capital (machine-hours), and L stands for labor (worker-hours). This non-linear relation is readily converted to a linear regression framework by taking logarithms of both sides and adding a stochastic error term.

$$\ln Q \;=\; \ln Q_0 \,+\, \alpha \ln K \,+\, \beta \ln L \,+\, u$$

The estimation procedure is to convert Q, K, and L into logarithms and use them in a linear regression framework. A similar specification is used in modeling a demand relation such as the following.

$$\ln Q \;=\; \beta_1 \,+\, \beta_2 \ln Y \,+\, \beta_3 \ln P \,+\, u$$

where Q is the quantity of the commodity demanded, Y is the income, and P is the price. The double-log model is often preferred because it has an interesting feature. Note that

$$\beta_2 = \frac{\partial \ln Q}{\partial \ln Y} = \frac{Y}{Q}\frac{\partial Q}{\partial Y}$$

which is the well-known **elasticity** of Q with respect to Y, that is, it is the **income elasticity**. Similarly, β_3 is the **price elasticity**. The double-log model has the important property that the regression coefficients have the interpretation of being the corresponding elasticities that are all constant.

Practice Problem

11.5 Derive expressions for the elasticity of Y with respect to X in linear, log-linear, and linear-log models.

Polynomial Curve Fitting

A simple extension of the linear specification is to include higher powers of dependent variables as in the following specification.

$$Y = \beta_1 + \beta_2 X + \beta_3 X^2 + \cdots + \beta_{k+1} X^k + u$$

The estimation procedure consists of generating the new variables X^2, X^3, and so on, and then regressing Y against a constant, X, and all these newly created variables. Quadratic formulations (with $k = 2$) are frequently used to model U-shaped relations (such as average cost functions). In general, polynomials of order greater than 2 are, however, not recommended. An obvious reason is that each term adds a parameter to be estimated, which in turn decreases the number of degrees of freedom. A stronger reason will be seen in Section 11.5 in which we discuss problems that highly correlated independent variables can cause.

Practice Problem

11.6 Suppose the basic formulation is the simple linear model $Y = \alpha + \beta X + u$. Describe step by step how the LM test could be used to test whether higher order polynomial terms should be added to the specification.

Interaction Terms

When discussing dummy variables, we used interaction terms in which the slope term of a model depended on dummy variables. This principle can be used in more general contexts also. To illustrate, let Y_t be the amount of electricity a household uses at time t and T_t be the temperature at that time. A simple linear relationship will give $Y_t = \alpha + \beta T_t + u_t$. Suppose that the marginal effect of temperature (β) is not constant but depends on the price of electricity (P_t). If electricity is expensive, then consumers might be willing to postpone turning on a fan or an air conditioner in order to save electricity. This means that β might decrease when P_t increases. A simple way to test for this is to assume that $\beta = \beta_0 + \beta_1 P_t$, which gives the following model.

$$Y_t = \alpha + \beta_0 T_t + \beta_1 (P_t T_t) + u_t$$

The term $P_t T_t$ is the interaction term and its coefficient measures the effect of price on the sensitivity of electricity demand to the temperature.

The Logit Model

When the dependent variable Y refers to a proportion so that the observed values are between 0 and 1, a standard regression model might result in unacceptable predictions. More specifically, suppose $Y_t = \beta' x_t + u_t$, where β is the vector of regression coefficients and x_t is the vector of observations on the independent variables. Given $x_t = x_0$, the predicted dependent variable is $\hat{Y}_0 = \hat{\beta}' x_0$. There is no assurance, however, that the predicted value will be between 0 and 1. In such a case, an often preferred specification is the **logit model** which is formulated as follows.

$$\ln\left[\frac{Y_t}{1 - Y_t}\right] = \beta' x_t + u_t$$

The estimation of the logit model is straightforward because we simply create a new dependent variable using the left-hand side and then regress it against the X's. Solving the above equation for Y_t, we can obtain the predicted dependent variable as

$$\hat{Y}_0 = \frac{1}{1 + e^{-\hat{\beta}' x_0}}$$

It is readily verified that $\hat{Y}_0$ will always lie between 0 and 1.

Practice Problem

11.7 Derive the marginal effect of x_0 on $\hat{Y}_0$ assuming for simplicity that β is a scalar.

11.3 NONLINEARITIES IN PARAMETERS

Although a variety of nonlinear specifications can be converted to a linear regression framework, there are cases where such a transformation may not be feasible. Consider, for example, the following model.

$$Y = \beta_1 + \beta_2 X^{\beta_3} + u$$

It is quite evident that there exists no transformation that would convert the above into a linear regression model in which Y or some transformation of it is expressed as a linear function of unknown parameters. Models that exhibit such nonlinearities in parameters require more complicated methods of estimation. In this section we present an introduction to such nonlinear estimation. Excellent references for this topic are Greene (1990), Judge *et al.* (1985), Amemiya (1985), Kmenta (1986), and Goldfeld and Quandt (1972).

The class of nonlinear models that is considered here is the following.

$$Y_t = g(x_t, \beta) + u_t \qquad \text{or} \qquad y = g(X, \beta) + u$$

where x_t is the $k \times 1$ vector of the tth observation on the X's and β is the $k \times 1$ vector of unknown coefficients. A linear approximation to the function g in the neighborhood of β^0 is given by (for simplicity, we have ignored the t-subscript)

$$g(x, \beta) \approx g(x, \beta^0) + \sum_{i=1}^{i=k} \left[\frac{\partial g}{\partial \beta_i} \right]_{\beta = \beta^0} (\beta_i - \beta_i^0)$$

For a given β^0, $g(x_t, \beta^0)$ and $\partial g / \partial \beta_i$ are known and hence we can rewrite the model as follows.

$$y - g(x, \beta^0) + \sum_{i=1}^{i=k} \left[\frac{\partial g}{\partial \beta_i} \right]_{\beta = \beta^0} \beta_i^0 = \sum_{i=1}^{i=k} \left[\frac{\partial g}{\partial \beta_i} \right]_{\beta = \beta^0} \beta_i + u$$

Here also the least squares procedure minimizes the error sum of squares $\Sigma \hat{u}_t^2$ with respect to β. First fix a value β^0 for the vector

β, evaluate at β^0 the partial derivatives of g with respect to the parameters, compute the transformed left-hand side variable in the above equation, and run that regression to obtain a second round estimate for β. This procedure can be iterated, in principle, until some convergence criterion is met. Although this procedure has generally been successful, sometimes convergence might be a problem because the initial choice of β^0 might be far away from the solution. Unfortunately, there is no rule or formula that would guide one to a better initial estimate. Trial and error with different starting values might lead to convergence and is often the method adopted.

Example 11.2

For the nonlinear model

$$Y = \beta_1 + \beta_2 X^{\beta_3} + u$$

we have

$$\frac{\partial g}{\partial \beta_1} = 1$$

$$\frac{\partial g}{\partial \beta_2} = X^{\beta_3}$$

$$\frac{\partial g}{\partial \beta_3} = \beta_2 X^{\beta_3} \ln X$$

The regression to run for each iteration is the following.

$$Y + \beta_3^0 \beta_2^0 X^{\beta_3^0} \ln X = \beta_1 + \beta_2 X^{\beta_3^0} + \beta_3 \beta_2^0 X^{\beta_3^0} \ln X + u$$

The CES and Translog Production Functions

A popular nonlinear specification is the **constant elasticity of substitution (CES)** production function which has the following form.

$$Y_t = \gamma [\delta K_t^{-\rho} + (1 - \delta) L_t^{-\rho}]^{-v/\rho} e^{u_t}$$

$$(\gamma > 0; \ 0 < \delta < 1; \ v > 0; \ \rho \geq 1)$$

Taking logarithms of both sides of the production function

$$\ln (Y_t) = \ln \gamma - \frac{v}{\rho} \ln [\delta K_t^{-\rho} + (1 - \delta) L_t^{-\rho}] + u_t$$

It is readily noted that this has the nonlinear structure specified earlier. The estimates of the parameters may be obtained by the nonlinear least squares procedure. However, by using a linear approximation around $\rho = 0$, we can obtain a simpler formulation that can be estimated directly by OLS. Kmenta (1986, p. 515) has derived the following quadratic approximation.

$$\begin{aligned}
\ln Y_t &= \ln \gamma + v\delta \ln K_t + v(1-\delta) \ln L_t \\
&\quad - \tfrac{1}{2}\rho v\delta(1-\delta)\,[\ln K_t - \ln L_t]^2 + u_t \\
&= \beta_1 + \beta_2 \ln K_t + \beta_3 \ln L_t + \beta_4[\ln K_t - \ln L_t]^2 + u_t
\end{aligned}$$

The term involving β_4 is the difference between the CES production function and the Cobb-Douglas production function encountered earlier. We can test that coefficient for significance using the standard t-test. The parameters of the original specification are easily derived in terms of the β's.

Practice Problem

11.8 Use a Taylor-series expansion of $\ln Y_t$ to verify the quadratic approximation. Also show that

$$\gamma = e^{\beta_1} \qquad\qquad \delta = \frac{\beta_2}{\beta_2 + \beta_3}$$

$$v = \beta_2 + \beta_3 \qquad\qquad \rho = \frac{-2\beta_4(\beta_2 + \beta_3)}{\beta_2\beta_3}$$

It has been noted in the CES specification that the term for β_4 is quadratic. A more general specification would be to expand that expression and assume that each of the terms has a different regression coefficient. This results in the following model, known as the **transcendental logarithmic (or translog) production function.**

$$\begin{aligned}
\ln Y_t &= \beta_1 + \beta_2 \ln K_t + \beta_3 \ln L_t + \beta_4(\ln K_t)^2 \\
&\quad + \beta_5(\ln L_t)^2 + \beta_6(\ln K_t)(\ln L_t) + u_t
\end{aligned}$$

The CES production function can be used to carry out a Lagrange multiplier test that the elasticity of substitution is equal to 1. In this production function, $1/(1-\rho)$ is the constant elasticity of substitution [see Ramanathan (1982), p. 15]. If $\rho = 0$, the production function reduces to the Cobb-Douglas form. An interesting test of hypothesis is therefore that $\rho = 0$. This can be carried out by an LM test which is based on testing whether the score $S = \partial \ln L / \partial \rho$ is zero. It can be shown [see Engle (1984)] that the score term is the same as the quadratic term in the Kmenta approximation. Therefore, the LM test for unitary elasticity of substitution is the same as a t–test on the quadratic term and hence is easy to carry out.

The Box-Cox Model

Another generalization of the linear specification makes use of the **Box-Cox transformation** given below.

$$z^{(\lambda)} = \frac{z^\lambda - 1}{\lambda}$$

The Box-Cox model is specified as

$$g(Y) = \alpha + \sum_{i=1}^{i=k} \beta_i X_i^{(\lambda)} + u$$

The dependent variable could be linear or logarithmic in Y. Since λ is generally unknown, the regression is nonlinear and can be estimated accordingly. A common procedure, however, is to fix a value of λ between -1 and $+1$, perform the Box-Cox transformation on each X, and estimate the resulting linear regression model. By searching over the interval $(-1, +1)$ for the value of λ that yields the smallest ESS, we can obtain an optimum value. It can be shown, however, that the standard errors will be underestimates (see Fomby *et al.*, pp. 426-431). The appropriate method is to estimate the asymptotic covariance matrix of the parameters using the approximation formula of Section 5.10.

If $\lambda = 1$, then the Box-Cox model becomes a linear regression model if $g(Y) = Y$ and log-linear if $g(Y) = \ln Y$. Using L'Hospital's rule we can show that for $\lambda = 0$, $z^{(\lambda)} = \ln z$. In this case, the Box-Cox model reduces to a linear-log or a double-log model.

A further variation of the Box-Cox model is to assume that $g(Y) = Y^{(\lambda)}$. For more details about this and other variations see Kmenta (1986) or Greene (1990).

Practice Problem

11.9 Derive an expression for the elasticity of Y with respect to X in the Box-Cox model $g(Y) = \alpha + \beta X^{(\lambda)} + u$. Examine the special cases, $g(Y) = Y$ and $g(Y) = \ln Y$.

11.4 SPECIFICATION ERRORS

The implicit presumption in Assumption 10.1 $(y = X\beta + u)$ is that the econometric model specified there is correct. A natural question to ask is "what are the consequences of misspecification?" In particular, suppose we omit an important variable from the model (for example, price in a demand equation). Would the parameter estimates still be unbiased, consistent, efficient, and BLUE? In contrast, suppose we had included a redundant explanatory variable, that is, one which has no effect on the independent variable. Would this jeopardize the statistical properties of the remaining parameter estimates? In the present section we explore these two types of model misspecification. Specification errors are also possible in the stochastic error term u and are examined in the next chapter.

Exclusion of Relevant Variables

Suppose the "true" DGP is

$$y = X\beta + u = X_1\beta_1 + X_2\beta_2 + u$$

but we excluded the variables in X_2 (that is, erroneously assumed that $\beta_2 = 0$) and estimated the model

$$y = X_1\beta_1 + v$$

The OLS estimator for β_1 is

$$\hat{\beta}_1 = (X_1'X_1)^{-1}X_1'y$$

Substituting for y from the true model into the above expression we obtain,

$$\hat{\beta}_1 = (X_1'X_1)^{-1}X_1'(X_1\beta_1 + X_2\beta_2 + u)$$
$$= \beta_1 + (X_1'X_1)^{-1}X_1'X_2\beta_2 + (X_1'X_1)^{-1}X_1'u$$

The last term has zero expectation because of Assumptions 10.2 and 10.3 that the X's are given and that $E(u) = 0$. Therefore,

$$E(\hat{\beta}_1) = \beta_1 + (X_1'X_1)^{-1}X_1'X_2\beta_2$$

It follows that unless $X_1'X_2 = 0$, $\hat{\beta}_1$ will be biased and inconsistent. The columns of the matrix $(X_1'X_1)^{-1}X_1'X_2$ are the OLS coefficients obtained when each of the excluded variables is regressed against the set of included variables. If the model is expressed as deviations from the respective means, then $X_1'X_2/T$ is the matrix of covariances between the two sets of variables. Thus, for unbiased-ness and consistency each of the omitted variables must be uncorre-lated with each of the included variables. As this condition is too stringent to be satisfied in practice, the OLS estimators of β_1 will be biased (the bias is known as the **omitted variable bias**) and incon-sistent. Omitted variables therefore destroy the properties of unbiasedness, consistency, and BLUE. For standard errors and hypothesis tests we need the estimate of σ^2, and hence it is impor-tant to examine the consequences of omitted variables on the covari-ance matrix of $\hat{\beta}_1$ estimated using the wrong model. The estimated residuals are

$$\hat{v} = y - X_1\hat{\beta}_1 = y - X_1(X_1'X_1)^{-1}X_1'y = M_1y$$

where

$$M_1 = I - X_1(X_1'X_1)^{-1}X_1'$$

which is symmetric, idempotent, and has the property that $M_1X_1 = 0$. Using the true model we have

$$\hat{v} = M_1(X_1\beta_1 + X_2\beta_2 + u) = M_1X_2\beta_2 + M_1u$$

Therefore

$$\hat{v}'\hat{v} = \beta_2'X_2'M_1X_2\beta_2 + u'M_1u + 2\beta_2'X_2'M_1u$$

Because the third term has zero expectation,

$$E(\hat{v}'\hat{v}) = \beta_2'X_2'M_1X_2\beta_2 + \sigma^2 \text{ trace } (M_1)$$

$$= \beta_2'X_2'M_1X_2\beta_2 + (T - k_1)\sigma^2$$

For the second term, we have used the method adopted in Sec-tion 10.2 on the estimation of the residual variance. The estimated error variance is $s^2 = \hat{v}'\hat{v}/(T - k_1)$, where k_1 is the number of regression coefficients in the estimated model (also equal to the number of columns in X_1). It is clear from the above derivation that s^2 is a biased estimator of σ^2 and it overestimates the true popula-tion error variance. This implies that statistical inferences are faulty and the tests of hypotheses are invalid. Note that this conclusion holds even if $X_1'X_2 = 0$. The consequences of omitted variables are thus drastically serious and hence it is important to pay a great deal of attention to the appropriate specification of a model.

Inclusion of Irrelevant Variables

The fear of the omitted variable bias in estimators and their standard errors often leads an investigator to overparametrize the model and add independent variables to the model even though they may not have significant effects on the dependent variable. Here we examine the consequences of this type of misspecification. The "true" DGP is now

$$y = X_1 \beta_1 + u$$

and the estimated model is

$$y = X\beta + v = X_1 \beta_1 + X_2 \beta_2 + v$$

The estimator and its expectation are as given below.

$$\hat{\beta} = (X'X)^{-1} X'y$$

$$E(\hat{\beta}) = (X'X)^{-1} X'(X_1 \beta_1 + u) = (X'X)^{-1} X'X_1 \beta_1$$

The matrix X_1 can be written as follows.

$$X_1 = [X_1 \quad X_2] \begin{bmatrix} I_{k_1} \\ O \end{bmatrix} = X \begin{bmatrix} I_{k_1} \\ O \end{bmatrix}$$

where I_{k_1} is an identity matrix of order k_1 and O is a $k_2 \times k_1$ matrix of zeros. Therefore,

$$E(\hat{\beta}) = (X'X)^{-1} X'X \begin{bmatrix} I_{k_1} \\ O \end{bmatrix} \beta_1 = \begin{bmatrix} \beta_1 \\ 0 \end{bmatrix}$$

It follows that $\hat{\beta}_1$ is unbiased even though redundant variables (with zero coefficients) were included in the model.

$$\hat{v} = y - X\hat{\beta} = My$$

where $M = I - X(X'X)^{-1} X'$. Hence the computed error sum of squares is given by

$$\hat{v}'\hat{v} = y'My = (u' + \beta_1'X_1') M (X_1 \beta_1 + u)$$

$$= u'Mu + \beta_1'X_1' M X_1 \beta_1 + 2\beta_1'X_1'Mu$$

But $X_1'M = [I \ O] X'M = 0$, and hence $\hat{v}'\hat{v} = u'Mu$. It follows from this that

$$E(\hat{v}'\hat{v}) = E(u'Mu) = \sigma^2 \, \text{trace}(M) = \sigma^2(T - k)$$

Therefore, $s^2 = \hat{v}'\hat{v} / (T - k)$ is still unbiased, which means that, unlike the case of omitting important variables, the statistical inferences are valid here. One might conclude that it is better to overparametrize the model because we would still obtain unbiased estimators for the parameters as well as the error variance. However, here too there is a cost and that is a loss in efficiency. The true estimators are BLUE and hence the estimators obtained using the wrong model have reduced precision. Also, the tests of hypotheses will not be as powerful as those using the correct specification. In spite of these costs, investigators often **overfit** the model by adding variables whose effects they may not be very sure of, simply to avoid the bias and faulty inferences caused by excluding relevant variables in the model.

11.5   MULTICOLLINEARITY

One of the assumptions made on the multiple regression model $y = X\beta + u$ was that the rank of X was full and less than the number of observations so that the cross-product matrix $X'X$ is nonsingular. If this assumption were violated, then the normal equations $X'X\hat{\beta} = X'y$ cannot be solved for $\hat{\beta}$ uniquely. The violation can occur if there is an exact linear relationship between one set of regressors and another. This situation is known as **exact multicollinearity** or **perfect collinearity**. To see this, consider the partitioning done earlier so that the model is written as

$$y = X_1\beta_1 + X_2\beta_2 + u$$

Suppose the second set of variables has an exact linear relationship with respect to the first set so that $X_2 = X_1A$, where A is a $k_1 \times k_2$ matrix of known values. Then the cross-product matrix is

$$X'X = \begin{bmatrix} X_1'X_1 & X_1'X_2 \\ X_2'X_1 & X_2'X_2 \end{bmatrix} = \begin{bmatrix} X_1'X_1 & X_1'X_2 \\ A'X_1'X_1 & A'X_1'X_2 \end{bmatrix}$$

which implies linear dependence and the singularity of $X'X$ that renders it impossible to obtain unique estimators of β.

In a practical situation, exact multicollinearity is not likely to occur and if it does, the solution is easy; simply exclude the variables causing the exact collinearity. For example, in the dummy variable section of this chapter, we alluded to the *dummy variable trap* which arises when all the dummy variables are included along with a constant term. In such a case, the dummies add up to 1, which is the constant term, thus causing perfect collinearity. The solution was to exclude one of the dummies or the constant term from the model.

Near Multicollinearity

Although perfect collinearity is rare, **near multicollinearity** in which there is at least one approximately linear relationship among the independent variables is quite common, especially when time series are used. As an illustration, consider the following double-log model of the demand for new housing in the year t in a certain community.

$$\ln H_t \; = \; \beta_1 + \beta_2 \ln P_t + \beta_3 \ln Y_t + \beta_4 \ln r_t + u_t$$

where H_t is the number of new housing units sold, P_t is the area's population, Y_t is the community's aggregate income, and r_t is the mortgage interest rate. Although population and income are both sensible variables to include in the specification, they usually increase steadily over time and hence one would expect the two variables to be highly correlated, implying a nearly exact relationship between $\ln P_t$ and $\ln Y_t$. In this case, the normal equations can be solved uniquely but the interpretation of the coefficients might be questionable. To explore this further, consider the partitioned regression model in which $k_1 = 1$, so that a single variable (whose observations are denoted by the vector x_1) is nearly linearly related to the X_2 variables. Using the partitioned matrix inversion formula given in Property A.8, the leading element of $(X'X)^{-1}$ can be written as the scalar

$$[\, x_1'x_1 - x_1'X_2(X_2'X_2)^{-1}X_2'x_1\,]^{-1} \; = \; [x_1'M_2x_1]^{-1}$$

where

$$M_2 \; = \; I - X_2(X_2'X_2)^{-1}X_2'$$

The variance of $\hat{\beta}_1$ is now given by

$$\mathrm{Var}\,(\hat{\beta}_1) \; = \; \frac{\sigma^2}{x_1'\,M_2\,x_1}$$

It is easy to show that the denominator is simply the error sum of squares obtained when x_1 is regressed against X_2 (verify it). If the relationship between x_1 and X_2 is near perfect, this error sum of squares will be small and hence the variance of $\hat{\beta}_1$ will be large. Thus, an important implication of high multicollinearity is that *the estimated standard errors of individual regression coefficients will be high, and this might make the corresponding t-statistic small and insignificant.* Now we see why polynomial curve fitting is not recommended; in a polynomial relation, X^2, X^3, and so on are likely to be highly correlated, making individual coefficients possibly insignificant.

Example 11.3

Table 11.3 has annual data for the United States on the number of new housing units and their determinants during the 23 years from 1963 to 1985. Housing units started are in thousands, POP is U.S. population in millions, GNP is gross national product in constant 1982 dollars (in billions), and INTRATE is new home mortgage yields in percent.

Table 11.4 has four double-log models (the prefix L denotes that the variable is in logarithms) estimated with the data. In Model A all the variables are present and we note that the t–statistics for LPOP and LGNP are very small, implying that the corresponding regression coefficients are not significantly different from zero. If strictly interpreted, this would seem to suggest that population and the gross national product do not significantly affect the number of new housing starts, clearly an unacceptable conclusion. The F-statistic for the joint significance of the POP and GNP elasticities is given by (using Model B as the restricted model)

$$F_c = \frac{(0.982 - 0.560) \div 2}{0.560 \div 19} = 7.16$$

Under the null hypothesis that the regression coefficients for LPOP and LGNP are zero, F_c has the F-distribution with d.f. 2 for the numerator and 19 for the denominator. The critical F^* for a 1 percent level of significance is 5.93, which is below F_c. Therefore we would strongly reject the null hypothesis that both coefficients are zero and conclude that at least one of them is nonzero. But this contradicts the earlier conclusion, based on the t-statistics, that each coefficient is significant. This apparent dilemma is due to the high correlation among the independent variables. The correlations are 0.998 between LPOP and LGNP, 0.946 between LPOP and LINTRATE, and 0.918 between LGNP and LINTRATE. Multicollinearity between LPOP and LGNP is almost perfect and accounts for higher standard errors and low t-statistics that make the coefficients insignificant. We note from Table 11.4 that if one of these variables is omitted from the model, then the significance of the other variable's coefficient immediately jumps up. If both LGNP and LPOP are omitted from the specification, we find that interest rate becomes insignificant also, and the adjusted R^2 becomes negative. This is a clear example of biased estimates due to the exclusion of important variables.

Table 11.3 Data on the Determinants of the Number of New Housing Units

YEAR	HOUSING	POP	GNP	INTRATE
1963	1634.9	189.242	1873.3	5.89
1964	1561	191.889	1973.3	5.82
1965	1509.7	194.303	2087.6	5.81
1966	1195.8	196.56	2208.3	6.25
1967	1321.9	198.712	2271.4	6.46
1968	1545.4	200.706	2365.6	6.97
1969	1499.5	202.677	2423.3	7.8
1970	1469	205.052	2416.2	8.45
1971	2084.5	207.661	2484.8	7.74
1972	2378.5	209.896	2608.5	7.6
1973	2057.5	211.909	2744.1	7.96
1974	1352.5	213.854	2729.3	8.92
1975	1171.4	215.973	2695	9
1976	1547.6	218.035	2826.7	9
1977	2001.7	220.239	2958.6	9.02
1978	2036.1	222.585	3115.2	9.56
1979	1760	225.055	3192.4	10.78
1980	1312.6	227.757	3187.1	12.66
1981	1100.3	230.138	3248.8	14.7
1982	1072.1	232.52	3166	15.14
1983	1712.5	234.799	3279.1	12.57
1984	1755.8	237.019	3489.9	12.38
1985	1745	239.283	3585.2	11.55

Source: *Economic Report of the President*, 1987.

Table 11.4 Estimated Models for Housing Starts[†]

Variable	Model A	Model B	Model C	Model D
Constant	−12.567	7.677	−20.728	−4.759
	(−0.97)	(22.22)	(−2.73)	(−1.45)
LPOP	2.695		5.842	
	(0.63)		(3.74)	
LGNP	1.075			1.873
	(0.78)			(3.79)
LINTRATE	−1.389	−0.147	−1.492	−1.229
	(−3.42)	(−0.94)	(−3.92)	(−3.96)
$\bar{R}^2$	0.366	−0.006	0.379	0.386
ESS	0.560	0.982	0.578	0.571

[†] Values in parentheses are the corresponding *t*-statistics.

In this particular example a caveat should be added. We will see in the next chapter that the housing model violates Assumption 10.5 that the error terms of different observations be uncorrelated, and such a violation would cast doubts on the validity of hypothesis tests. Nevertheless, as will be seen later, the conclusion reached here is valid even if we modify the analysis to take account of the violation.

In addition to increasing the standard errors of regression coefficients, multicollinearity makes the interpretation of individual parameter estimates difficult. This can be illustrated with a simple model in which all the variables are expressed as deviations from the mean and which has only two independent variables. The model is thus

$$y = \beta_1 x_1 + \beta_2 x_2 + u$$

where $y_t = Y_t - \overline{Y}$, and similarly for the X's. The covariance matrix of the estimated coefficients is given by $\sigma^2(X'X)^{-1}$, which now takes the form of the 2×2 matrix

$$\sigma^2(X'X)^{-1} = \sigma^2 \begin{bmatrix} x_1'x_1 & x_1'x_2 \\ x_2'x_1 & x_2'x_2 \end{bmatrix}^{-1} = \sigma^2 \begin{bmatrix} a_{11} & a_{12} \\ a_{12} & a_{22} \end{bmatrix}$$

The covariance between $\hat{\beta}_1$ and $\hat{\beta}_2$ is given by

$$\text{Cov}(\hat{\beta}_1, \hat{\beta}_2) = \sigma^2 a_{12} = \frac{-\sigma^2 x_1'x_2}{(x_1'x_1)(x_2'x_2) - (x_1'x_2)^2}$$

Using the fact that the correlation coefficient between x_1 and x_2 can be expressed as (see Section 5.1)

$$r = \frac{x_1'x_2}{\sqrt{(x_1'x_1)(x_2'x_2)}}$$

the covariance becomes

$$\text{Cov}(\hat{\beta}_1, \hat{\beta}_2) = \frac{-\sigma^2 r}{\sqrt{(x_1'x_1)(x_2'x_2)} (1 - r^2)}$$

Suppose there is high collinearity so that r is close to ± 1. It is evident from the above derivation that the covariance between the regression coefficients will be very high also (numerically). This means that each coefficient is capturing part of the effect of the other variable and hence it is difficult to obtain the *separate* effects of X_1 and X_2 on Y. In other words, we cannot hold X_1 constant and increase X_2 alone, because X_1, being correlated with X_2, will also change as a result. In a model with several explanatory variables, the chances of multicollinearity are much higher and therefore the interpretation of the individual coefficients may be even more difficult, and the magnitudes of coefficients might change drastically as variables are added or deleted. This point can be seen from Table 11.4, in which the coefficients of LPOP and LGNP differ considerably from one model to another. The danger of multicollinearity suggests that indiscriminate use of explanatory variables should be avoided; instead, theory and an understanding of the underlying behavior should be used in formulating appropriate models.

Does the presence of multicollinearity destroy any of the desirable properties of the OLS estimators? The answer is no because multicollinearity is a data problem and does not affect any of the statistical properties of the estimators, provided, of course, the model is correctly specified.

The effects of multicollinearity and the means of identifying them are summarized below.

Estimates of parameters are still unbiased, consistent, and BLUE.

The standard errors of the coefficients might be high, making individual coefficients insignificant.

The model might exhibit high R^2 but low t-values.

A joint F-test might reject the null hypothesis that several coefficients are zero, but individual t-tests might accept them, leading to an apparent contradiction.

Coefficients may change considerably when variables are added or deleted, thus making the interpretation of individual coefficients difficult.

Example 11.4

Ramanathan (1992, Example 5.2, p. 235) has an interesting empirical example in which a change in model specification so drastically altered the results that the sign of a regression coefficient was reversed. The models estimated were the following.

$$\text{Model A:} \quad E_t = \alpha_0 + \alpha_1 A_t + u_{1t}$$
$$\text{Model B:} \quad E_t = \beta_0 + \beta_1 M_t + u_{2t}$$
$$\text{Model C:} \quad E_t = \gamma_0 + \gamma_1 A_t + \gamma_2 M_t + u_{3t}$$

Table 11.5 Estimated Models for Auto Expenditure [†]

Variable	Model A	Model B	Model C
Constant	−626.24	−796.07	7.29
	(−5.98)	(−5.91)	(0.06)
A	7.35		27.58
	(22.16)		(9.58)
M		53.45	−151.15
		(18.27)	(−7.06)
$\bar{R}^2$	0.90	0.86	0.95
d.f.	55	55	54

[†] The values in parentheses are the corresponding t-statistics.

where E is cumulative expenditure on the maintenance of a particular automobile, M is the cumulative mileage in thousands, and A is the car's age in weeks since the original purchase. The more a car is driven, the greater the maintenance cost and, similarly, the older the car the greater the cost of maintaining it. Between two cars with the same age, the one driven more is likely to incur higher costs. It therefore makes sense to include both variables in the model. However, we see from Table 11.5 (in which the values in parentheses are the corresponding t-statistics) that in Model C, which has both variables, the coefficient of miles is significantly negative and counterintuitive, whereas it was significantly positive in Model B, which does not have age. The coefficient of correlation between age and miles is 0.996, which implies near perfect collinearity. We see from this example that severe multicollinearity can cause a reversal of signs.

Identifying Multicollinearity

We have seen in the two examples that the surest sign of multicollinearity is a drastic change in the numerical values of regression coefficients when variables are added or deleted. Thus, if coefficients are quite sensitive to the model specification then we can strongly suspect that multicollinearity is the source of the problem. It is also useful to compute the correlation matrix for the explanatory variables in a model and look for high values. It should be pointed out however that, although we have illustrated multicollinearity with high pairwise correlations between independent variables, collinearity often occurs in the form of close linear relationships among three or more variables, even though pairwise correlation coefficients may not seem very high. Kmenta (1986, p. 434) has an example in which three variables are exactly linearly related, but the correlations between any pair of them is no higher than 0.5.

Belsley, Kuh, and Welsch (1980) suggest computing the **condition number** of the cross-product matrix $X'X$ defined as

$$\kappa = \sqrt{\lambda_{max}} / \sqrt{\lambda_{min}}$$

where the λ's are the maximum and minimum eigenvalues of $X'X$. To avoid dependence on the scaling of the data, the ith column of the matrix is first divided by $\sqrt{x'_i x_i}$. A condition number in excess of 20 is deemed to be indicative of multicollinearity problems.

Although these and other methods proposed are helpful in detecting multicollinearity, they are not that prevalent. Many econometricians argue that formal tests or measures are either meaningless or are not fruitful [see Maddala (1977, p. 186)].

Remedies to the Multicollinearity Problem

The surest way to reduce multicollinearity is to eliminate one or more variables from the model such as, for example, the ones with the least significant coefficients. This often improves the standard errors of the remaining coefficients and makes formerly insignificant coefficients significant. The danger in this, however, is that of excluding a relevant variable. If collinearity were severe enough, the bias from eliminating the offending variable may be negligible because the remaining variables will be able to capture its effects.

In the rest of the section we briefly discuss a number of proposed solutions, all of which have serious problems. Perhaps a careful specification of the model, elimination of redundant variables, and "benign neglect" is as good as any other procedure.

Extraneous information: In some situations extraneous information is used to take care of multicollinearity problems. For instance, time series data on income and price often exhibit high correlation, which makes the income and price elasticities unreliable. To take care of the problem, one estimates the income elasticity from cross-section studies and uses that in the time series regression. A serious problem with this procedure is that income elasticity from cross-section data might be measuring something quite different from that in time series data [see Meyer and Kuh (1957)].

Ridge regression: This is a mechanical procedure to handle the problem [see Hoerl and Kennard (1970), Judge *et al.* (1985, p. 912)]. A ridge estimate is defined as

$$b_R = [X'X + \gamma I]^{-1} X'y$$

where γ is an arbitrary positive constant which can be chosen so as to yield "stable" estimates when γ is changed slightly. Alternatively, one can minimize the mean squared error $E(b_R - \beta)'(b_R - \beta)$ with respect to γ. Because ridge estimators are biased and make statistical inference difficult (conventional distribution theory is not applicable here), this method is not very popular in econometrics.

Principal component analysis: This is another purely statistical procedure suggested as a solution to the problem of multicollinearity. Suppose we obtain a linear combination of the X's

$$z_1 = a_1 x_1 + a_2 x_2 + \cdots + a_k x_k$$

where we choose the a's so as to maximize the variance of z_1 subject to the normalization condition $\Sigma_i a_i^2 = 1$. Then z_1 is called the first principal component. Next obtain a similar linear combination

$$z_2 = b_1 x_1 + b_2 x_2 + \cdots + b_k x_k$$

but add the condition that z_1 and z_2 are orthogonal. This is the second principal component. If we use k principal components and regress y against the z's, it is equivalent to regressing y against the X's and therefore there is no gain in this procedure. Therefore, a common procedure is to choose a smaller subset of principal components and regress y against them. Because the z's are orthogonal to each other there is no multicollinearity problem. However, linear combinations have no economic interpretation that can be meaningfully attached. Also, the first principal component need not be the one most correlated with the dependent variable. In fact the order of the principal components has no relation to the order of the corresponding correlations with y.

EXERCISES

11.1   Suppose you have data for 100 sample families on the following variables:

        C = Annual consumption expenditure
        Y = Annual income
        W = Net worth of the family
        N = Family size

In addition, the age group to which the head of the household belongs is known as ≤ 30, 31–55, ≥ 56 (note that the actual age is not known, just the group).

(a) Formulate an econometric model that will enable you to test *all* of the following hypotheses; (i) the marginal propensity to consume out of income (MPC) decreases as income increases, (ii) other things being equal, the higher the net worth the higher the MPC, (iii) the *additional* consumption for an *extra* person added to the family decreases as the family size increases because of economies of scale in cooking, housekeeping, and so on, and (iv) other things being equal, the MPC is different across families whose heads belong to different age groups.

(b) Describe with respect to your model how you would go about testing all of the above hypotheses jointly; that is, write down the complete model, the null hypotheses, describe the regressions to run, the test statistic, its distribution and degrees of freedom, and the criterion for acceptance or rejection of the null. (*Note*: Matrix notation is not allowed here.)

11.2 You want to study the determinants of commercial television market shares and have the following basic specification (omitting the t subscript).

$$\text{SHARE} = \alpha + \beta \, \text{NSTAT} + \gamma \, \text{CABLE} + u$$

where SHARE is the market share (in percent) of a station, NSTAT is the number of competing television stations in the area, and CABLE is the percentage of households wired for cable television. We also know whether the station is VHF or not and with which network it is affiliated (ABC, CBS, NBC, others). Carefully describe how you will go about testing whether these qualitative variables affect the above relationship. More specifically,

(a) Formulate another econometric model that will enable you to test the hypotheses (clearly define what additional variables should be generated).

(b) State the null and alternative hypotheses.

(c) For the Wald test describe the regressions to run, test statistic to compute, and the decision rule.

(d) Describe step by step how the LM test can be used instead of the Wald test.

11.3 Use the data on home prices and their determinants presented in Table 11.1 to test (by Wald and LM tests) whether logarithms of X2, X3, X4, X5, and X6 should be added to Model A. Estimate a more general model including these variables, eliminate redundant variables, and reduce the model appropriately. Does sale price indicate significant nonlinear responses to any of the variables?

11.4 Consider the two-variable double-log model

$$\ln Y_t = \alpha + \beta \ln X_t + u_t$$

X_t is nonrandom and $u_t \sim N(0, \sigma^2)$ and is independent of u_s $(t \neq s)$.

(a) Derive the expected value of Y_t and show that it is NOT equal to $e^{\alpha} X_t^{\beta}$.

(b) Suppose we obtain the estimates $\hat{\alpha}$, $\hat{\beta}$, and $\hat{\sigma}^2$. Describe how we can obtain a forecast of Y_t, for a given X_t, that takes into account a correction for the bias shown in (a).

11.5 Table 11.6 has cross-section data by States on the women's labor force participation rate (percentage of the female population that is either employed or is seeking employment) and its determinants. The definitions of the variables are given below:

WLFP = participation rate (%) of all women over 16
YF = median earnings ($) by females working 50 to 52 weeks in 1979
YM = median earnings by males
EDUC = percentage of female high school graduates over 24 years of age
UE = unemployment rate in percent
WH = percentage of females over 16 who are white
DR = divorce rate (averaged over 1970-1979)
MR = marriage rate of women who are at least 15
URB = percentage of urban population

Table 11.6 Data on the Determinants of Women's Labor Force
Participation Rates[†]

WLFP	YF	YM	EDUC	UE	WH	DR	MR	URB
45.4	2922	7420	53.5	7.1	74	7	56.4	58.4
59.7	4956	13446	78.1	9.2	78.5	8.6	61.9	48.4
47.8	3187	8770	69.7	5.1	85.9	8.2	58.5	79.6
44.6	2569	6272	52.5	6.2	83.9	9.3	59.6	50
52.6	3915	9750	71.8	6.2	79.3	6.1	53	90.9
55.3	3494	9817	76.5	4.8	90	6	57.2	78.5
53.6	3681	10103	67.3	5.1	91.6	4.5	53.9	77.4
51.6	3433	10329	66.9	8	83.6	5.3	54	72.2
45.8	3379	7619	61.6	6	86	7.9	56.3	80.5
52.3	3299	7200	53.7	5.1	73.4	6.5	61.3	60.3
57.8	4082	9489	66.5	7.7	33	5.5	57.3	83.1
49	2662	8389	69.2	5.7	96.6	7.1	64.2	54.1
51.6	3825	10322	62.4	5.5	82.7	4.6	53.6	83
50.4	3303	9847	63.2	6.4	91.8	7.7	58.1	64.9
50.1	3096	9223	71.1	4.1	97.9	3.9	59.2	57.2
51	3427	8747	69.8	3.4	92.8	5.4	59.8	66.1
43.8	3060	6932	51.3	5.6	92.6	4.5	59.1	52.3
44.2	2446	7553	53	6.7	70.5	3.8	54.6	66.1
47.9	2939	7364	67.7	7.2	99.1	5.6	56.9	50.8
54.6	4402	10803	66.4	5.6	76.2	4.1	52.6	76.6
52.9	3795	9421	69.7	6.1	94.4	3	49.5	84.6
48.8	3550	10364	67.3	7.8	86.1	4.8	55.2	73.8
54	3215	9005	70.9	4.2	97.3	3.7	57	66.4
46.2	2366	5842	50.6	5.8	66.4	5.6	54.8	44.5
49.3	3348	8149	60.8	4.5	89.1	5.7	56.6	70.1
49	2545	8161	71.8	5.1	95.1	6.5	61	53.4
51.3	3101	9045	71.7	3.2	95.7	4	58.9	61.5
60.1	4295	9636	73	5.1	88.9	16.8	57	80.9
54.5	3421	8785	69.6	3.1	99	5.9	56.7	56.4
50.6	3855	10362	62.9	6.9	84.9	3.2	53.2	88.9
46.5	2991	7485	63	6.6	78.4	8	58.4	69.8
48.2	4246	9763	62.6	7.1	80.9	3.7	49.2	88.9
53.9	3602	6975	53.4	4.8	77.2	4.9	56.5	45
47.3	2478	8012	66.4	3.7	96.8	3.2	60.9	44.3
48	3367	10003	65.7	5.9	89.4	5.5	56.3	75.8
48	2977	7716	61	3.4	87.5	7.9	58.5	68
50.2	3143	8920	74.6	6.8	95.6	7	60.6	67.1
45.7	3375	9256	60.5	6.9	90.4	3.4	53.8	71.5
52.4	3556	8737	56.1	6.6	95.7	3.9	51.3	87.1

(Continued)

Table 11.6 (Continued)

WLFP	YF	YM	EDUC	UE	WH	DR	MR	URB
52.9	3022	7373	53	5	70.2	4.7	55.2	47.6
49.3	2500	7170	67.8	3.5	96.2	3.9	59.5	44.6
48.9	3246	6817	50.7	5.8	84.3	6.8	57.3	58.7
51	3214	8542	59.7	4.5	81.1	6.9	58.6	79.7
49.5	2899	9535	77.7	4.3	91.7	6.9	62.4	80.4
51.8	3194	7443	69.6	5.1	99.2	4.6	54.4	32.2
52.4	3554	9176	61.9	4.7	80.1	4.5	55.9	63.1
50.6	3173	10070	75.6	6.8	92.1	6.9	58.1	72.6
36.5	2893	7989	49.8	6.7	96	5.3	59.7	39
52.7	3218	9346	68.3	4.5	95.2	3.6	56.8	65.9
51.6	2791	9420	73.7	2.8	95.6	7.8	65.2	60.5

[†]   Data compiled by Jennifer Whisenand from the 1980 Census.

(a)   What signs would you expect for the regression coefficients if you related WLFP to the other explanatory variables?

(b)   Because the dependent variable (WLFP) is a percentage, a linear model specification might give inadmissible predictions and hence a logit model would be appropriate. First generate $Y = \ln[\text{WLFP}/(100 - \text{WLFP})]$ and then regress Y against all the other variables.

(c)   Test the model for overall significance and test each regression coefficient for significance (ignore the constant term).

(d)   Next eliminate variables with insignificant coefficients one at a time until the coefficients (except the constant) are significant at 10 percent or lower levels. In this final model derive the marginal effects of the remaining explanatory variables and interpret their meaning.

11.6   Consider the partitioned regression model

$$y = X_1\beta_1 + X_2\beta_2 + u$$

(a)   Estimate β_2 in an indirect way using the steps of an LM test. First estimate β_1 from the model $y = X_1\beta_1 + v$ and get $\hat{v}$. Then estimate the model $\hat{v} = X_2\gamma + w$. Show that $\hat{\gamma}$ is a biased estimator of $\hat{\beta}_2$. Derive the conditions under which it will become unbiased.

(b)   Yet another method is to proceed as follows. Estimate the model $X_2 = X_1\delta + \theta$ by regressing X_2 against X_1. Next obtain the residuals $\hat{\theta}$. Then regress $\hat{v}$ against $\hat{\theta}$, that is, estimate $\hat{v} = \hat{\theta}\pi + $ error, and get $\hat{\pi}$. Show that $\hat{\pi}$ is an unbiased estimator of β_2. Would that estimator be the same or different from the OLS estimator applied on the original model?

11.7   Consider the Cobb-Douglas production function

$$Q_t = Q_0 e^\alpha K_t^\beta L_t^\gamma$$

You have time series data on Q_t, K_t, and L_t, and the goal is to estimate the unknown parameters α, β, and γ.

(a)   Formulate an estimable econometric model and describe how the parameters may be estimated.

(b)   I believe that the parameters are not constant but **time-varying**, for example, $\alpha = f(t)$ and similarly for the other parameters. Assume a simple form for $f(t)$ and formulate a second model with which to test the null hypothesis that the parameters are constant and not time-varying. Be sure to describe the steps for both the Wald test and the LM test. (*Note*: Matrix notation is not allowed here.)

11.8   Consider the model $y = X_1\beta_1 + X_2\beta_2 + u$. Suppose we have an extraneous estimator of β_1 (call it b_1) with the covariance matrix V. Using the relation

$$y^* = y - X_1 b_1 = X_2\beta_2 + u$$

we obtain the OLS estimator $\hat{\beta}_2$ of β_2. Is the estimator of β_2 that is obtained in this way biased or not? Show that the covariance matrix of $\hat{\beta}_2$ is given by

$$E[(\hat{\beta}_2 - \beta_2)(\hat{\beta}_2 - \beta_2)'] = \sigma^2(X_2'X_2)^{-1} + Z V Z'$$

where $Z = (X_2'X_2)^{-1}X_2'X_1$.

12

NONSPHERICAL DISTURBANCES

The multiple regression model $y = X\beta + u$ used in the previous two chapters has made a number of assumptions about the random residual vector u. One of the assumptions was that u is a random vector with the covariance matrix $\sigma^2 I$. This assumption, known as **sphericalness**, may be unrealistic in many practical situations. In the present chapter we study **nonsphericalness** by replacing the above assumption by the more general assumption that the covariance matrix is $\sigma^2 \Omega$, where Ω is a symmetric positive definite nonsingular matrix with possibly unknown parameters, the order of the matrix being equal to the number of observations (T). Two specific cases of nonsphericalness are also examined. The first is **heteroscedasticity (HSK)**, which arises when the error variances are different across observations. The second case examined is **serial correlation** (or **autocorrelation**), which arises when the error term from one observation is correlated with that from another observation. In each case we examine the consequences of ignoring the violations of the basic assumptions, develop tests, and suggest new estimation procedures.

12.1 GENERAL AITKEN ESTIMATION

Denote the covariance between u_t and u_s by $\sigma^2 \rho_{ts}$, so that $E(u_t u_s) = \sigma^2 \rho_{ts}$ (note that the mean of u_t is zero). We then have (because $\rho_{tt} = 1$ and Ω is symmetric),

$$E(uu') = \sigma^2 \Omega = \sigma^2 \begin{bmatrix} 1 & \rho_{12} & \cdot & \cdot & \rho_{1T} \\ \rho_{12} & 1 & \cdot & \cdot & \rho_{2T} \\ \cdot & \cdot & \cdot & \cdot & \cdot \\ \rho_{1T} & \rho_{2T} & \cdot & \cdot & 1 \end{bmatrix}$$

where we assume, for the present, that the ρ's are known.

The Consequences of Using OLS

Suppose we ignore nonsphericalness, apply the OLS procedure, and estimate the model by $\hat{\beta} = (X'X)^{-1}X'y$. What are the statistical properties of the estimator and its variance? First, let us check the unbiasedness.

$$\hat{\beta} = (X'X)^{-1}X'(X\beta + u) = \beta + (X'X)^{-1}X'u$$

$$E(\hat{\beta}) = E[\beta + (X'X)^{-1}X'u] = \beta$$

provided, as before, $E(X'u \mid X) = 0$. Therefore the unbiasedness property is preserved. The covariance matrix of the estimator is given by

$$\text{Var}(\hat{\beta}) = E[(\hat{\beta} - \beta)(\hat{\beta} - \beta)'] = E[(X'X)^{-1}X'uu'X(X'X)^{-1}]$$
$$= (X'X)^{-1}X'(\sigma^2 \Omega)X(X'X)^{-1} = \sigma^2 (X'X)^{-1}X'\Omega X(X'X)^{-1}$$

If the u's are normally distributed, we have

$$\hat{\beta} \sim MVN[\beta, \sigma^2 (X'X)^{-1}X'\Omega X(X'X)^{-1}]$$

It follows that the conventional measure $\sigma^2 (X'X)^{-1}$ no longer reflects the true covariance matrix *even if σ^2 is estimated correctly.* Furthermore, the t- and F-distributions derived for test statistics are no longer appropriate and hence hypothesis tests based on OLS estimators are no longer valid. It will be shown in the next section that the BLUE property also is invalid. To check whether $\hat{\beta}$ is consistent, we write the variance of the estimator as

$$\text{Var}(\hat{\beta}) = \frac{\sigma^2}{T} \left[\frac{X'X}{T}\right]^{-1} \frac{X'\Omega X}{T} \left[\frac{X'X}{T}\right]^{-1}$$

If $\text{plim}(X'X/T)$ and $\text{plim}(X'\Omega X/T)$ are both finite positive definite matrices, then the variance of $\hat{\beta}$ converges to zero, which means that $\hat{\beta}$ converges in probability to β (that is, it is consistent). To summarize, *ignoring nonsphericalness and applying OLS yields estimators that are unbiased and consistent but not BLUE (that is, not efficient), and hypothesis tests that are not valid.*

Generalized Least Squares (GLS) Estimation

We now prove that the OLS estimators of β are not BLUE by deriving more general estimators. First note that since Ω is a symmetric positive definite matrix, it is possible to find a nonsingular matrix P such that $\Omega = PP'$ (refer to Property A.16 of Appendix A). Premultiplying the equation $y = X\beta + u$ by P^{-1}, the model becomes

$$P^{-1}y = P^{-1}X\beta + P^{-1}u \qquad \text{or} \qquad y^* = X^*\beta + u^*$$

We observe that $E(u^*) = 0$ and

$$\text{Var}(u^*) = E(u^*u^{*\prime}) = P^{-1}\sigma^2\Omega(P^{-1})' = \sigma^2 I$$

Therefore the transformed errors u^* have all the nice properties and hence we can apply OLS to the transformed equation and obtain estimates that are BLUE. This amounts to minimizing the generalized sum of squares $u^{*\prime}u^* = u'\Omega^{-1}u = (y - X\beta)'\Omega^{-1}(y - X\beta)$.

$$\hat{\beta}^* = (X^{*\prime}X^*)^{-1}X^{*\prime}y^* = [X'(P^{-1})'P^{-1}X]^{-1} X'(P^{-1})'P^{-1} y$$

$$= (X'\Omega^{-1}X)^{-1}X'\Omega^{-1}y$$

The above estimator, known as the **generalized least squares (GLS) estimator** or as the **generalized Aitken estimator**, is BLUE and hence the OLS estimator of β is inefficient. Because the transformed model has "well-behaved" errors, we can perform the t- and F-tests on the transformed model.

Maximum Likelihood Estimation

If the original disturbance vector u has the multivariate normal distribution, then the likelihood function is (refer to Section 5.10 for the general multivariate normal density function)

$$\frac{1}{(2\pi)^{n/2}} \frac{1}{|\sigma^2\Omega|^{1/2}} \exp[-u'(\sigma^2\Omega)^{-1}u/2]$$

Setting $u = y - X\beta$ and taking logarithms, the log-likelihood function is

$$\ln L = -\frac{n}{2}\ln(2\pi) - \frac{n}{2}\ln(\sigma^2) - \frac{1}{2}\ln|\Omega|$$
$$- \frac{1}{2\sigma^2}(y - X\beta)'\Omega^{-1}(y - X\beta)$$

Since the Ω matrix does not involve β, we readily see that maximizing $\ln L$ is equivalent to minimizing $(y - X\beta)'\Omega^{-1}(y - X\beta)$, which is the same as the GLS estimator derived earlier. Thus GLS is also maximum likelihood. Minimization of the log-likelihood with respect to σ^2 would yield the MLE for that parameter. An unbiased estimator for it is given by

$$s^2 = \frac{(y - X\hat{\beta}^*)'\Omega^{-1}(y - X\hat{\beta}^*)}{T - k}$$

If the elements of Ω are unknown but are consistently estimated by $\hat{\Omega}$, then we can obtain estimates of the parameters [known as the **estimated generalized least squares (EGLS)** estimates] as

$$\hat{\beta}^* = (X'\hat{\Omega}^{-1}X)^{-1}X'\hat{\Omega}^{-1}y \qquad\qquad s^2 = \frac{(y - X\hat{\beta}^*)'\hat{\Omega}^{-1}(y - X\hat{\beta}^*)}{T - k}$$

Because Ω is consistently estimated from the sample, EGLS estimates are consistent and asymptotically efficient, but not unbiased or BLUE.

12.2 HETEROSCEDASTICITY

A special case of nonsphericalness of disturbances arises when the error variance is not constant across observations. This is the case of *heteroscedasticity* (which means unequal scatter) mentioned earlier and is very prevalent when cross section data are used. For example, suppose we relate the consumption expenditures of a family to the family's income. Lower income households have little flexibility in the expenses because most of their income will go for food, clothing, shelter, and medical care. The difference between average consumption and an individual family's consumption is not likely to be large, which makes the variation in the error term low. In contrast, a high income family might be a big saver or a big spender, which makes the difference between mean and actual expenditures large. Thus the variance of u_t is likely to be larger for high income households and smaller for low income households. This point is illustrated in Figure 12.1 in which the scatter diagram spreads out as income increases, and also in Figure 10.3. When disturbances are heteroscedastic, the covariance matrix of the error terms has the form

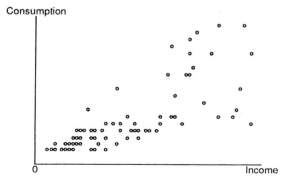

Figure 12.1 An example of heteroscedasticity

$$E(uu') = \begin{bmatrix} \sigma_1^2 & 0 & . & . & 0 \\ 0 & \sigma_2^2 & . & . & 0 \\ . & . & . & . & . \\ 0 & 0 & . & . & \sigma_T^2 \end{bmatrix}$$

We have seen earlier that if heteroscedasticity is ignored and the OLS procedure is applied, we get estimates that are unbiased but inefficient. The covariance matrix was shown to be

$$\text{Var}(\hat{\beta}) = \sigma^2 (X'X)^{-1} X' \Omega X (X'X)^{-1}$$

White (1980) proposed a **heteroscedasticity consistent covariance matrix estimator** for this by using a consistent estimator for Ω. A simple example of this is to use $\hat{u}_t^2$ as an estimator of σ_t^2, so that the estimated covariance matrix is

$$\text{Var}(\hat{\beta}) = \sigma^2 (X'X)^{-1} X' \hat{\Omega} X (X'X)^{-1}$$

where $\hat{\Omega}$ is the diagonal matrix $\text{diag}[\hat{u}_t^2]$.

A modified estimator suggested by Hinkley (1977) makes a correction for degrees of freedom by multiplying the above by $T/(T-k)$. MacKinnon and White (1985) proposed a further modification to this (known as a **jacknife** estimator). The procedure is to first estimate the model T times, each time excluding one observation. The resulting variability in estimates is then exploited to construct an estimate of Ω that is consistent and has better small sample properties.

Heteroscedasticity Known up to a Proportional Factor

In some situations, the diagonal elements may be known except for an unknown proportional factor. Consider, for instance, the case in which we have access only to data averaged within groups. The model will now take the form

$$\bar{y}_t = \bar{X}_t'\beta + \bar{u}_t \qquad\qquad \text{Var}(\bar{u}_t) = \sigma_t^2 = \sigma^2/n_t$$

where n_t is the number of observations in the tth group. We readily note that this gives rise to HSK.

As another example, consider the relationship between a company's sales and its expenditures on advertising. The variance of the error term might be small for small companies and large for big firms. A plausible assumption is therefore $\sigma_t = \sigma N_t$, where N_t is the number of employees. In the household consumption expenditure example, HSK might take the form $\sigma_t = \sigma Z_t$, where Z_t is the tth household's income, so that the disturbance variance increases with income.

Cross section data across states or cities often exhibit HSK. To illustrate, consider the relationship between the number of burglaries in a city and its determinants. For small cities the error variance may be low and for large cities the variance may be large. We may thus have $\sigma_t = \sigma P_t$, where P_t is the population of the tth city.

In the cases illustrated above in which HSK is of the form $\sigma_t = \sigma Z_t$, for $t = 1, 2, \ldots, T$ with known Z's, the Ω matrix is of the form diag $[Z_t^2]$. We readily see that the P matrix that makes $\Omega = PP'$ is diag $[Z_t]$ and hence $P^{-1} = $ diag $[1/Z_t]$. The transformed model reduces to the simple form

$$\frac{Y_t}{Z_t} = \left[\frac{X_t}{Z_t}\right]' \beta + \frac{u_t}{Z_t}$$

OLS applied to this equation gives estimates that have all the desirable properties, that is, they are unbiased, consistent, efficient, and BLUE. The above transformed model and estimation can be interpreted in another way also. To see this, define $w_t = 1/Z_t$ and rewrite the model as

$$w_t Y_t = (w_t X_t)' \beta + w_t u_t$$

OLS is equivalent to minimizing the sum of squares $\Sigma (w_t u_t)^2$ with respect to the β's. This is known as **weighted least squares (WLS)** where the weights are proportional to the inverse of the standard deviation of the error variances. WLS is therefore equivalent to estimating the parameters by OLS after multiplying every variable by w_t.

Lagrange Multiplier Tests for Heteroscedasticity

Before applying GLS or WLS, one needs to test whether HSK is present in the model. In this section we present three tests for HSK all of which are based on the Lagrange multiplier (LM) test principle introduced in Chapter 8 and applied to a regression model in Chapter 10. The tests are the **Breusch-Pagan test** (1979, 1980), **White's test** (1980), and **Engle's ARCH test** (1982, 1983). Other tests by Goldfeld and Quandt (1965), Glesjer (1969), and others are not presented here as they are not as general as the tests listed above. For details on those tests, see Greene (1990), Johnston (1984), and Judge *et al* (1985).

The Breusch-Pagan test: This test is conducted on the model given by

$$Y_t = X_t' \beta + u_t \qquad \sigma_t^2 = h(\alpha_0 + \alpha' z_t)$$

where z_t is a set of variables with known observations (some of which may be from the explanatory variables X_t), $h(\cdot)$ is of known form with first and second derivatives, and the α's are unknown parameters. The null hypothesis of homoscedasticity corresponds to the vector α being zero. The actual derivation of the LM test statistic is quite tedious and is not carried out here. Details may be found in the papers by Breusch and Pagan (1979, 1980) and Engle (1978, 1984). The steps for the test are as follows:

Step 1   Regress Y_t against X_t and obtain the residuals $\hat{u}_t = Y_t - X_t' \beta$.

Step 2   Compute $f_t = (\hat{u}_t^2 / \hat{\sigma}^2)$, where $\hat{\sigma}^2$ is the estimated variance under the null hypothesis obtained in Step 1.

Step 3   Estimate the auxiliary regression of f_t on the variables z_t and compute TR^2, which is the product of the number of observations and the unadjusted R^2 of the auxiliary regression.

Step 4   Under the null hypothesis of no HSK, TR^2 is asymptotically distributed as χ_p^2, where p is the number of α's in the hypothesis restriction. Therefore the test criterion would be to reject the null if the test statistic is "large" or if the corresponding p-value is "small."

Breusch and Pagan presented the test using one-half of the explained sum of squares of the auxiliary regression, but the TR^2 statistic is easier to compute and, as shown by Engle (1984), is asymptotically equivalent. Furthermore, Koenker (1981) has shown that this form is more robust under departures from normality.

White's test: The LM test proposed by White (1980) covers more general forms of heteroscedasticity and in particular is suited to the situation when the normality assumption for the disturbance term may not hold. He specifies the residual variance structure as follows.

$$\sigma_t^2 = \alpha_0 + \sum_{j=1}^{j=k} \sum_{m=j}^{m=k} \alpha_s X_{tj} X_{tm}$$

This includes the squares and cross-products of the X's. The steps for carrying out White's test are given below.

Step 1 Regress Y_t against X_t and obtain the residuals $\hat{u}_t = Y_t - X_t'\beta$.

Step 2 Estimate the auxiliary regression of $\hat{u}_t^2$ on the variables in X_t, their squares, and cross-product terms. For example, if the original model were

$$Y_t = \beta_0 + \beta_1 X_{1t} + \beta_2 X_{2t} + u_t$$

the auxiliary equation will be to regress $\hat{u}_t^2$ against a constant, X_1, X_2, X_1^2, X_2^2, and $X_1 X_2$.

Step 3 Compute TR^2, which is the product of the number of observations and the unadjusted R^2 of the auxiliary regression.

Step 4 Under the null hypothesis of no HSK, TR^2 is asymptotically distributed as χ_p^2, where p is the number of α's in the hypothesis restriction. Therefore the test criterion would be to reject the null if the test statistic is "large" or if the corresponding p-value is "small."

Example 12.1

Using cross section data for the 50 states of the United States, Ramanathan (1992) has estimated the following simple model relating expenditures on travel (E) to personal income (Y).

$$E_t = \alpha + \beta Y_t + u_t$$

One can expect that large states will have a much larger variance of the disturbance term than small states. Thus, we might expect $\sigma_t = \sigma_0 + \sigma P_t$, where P_t is the population of the tth state. One-half of the regression sum of squares was 10.5, which is higher than the critical χ_1^2 for a 1 percent test. Therefore the Breusch-Pagan test rejects homoscedasticity. The auxiliary regression for White's test is to regress $\hat{u}_t^2$ against a constant, P_t, and P_t^2. TR^2 for this regression has a χ_2^2 distribution under the null hypothesis. Its value was 5.36, which is slightly below the 5 percent critical value. Using the p-value approach, the hypothesis would be rejected at 7 percent. To apply the WLS procedure under the assumption that $\sigma_t = \sigma P_t$, we divide the model by P_t and estimate the transformed equation

$$(E_t/P_t) = \alpha (1/P_t) + \beta (Y_t/P_t) + (u_t/P_t)$$

It should be noted that in the transformed model we relate per capita expenditure to per capita personal income. This kind of normalization by population often reduces HSK problems. We can also use the WLS procedure under White's general specification. In this case we estimate the residual variances as

$$\hat{\sigma}_t^2 = \hat{\alpha}_0 + \hat{\alpha}_1 P_t + \hat{\alpha}_2 P_t^2$$

then compute the weights $w_t = 1/\hat{\sigma}_t$ and then apply WLS. This method might break down, however, if any of the predicted $\hat{\sigma}_t^2$ values is not positive, because then we cannot take the reciprocal of the square root. If such a situation arises, a regression of the logarithm of $\hat{u}_t^2$ against a constant, P_t, and P_t^2 is often used to generate

$$\hat{\sigma}_t^2 = \exp(\hat{\alpha}_0 + \hat{\alpha}_1 P_t + \hat{\alpha}_2 P_t^2)$$

which is always positive. This approach, however, might lead to large negative values for $\ln(\hat{u}_t^2)$ when $\hat{u}_t^2$ is small.

Because we are estimating the Ω matrix when using White's approach, WLS is the same as EGLS. The estimates will generally be more efficient than OLS estimates but we cannot be assured that they are unbiased and BLUE.

Engle's ARCH test: Although HSK is most prevalent in cross section data, time series data can also exhibit HSK. Some forecasters, especially those analyzing the securities market, have observed that the variance of prediction errors is not constant but varies with time. To accommodate this type of HSK, Engle (1982) formulated the following auxiliary equation called **autoregressive conditional heteroscedasticity (ARCH)**.

$$\sigma_t^2 = \alpha_0 + \alpha_1 \sigma_{t-1}^2 + \cdots + \alpha_p \sigma_{t-p}^2 + \varepsilon_t$$

This equation is known as the pth-order ARCH process. The testing and estimation procedures are identical to those done earlier for White's test. The only difference is that lagged values of $\hat{u}_t^2$ are used in the auxiliary regression instead of squares and cross-products of the X's.

12.3   AUTOCORRELATION

The assumption that the covariance matrix of the disturbance terms in a regression model is $\sigma^2 I$ implies that the error terms u_t and u_s $(t \neq s)$ are uncorrelated. This property is called **serial independence**. When analyzing time series data, this assumption is frequently violated. Error terms from successive or nearby time periods are often correlated. This property is known as **autocorrelation** (or **serial correlation**) and can often be identified by the **residual plot** that graphs the OLS residuals $\hat{u}$ against time (see Figure 12.2 for an illustration).

We note in the graph that the residuals are clustered together, thus violating the assumption of independence. If the disturbances are autocorrelated, then the Ω matrix will have nonzero elements in the off-diagonal entries. As in the case of HSK, applying OLS in the presence of serial correlation will result in unbiased and consistent but inefficient estimators. The common method of treating autocorrelation is to assume that the error terms u_t follow an **autoregressive process** specified as follows.

$$u_t = \rho_1 u_{t-1} + \rho_2 u_{t-2} + \cdots + \rho_p u_{t-p} + \varepsilon_t$$

where the ρ's are known as the autoregression coefficients, the above pth-order autoregressive process is referred to as **AR** (p), and the ε's are assumed to have zero means, constant variance, and zero covariance. The order of the AR process is often dictated by the periodicity of the data series. Thus, for example, quarterly data might exhibit fourth-order serial correlation and monthly data might have AR (12).

Lagrange Multiplier Test for AR (p)

The LM test procedure is applicable here, also in a manner that is easily carried out. To motivate the procedure, the model is rewritten as follows.

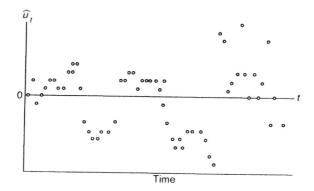

Figure 12.2 An illustration of serial correlation

$$Y_t = X_t'\beta + \rho_1 u_{t-1} + \rho_2 u_{t-2} + \cdots + \rho_p u_{t-p} + \varepsilon_t$$

The null hypothesis is that each of the ρ's is zero, that is, $\rho_1 = \rho_2 = \cdots = \rho_p = 0$. The test procedure is very similar to the LM test presented in Chapter 10 (Section 10.5) for the addition of variables. Here the added variables are u_{t-1}, u_{t-2}, and so on. But because these are unobserved variables we use estimates for them. The steps of the procedure are given below [see Godfrey (1978) and Engle (1984) for detailed justification].

Step 1   Estimate the model $Y_t = X_t'\beta + u_t$ by OLS and save the residuals $\hat{u}_t = Y_t - X_t'\beta$.

Step 2   Lag the residuals and generate the variables $\hat{u}_{t-1}$, $\hat{u}_{t-2}, \ldots, \hat{u}_{t-p}$.

Step 3   Regress $\hat{u}_t$ against all the independent variables in the model plus $\hat{u}_{t-1}, \hat{u}_{t-2}, \ldots, \hat{u}_{t-p}$. Note that the effective number of observations in this auxiliary regression is $T - p$ because $\hat{u}_{t-p}$ is defined only for the period $p + 1$ through T.

Step 4   Compute $(T - p)R^2$ from Step 3. Under the null hypothesis of serial independence, it has a χ_p^2 distribution (asymptotically). Reject the null if the LM test statistic is "large" or the corresponding p-value is "small."

The Special Case of AR (1)

The case of the first-order autocorrelation is the most widely used form for capturing serial correlation. We have $u_t = \rho u_{t-1} + \varepsilon_t$ with $|\rho| < 1$. Repeated substitution of this gives

$$u_t = \varepsilon_t + \rho \varepsilon_{t-1} + \rho^2 \varepsilon_{t-2} + \cdots$$

A measure of ρ is given by (using the OLS residuals $\hat{u}_t$)

$$\hat{\rho} = \frac{\sum\limits_{t=2}^{t=T} \hat{u}_t \hat{u}_{t-1}}{\sum\limits_{t=1}^{t=T} \hat{u}_t^2}$$

Because each of the ε's has zero mean, $E(u_t) = 0$.

$$\mathrm{Var}(u_t) = \sigma_\varepsilon^2 (1 + \rho^2 + \rho^4 + \cdots) = \frac{\sigma_\varepsilon^2}{1 - \rho^2}$$

Using the fact that $E(\varepsilon_t \varepsilon_{t-s}) = 0$ for $s \neq 0$ we can show that the covariance between u_t and u_{t-s} is given by (verify it)

$$E(u_t u_{t-s}) = \rho^s \frac{\sigma_\varepsilon^2}{1-\rho^2} = \sigma^2 \rho^s$$

The Ω matrix therefore has the form

$$\Omega = \begin{bmatrix} 1 & \rho & \rho^2 & \cdot & \cdot & \rho^{T-1} \\ \rho & 1 & \rho & \cdot & \cdot & \rho^{T-2} \\ \cdot & \cdot & \cdot & \cdot & \cdot & \cdot \\ \rho^{T-1} & \rho^{T-2} & \rho^{T-3} & \cdot & \cdot & 1 \end{bmatrix}$$

The most widely used test for AR(1) is the **Durbin-Watson test**. First compute the **Durbin-Watson (DW) statistic**

$$d = \frac{\sum_{t=2}^{t=T} (\hat{u}_t - \hat{u}_{t-1})^2}{\sum_{t=1}^{t=T} \hat{u}_t^2}$$

As seen below, for a large sample, $d \approx 2(1 - \hat{\rho})$.

$$d = \frac{\sum_{t=2}^{t=T} \hat{u}_t^2 + \sum_{t=2}^{t=T} \hat{u}_{t-1}^2 - 2 \sum_{t=2}^{t=T} \hat{u}_t \hat{u}_{t-1}}{\sum_{t=1}^{t=T} \hat{u}_t^2}$$

Because the sample size is large and the residuals are generally small, we have

$$\sum_{t=2}^{t=T} \hat{u}_t^2 \approx \sum_{t=2}^{t=T} \hat{u}_{t-1}^2 \approx \sum_{t=1}^{t=T} \hat{u}_t^2$$

We note from this that the first two terms of d will approximately cancel with the denominator, and the third term is $2\hat{\rho}$. Therefore, $d \approx 2(1 - \hat{\rho})$. Because ρ ranges from -1 to $+1$, the range for d is 0 to 4. Positive serial correlation corresponds to $0 < d < 2$ and negative serial correlation corresponds to $2 < d < 4$. Serial independence implies that $\rho = 0$ and hence $d \approx 2$. We would thus reject the null hypothesis of serial independence if d is nearly 2. However, calculating the exact critical value is cumbersome because the statistical distribution of the DW statistic depends on the data matrix X. This is a major drawback of the DW test. Durbin and Watson's solution to this problem was to derive limiting distributions between which the true distribution lies. The test is carried out with these values. The steps for the DW test are listed below.

Step 1   Estimate the model by OLS and save the residuals $\hat{u}_t$.

Step 2   Compute the Durbin-Watson statistic d (most computer programs provide this automatically).

Step 3   Look up either Table B.7a or B.7b from Appendix B (depending on whether the alternative hypothesis is $\rho \neq 0$ or $\rho > 0$) and obtain the critical values d_L and d_U. Reject serial independence if $d \leq d_L$. If $d \geq d_U$, we cannot reject it. If $d_L < d < d_U$, the test is inconclusive. If $d > 2$, then use $4-d$ instead of d in the above test (the alternative in this case is $\rho < 0$).

The inconclusiveness is a serious shortcoming of the DW test procedure. Also, the test is not applicable for higher order autocorrelations or when the model has lagged dependent variable terms such as Y_{t-1}, Y_{t-2}, and so on as explanatory variables. Finally, the tables of critical values are not applicable if the number of regressors is large [see Savin and White (1977) and Farebrother (1980) for tables extending to 20 regressors]. For these reasons, the LM test is recommended as the preferred method of testing for serial correlation. It does not have the inconclusiveness, it can be applied for any order of AR, and it is applicable for any number of regressors and when lagged dependent variables are present. The LM test, however, is a large sample test and requires at least 30 observations, preferably many more.

Treatment of Autocorrelation

Because the nature of the error structure in a model is unknown, no estimation procedure can guarantee the elimination of autocorrelation. Some general approaches might be worth trying before using alternative estimation procedures. For instance, if a model excluded some variables that had increasing or decreasing time trends, then this misspecification will manifest itself as an apparent serial correlation. A careful specification of the model might eliminate the autocorrelation. In some cases, incorporating nonlinearities or using a double-log model might take care of serial dependence of the error terms. The inclusion of lagged dependent variables (that is, Y_{t-1}, Y_{t-2}, and so on) as explanatory variables often eliminates autocorrelation. Another alternative is to specify the model in the following **first difference** form.

$$\Delta Y_t \;=\; \Delta X_t' \beta \;+\; v_t$$

where $\Delta Y_t = Y_t - Y_{t-1}$ is the first difference of Y_t and similarly for X_t.

Estimating the Parameters in the AR (1) Case

If significant autocorrelation persists in spite of alternative functional forms, then we know that OLS estimators are not efficient and tests of hypothesis are invalid. It would therefore be useful to have alternative estimation procedures in such situations. Before developing a procedure, we rewrite the model in a transformed way. We have,

$$u_t \;=\; Y_t - X_t' \beta \qquad\qquad u_{t-1} \;=\; Y_{t-1} - X_{t-1}' \beta$$

Using these in $u_t = \rho u_{t-1} + \varepsilon_t$, we have

$$Y_t - X_t' \beta \;=\; \rho \left(Y_{t-1} - X_{t-1}' \beta \right) + \varepsilon_t$$

which can be written as follows.

$$Y_t - \rho Y_{t-1} \;=\; (X_t - \rho X_{t-1})' \beta + \varepsilon_t \qquad \text{or} \qquad Y_t^* \;=\; X_t^{*\prime} \beta + \varepsilon_t$$

The above transformation of the variables is known as **quasi-differencing**. If ρ were known, we could transform the variables and obtain Y^* and X^* and then apply OLS to the transformed model. Since ρ is unknown, we have to devise some method of estimating it first. The two most common procedures for estimating the parameters are the **Hildreth-Lu (HILU) search procedure** and the **Cochrane-Orcutt (CORC) iterative procedure**. In this section we discuss only the HILU technique. The CORC method is applicable for general AR also and is discussed later. The HILU procedure [see Hildreth and Lu (1960)] would first choose a value of ρ between -1 and $+1$ (exclusive of those two values), then transform the variables to obtain Y^* and X^*, apply OLS to the transformed model, and compute the error sum of squares $\hat{\varepsilon}_t^2$. By proceeding similarly for different values of ρ we can choose the one for which the sum of squared residuals is the lowest. The corresponding $\hat{\beta}$ will be the HILU estimate of the parameters. Because the search procedure must cover the range -1 to $+1$ quite thoroughly, this method would be very computer intensive. The application at the end of this chapter illustrates the tests and estimation procedures discussed here.

Estimating the Parameters in the AR (*p*) Case

If a higher order autocorrelation structure is specified for the disturbance terms, the HILU technique is cumbersome because we have to search over too many parameters. Cochrane and Orcutt (1949) proposed an iterative procedure that is usually quick to converge. The *p*th-order generalization of the quasi-differenced model is given by

$$Y_t - \rho_1 Y_{t-1} - \rho_2 Y_{t-2} - \cdots - \rho_p Y_{t-p}$$
$$= (X_t - \rho_1 X_{t-1} - \rho_2 X_{t-2} - \cdots - \rho_p X_{t-p})' \beta + \varepsilon_t$$

The disturbance term is specified as

$$u_t = \rho_1 u_{t-1} + \rho_2 u_{t-2} + \cdots + \rho_p u_{t-p} + \varepsilon_t$$

The generalized Cochrane-Orcutt algorithm alternates between these two equations until a convergent solution is obtained. The specific steps are as follows.

Step 1   Estimate the original model $Y_t = X_t'\beta + u_t$ by OLS and save the residuals $\hat{u}_t = Y_t - X_t'\hat{\beta}$.

Step 2   Regress $\hat{u}_t$ against $\hat{u}_{t-1}$, $\hat{u}_{t-2}$, ... , $\hat{u}_{t-p}$ (with no constant term) and obtain estimates of the ρ's (note that the effective range of observations is from $p + 1$ through *T*.

Step 3   Using these estimates construct the quasi-differenced variables. Then estimate the transformed model and obtain the next round of estimates of β.

Step 4   From these revised $\hat{\beta}$'s compute the new $\hat{u}_t$ and its lags. Then go back to step 2 and iterate until the error sum of squares of the quasi-differenced model does not change by more than some percentage such as 0.1.

The CORC procedure often converges in 5 to 10 iterations and hence is computationally much faster than the HILU procedure for AR(1). However, a drawback of the procedure is that if the error sum of squares as a function of ρ has more than one minima, then the method might choose a local minimum rather than the global minimum (see Ramanathan, 1992, Example 9.8 for an illustration). For AR(1), a hybrid procedure that is superior to both of these methods is to search by the HILU technique at broad steps of length 0.1 between −1 and +1, choose the ρ that minimizes the error sum of squares in this first pass as the starting point, and apply the CORC iterative procedure. If a general order AR structure is specified, then the CORC procedure is superior to the HILU technique, which is too tedious to be practical.

If the LM test for serial correlation determines that the error structure is AR(1) or AR(2), then the full maximum likelihood method is computationally practical and yields asymptotically efficient estimators. For details on this see Beach and MacKinnon (1978a, 1978b), Judge *et al* (1985, Section 8.2), Greene (1990, Section 15.6), and Johnston (1984, p. 325).

Forecasting in the Presence of Serial Correlation

If there is serial correlation in the error terms, then the usual forecast of the dependent variable as $X'_t\hat{\beta}$ is generally inefficient. It is possible to exploit the autoregressiveness of the disturbance terms and obtain improved forecasts. To see this, consider the case of the first-order autocorrelation with

$$Y_t = X'_t\beta + u_t \qquad \text{and} \qquad u_t = \rho u_{t-1} + \varepsilon_t$$

If $\hat{\rho}$ is an estimate of the autocorrelation coefficient, we have

$$\hat{u}_{T+1} = \hat{\rho}\hat{u}_T, \qquad \hat{u}_{T+2} = \hat{\rho}^2\hat{u}_T, \qquad \hat{u}_{T+h} = \hat{\rho}^h\hat{u}_T$$

Because the residual term $\hat{u}_T$ can be estimated from the sample, a better measure of the forecast error for h steps ahead can be obtained and hence we can generate the following "h-step ahead" forecast of the dependent variable.

$$\hat{Y}_{T+h} = X'_t\hat{\beta} + \hat{\rho}^h\hat{u}_T$$

This idea is easily extended to a higher order error structure.

An Application

We illustrate the testing and estimating procedures by applying them to a model of the weekly sales of a grocery store. Table 12.1 has data for 69 weeks of operator for a store in the coastal town of Del Mar, California. The variable definitions are given below (Y is the dependent variable).

Y = Weekly sales (in thousands) from January 1991 to April 1992

X1 = Labor dollars for the week (in thousands)

X2 = Sales in same week of previous year (in thousands)

X3 = 1 if a holiday week, 0 otherwise

X4   =   1 if summer time, 0 otherwise
X5   =   1 if race season at Del Mar race track
X6   =   1 if a top five holiday week
X7   =   1 if the Del Mar Fair is going on
X8   =   1 if adjacent stores are closed
X9   =   1 if construction on or closure of area streets
X10 =   Percentage of reduction in the price of items
         in the weekly advertisement
X11 =   1 if there was a special promotion for the week
X12 =   Average weekly water temperature at Del Mar
X13 =   Average weekly high air temperature at Del Mar
X14 =   Average weekly low air temperature at Del Mar
X15 =   Total rainfall in Del Mar for the week

The estimation and testing was carried out using the ECSLIB program. Readers are strongly encouraged to use the data provided here, reproduce the results, and carry out further analyses.

Besides being a coastal town, Del Mar hosts the annual fair and has a race track. Dummy variables are included to capture differences in sales patterns during the weeks when such activities take place. Initially an AR(1) error structure was tested using both the DW and LM tests (the lagged dependent variable X2 was excluded here as otherwise the DW test is invalid). The DW statistic was 1.486. From Table III of Savin and White (1977), we note that $d_L = 1.172$ and $d_U = 2.106$ for a 5 percent level of significance. We readily see that the DW test is inconclusive. Using the ECSLIB program the LM test statistic was computed as 5.17, which has the p-value 0.023. Thus the null hypothesis of serial independence is rejected at the 2.3 percent level. The model was then estimated by the CORC and the HILU-CORC hybrid methods and the results were very close. They are not presented here, however, because a higher order AR specification is more appropriate for a weekly data series. Furthermore, information was available for the sales values during the same week the year before (X2). As we can expect that including last year's sales would improve the predictability of current sales, a new model was estimated that included X2. The LM test statistic for a fourth-order autoregressive model was 10.467 with a p-value of 0.033, which indicates significant serial correlation. The model was then estimated by the generalized CORC procedure described earlier. As can be expected, many of the regression coefficients were not statistically significant, including some of the AR coefficients. These were then deleted, a few at a time, and the model reestimated until the model selection statistics did not improve further.

Table 12.1   Determinants of Weekly Sales in a Grocery Store †

Y	X1	X2	X3	X4	X5	X6	X7	X8	X9	X10	X11	X12	X13	X14	X15
328.236	31.418	315.243	1	0	0	1	0	0	0	46.52	1	57.43	63.71	46	1.895
306.577	26.546	319.264	0	0	0	0	0	0	0	53.34	0	58.71	64.86	45	0.26
287.755	26.242	314.373	0	0	0	0	0	0	0	37.05	0	58.71	71.57	40.86	0
308.18	26.371	318.37	0	0	0	0	0	0	0	39.08	0	57.57	65.57	41.29	0
295.343	25.349	322.897	0	0	0	0	0	0	0	51.17	1	57	66	40.86	0
295.525	25.629	309.487	0	0	0	0	0	0	0	37.21	0	57.29	67.14	46.29	0
303.966	25.687	332.638	0	0	0	0	0	0	0	48.21	1	58.71	67.43	49.57	0.005
300.959	30.665	320.839	1	0	0	0	0	0	0	45.83	1	59.14	68.71	45.14	0
301.385	25.926	312.496	0	0	0	0	0	0	0	50.66	1	59.86	65.14	51.14	2.935
301.824	26.246	305.747	0	0	0	0	0	0	0	41.79	0	59.43	68.29	47.88	0.005
313.037	27.751	323.583	0	0	0	0	0	0	0	28.81	0	58.86	63.86	44.43	0.295
315.619	27.079	319.882	0	0	0	0	0	0	0	43.07	1	58	63.43	48.71	2.63
336.274	28.229	347.282	0	0	0	0	0	0	0	35.87	0	59.29	64	47.86	2.19
291.238	25.56	304.681	0	0	0	0	0	0	0	51.94	1	59	69.86	51.57	0.005
290.1	25.88	305.675	0	0	0	0	0	0	0	41.64	0	59.71	71.14	46.86	0
296.384	25.643	311.886	0	0	0	0	0	0	0	49.41	0	60.71	68.14	53.57	0
296.509	26.016	308.914	0	0	0	0	0	0	0	34.57	1	62	69.43	56.43	0.09
311.603	26.368	313.424	0	0	0	0	0	0	0	45.2	0	63	72.29	53.57	0.005
303.753	25.501	316.658	0	0	0	0	0	0	0	36.71	1	62.43	73.57	53.71	0
293.053	25.609	311.743	0	0	0	0	0	0	0	37.74	0	61.86	71.29	50	0.005
304.624	27.485	325.193	0	0	0	0	1	0	0	35.76	1	62	69.71	56.57	0
303.588	30.774	321.197	0	1	0	0	1	0	0	57.27	1	61.57	70.57	54.43	0.005
306.095	27.389	307.771	0	1	0	0	0	0	0	43.61	1	62.29	71.14	59.14	0
310.912	26.538	348.037	0	1	1	0	0	0	0	40.2	0	63.14	71.71	62.14	0
286.562	25.354	303.262	0	0	0	0	0	0	0	34.9	1	64.29	71.86	59.86	0

(Continued)

Table 12.1 (Continued)

Y	X1	X2	X3	X4	X5	X6	X7	X8	X9	X10	X11	X12	X13	X14	X15
287.975	24.255	339.008	0	1	0	0	0	0	0	48.49	1	66	73.14	60.14	0.005
328.568	31.843	336.659	1	1	0	0	0	0	0	41.02	0	66.14	73.42	63.14	0.01
297.911	25.209	340.503	0	1	1	0	0	0	0	45.96	0	67	75.43	63.71	0
298.536	25.764	315.59	0	1	1	0	0	0	0	40.84	0	67	75	64.29	0.005
318.003	26.943	367.525	0	1	1	0	0	0	0	53.6	1	67	75.14	64.71	0
319.498	27.738	366.888	0	1	1	0	0	0	0	28.28	0	67	74.86	64.86	0.36
328.475	28.017	367.867	0	1	1	0	0	0	0	37	0	66.71	76	63.86	0
316.474	28.316	360.422	0	1	1	0	0	1	0	44.74	1	67.14	77.86	67	0
306.611	28.242	357.401	0	1	1	0	0	0	0	36.9	0	67	77.29	66.57	0
328.398	29.319	366.549	0	1	1	0	0	0	0	44.56	0	67.86	77.57	62.43	0
304.56	32.499	334.637	0	1	0	0	0	0	0	43.34	0	67.71	76.29	65.57	0.35
266.921	26.624	320.041	0	0	0	0	0	0	0	43.53	0	66.43	74.43	62.14	0.02
264.511	24.731	307.992	0	0	0	0	0	1	0	42	1	67	76.43	63	0.015
260.543	23.283	304.149	0	0	0	0	0	1	0	42.84	0	67.71	79.33	63.33	0.005
279.775	23.653	300.041	0	0	0	0	0	1	0	45.21	0	66.71	78.86	63.14	0
270.141	23.871	301.593	0	0	0	0	0	1	0	35.18	0	66.14	78.57	60.71	0
274.723	22.064	306.825	0	0	0	0	0	0	0	46.52	0	66.14	73	60	0
266.309	23.004	289.733	0	0	0	0	0	0	0	45.72	0	65.86	73.43	60.29	0.095
265.058	23.189	315.322	0	0	0	0	0	0	0	51.25	1	63.29	73.43	43.29	0.005
268.419	23.822	315.529	0	0	0	0	0	0	0	48.49	0	62	75	49.14	0
271.993	28.49	314.992	0	0	0	0	0	0	0	41.96	0	61.71	73.71	48.43	0
269.413	24.788	297.365	0	0	0	0	0	0	0	32.65	0	60.86	74.43	44.57	0.05
313.84	31.546	352.235	1	0	0	1	0	0	0	36.89	0	59.14	68.71	38.43	0.005
264.649	24.021	304.853	0	0	0	0	0	0	0	46.89	0	57	65.14	41.14	0.005
286.44	24.377	327.378	0	0	0	0	0	0	0	49.6	0	58	69.14	46.14	0.235

(Continued)

Table 12.1 (Continued)

Y	X1	X2	X3	X4	X5	X6	X7	X8	X9	X10	X11	X12	X13	X14	X15
330.869	27.643	407.501	1	0	0	1	0	0	0	41.29	0	57.43	68.14	48	0.3
319.836	28.239	306.491	1	0	0	1	0	0	0	55.67	1	57	66.57	42.57	0.28
313.346	30.5	328.236	1	0	0	1	0	0	0	51.93	0	57	66.43	45.57	1.3
281.598	24.282	306.577	0	0	0	0	0	0	0	44.44	0	57.71	66.14	42.43	1.15
266.501	24.616	287.755	0	0	0	0	0	0	0	51.63	0	57.28	69	38	0
277.513	24.129	308.18	0	0	0	0	0	0	0	46.38	0	56.57	69.29	38.43	0
257.803	23.161	295.343	0	0	0	0	0	0	0	41.73	0	57.14	70.29	43.14	0
263.12	24.014	295.525	0	0	0	0	0	0	0	34.6	0	57.86	69.14	47.57	1.85
286.616	24.753	303.966	0	0	0	0	0	0	0	43.56	1	58	65.57	53.14	2.005
270.022	28.94	300.959	0	0	0	0	0	0	0	51.85	0	59.14	70.57	50.71	0
289.303	24.403	301.385	0	0	0	0	0	0	0	38.89	0	59.29	70.57	46	0
268.89	24.637	301.824	0	0	0	0	0	0	0	42.34	0	59.29	65.43	51.43	1.87
271.621	23.862	313.037	0	0	0	0	0	0	0	46.56	0	60.57	68.43	54.14	0.02
281.666	24.09	315.619	0	0	0	0	0	0	0	43.14	0	60.43	69.43	50.86	0.56
278.57	25.058	336.274	0	0	0	0	0	0	0	52.13	0	60.86	69.71	54.86	1.405
263.262	24.9	291.238	0	0	0	0	0	0	0	37.11	0	61.71	70.14	56	0.195
260.512	24.323	290.1	0	0	0	0	0	0	1	41.9	0	62.57	71.71	59	0
315.233	26.587	296.384	0	0	0	0	0	0	1	33.13	0	63.86	74.29	57.57	0
303.663	25.871	296.509	0	0	0	0	0	0	1	47.86	0	66.43	79.43	55.71	0

† Data for this table was compiled by James McMillen.

Table 12.2   Final Model of Weekly Sales in a Grocery Store

Variable	Coefficient	Std. err.	t-stat	p-value	
Constant	81.907	22.119	3.703	0.0005	***
X2	0.413	0.063	6.528	< 0.0001	***
X1	2.742	0.544	5.040	< 0.0001	***
X5	11.096	5.030	2.206	0.0314	**
X6	17.507	5.155	3.396	0.0012	***
X7	9.120	5.379	1.696	0.0953	*
X9	40.327	6.747	5.977	< 0.0001	***
X15	2.644	1.444	1.832	0.0721	*
$\hat{\rho}_2$	0.536	0.100	5.357	< 0.0001	***
$\hat{\rho}_3$	0.263	0.100	2.629	0.0107	**

Table 12.2 presents the results of this "final" model for which all the regression coefficients are significant at levels below 10 percent. A single asterisk indicates significance at levels between 5 and 10 percent, two asterisks imply significance between 1 and 5 percent, and three asterisks indicate significance at levels below 1 percent. In this final model the disturbance equation was $u_t = \rho_2 u_{t-2} + \rho_3 u_{t-3} + \varepsilon_t$ and excluded the first- and fourth-order terms because they were insignificant.

The signs of all the regression coefficients agree with prior intuition except for those of X9 and X15. The adjusted R^2 for the final model was 0.806, which is quite good for weekly sales because they are usually quite volatile. As can be expected, the previous year's sales in the same week is a very good predictor of current sales.

EXERCISES

12.1   Consider the equations

$$Y_1 = X_1 \beta + u_1 \qquad Y_2 = X_2 \beta + u_2$$

where $E(u_1) = E(u_2) = E(u_1 u_2') = 0$, $E(u_1 u_1') = \sigma^2 I$, and $E(u_2 u_2') = \lambda \sigma^2 I$; X_i is $T \times k$; and the other vectors have appropriate dimensions. Note that β is the same in both equations.

(a)   Derive the GLS estimator of β assuming that λ is known.

(b)   Suppose OLS is applied to each equation separately and then one takes the arithmetic average. Will that be unbiased? How does its variance compare with that of the GLS estimator? Carefully justify all your answers.

(c)   Suppose λ is unknown. Suggest a way of estimating it and then applying GLS. What can you say about the statistical properties of this estimator as compared to the one obtained in (a), in terms of consistency and unbiasedness?

12.2   Example 10.2 is a cross section model of the demand for cable television. Perform a White's test for heteroscedasticity on that and, if HSK is present, obtain the EGLS estimates and compare the results with those in Example 10.3.

12.3   Using the data in Table 10.3, perform an analysis similar to Exercise 12.2.

12.4   Using the data in Table 11.1, perform an analysis similar to Exercise 12.2.

12.5   Using the data in Table 11.6, perform an analysis similar to Exercise 12.2.

12.6   Example 11.3 uses time series data but ignores serial correlation. With the data provided in Table 11.3, test the model in Example 11.3 for AR(1) using both the DW and LM tests. If there is significant serial correlation what can you say about the estimates obtained in Example 11.3? Estimate the parameters using the HILU and the CORC methods. How do they compare with each other? Next test the model for an AR(3) residual structure. If significant higher order serial correlation is present, then estimate the parameters using the generalized CORC and interpret the results. Finally, conduct an ARCH test and, if significant, obtain the appropriate EGLS estimates.

12.7   Using a regression program redo the analysis of the weekly sales application presented in this chapter, but use only the data for the first 52 weeks. Use your "best" model to forecast sales for the period of week 53 through week 69. Then compare them with the actual sales. In particular, regress the actual sales against a constant and forecast sales and test whether the constant term is zero and the slope term is 1 (which would be the case for a perfect forecast).

APPENDIX A: MATRIX ALGEBRA

This appendix summarizes the basic results of matrix algebra used in the book. When dealing with multiple variables, matrices (which are simply rectangular arrays of numbers) provide a compact framework of analysis. Here we review only those concepts that are actually used in the book. No proofs are provided, just the results. Readers should refer to Bellman (1970), Dhrymes (1970), Maddala (1974), Hadley (1973), and Strang (1976) for more elaborate treatments.

Vectors and Matrices

A **matrix** is a rectangular array of numbers written as

$$A = [a_{ij}] = \begin{bmatrix} a_{11} & a_{12} & \cdots \cdots & a_{1n} \\ a_{21} & a_{22} & \cdots \cdots & a_{2n} \\ \cdots & \cdots & \cdots & \cdots \\ a_{m1} & a_{m2} & \cdots \cdots & a_{mn} \end{bmatrix}$$

The above matrix is said to have **dimensions** $m \times n$. A single row or column of the matrix is known as a **vector**. A **column vector** has its elements arranged in a column as in

$$b = \begin{bmatrix} b_1 \\ b_2 \\ \vdots \\ b_n \end{bmatrix}$$

A **row vector** has elements arranged in a row, as in $(r_1, r_2, \ldots, r_n)$, and is denoted by a lowercase letter with a prime (for example, r'). A matrix is thus a column of row vectors or a row of column vectors. If $m = n$, we have a **square matrix**. A **symmetric matrix** has the property that $a_{ij} = a_{ji}$ for all i, j. A **triangular** matrix has only zeros, either below a diagonal from upper left to lower right (**upper triangular**) or above it (**lower triangu-lar**). A **diagonal matrix** is both upper and lower triangular and hence has all entries are zero both above and below the diagonal. It is often written as diag $[a_{ii}]$. If a diagonal matrix has only ones along the diagonal, that is, $a_{ii} = 1$ for $i = 1, 2, \ldots, n$, and $a_{ij} = 0$ for $i \neq j$, then it is called the **identity matrix**, denoted by I.

Matrix Manipulations

We have $A = B$ if and only if they have the same dimensions and $a_{ij} = b_{ij}$ for all i and j. The matrix $[a_{ji}]$ in which rows and columns of $A = [a_{ij}]$ are interchanged is called the **transpose** and is written as A'. Thus,

$$A' = \begin{bmatrix} a_{11} & a_{21} & \cdots & a_{m1} \\ a_{12} & a_{22} & \cdots & a_{m2} \\ \cdots & \cdots & \cdots & \cdots \\ a_{1n} & a_{2n} & \cdots & a_{mn} \end{bmatrix}$$

has dimensions $n \times m$. Note that the transpose of a column vector is a row vector and vice versa. If two matrices have the same dimensions, they can be added using the rule, $C = [c_{ij}] = A + B = [a_{ij} + b_{ij}]$. It is readily seen that $A + B = B + A$, $(A + B) + C = A + (B + C)$, and $(A + B)' = A' + B'$.

Consider two vectors of the same dimension.

$$a = \begin{bmatrix} a_1 \\ a_2 \\ \vdots \\ a_n \end{bmatrix} \quad \text{and} \quad b = \begin{bmatrix} b_1 \\ b_2 \\ \vdots \\ b_n \end{bmatrix}$$

The scalar sum $a_1 b_1 + a_2 b_2 + \cdots + a_n b_n$ is known as the **inner product** of a and b and is denoted as $a'b$ (also the same as $b'a$). If $a'b = 0$, then a and b are said to be **orthogonal**.

Let A be $m \times n$ and B be $n \times p$, $a'_i = (a_{i1}, a_{i2}, \ldots, a_{in})$ be the ith row of A, and $b'_j = (b_{j1}, b_{j2}, \ldots, b_{jn})$ be the jth column of B written as the row vector b'_j. Then the inner product

$$c_{ij} = a'_i b_j = \sum_{k=1}^{k=n} a_{ik} b_{kj}$$

is well defined for each i and j. These inner product values can be arranged in a matrix $C = [c_{ij}]$. This operation is known as **matrix multiplication**. Note that for a matrix multiplication to be valid, the matrices must be **conformable**, that is, the number of columns in the first matrix must be the same as the number of rows in the second matrix. An example of matrix multiplication is given below.

$$C = AB = \begin{bmatrix} 1 & 2 & 3 \\ -4 & 5 & -6 \end{bmatrix} \begin{bmatrix} -2 & 0 & 5 \\ 7 & 8 & 5 \\ 3 & 4 & 5 \end{bmatrix}$$

$$= \begin{bmatrix} 1(-2) + 2(7) + 3(3) & 1(0) + 2(8) + 3(4) & 1(5) + 2(5) + 3(5) \\ -4(-2) - 5(7) - 6(3) & -4(0) - 5(8) - 6(4) & -4(5) - 5(5) - 6(5) \end{bmatrix}$$

$$= \begin{bmatrix} 21 & 28 & 30 \\ -45 & -64 & -75 \end{bmatrix}$$

In the above example, the matrix product BA is undefined because the number of columns in B does not equal the number of rows in A and hence they are not conformable. Even if AB and BA exist, it is generally *not true* that $AB = BA$. Thus, matrix multiplication is not commutative. If a square matrix A has the property that $A^2 = A \cdot A = A$, then A is said to be **idempotent**.

It is possible to multiply a matrix and a vector, provided they are conformable. For instance, if A is $m \times n$ and b is $n \times 1$, then we obtain an $m \times 1$ vector. In the example given above, let b be the first column of B. Then Ab is the first column of C.

Given two $n \times n$ matrices A and B, if $AB = I = BA$, then B is said to be the **inverse** of A and is written as $B = A^{-1}$. If A^{-1} exists, A is said to be **nonsingular**, otherwise it is **singular**. Several properties are worth noting.

Property A.1

If A^{-1} exists, then it is unique, that is, if $AB = BA = I = AC = CA$, then $B = C = A^{-1}$.

Property A.2

If O is a conformable matrix of zeros, then AO has only zeros.

Property A.3 : $AI = IA = A$.

Property A.4 : $A(BC) = (AB)C$ and $A(B + C) = AB + AC$.

Property A.5 : $(AB)' = B'A'$.

Property A.6 : *If A is nonsingular, then* $(A')^{-1} = (A^{-1})'$.

Property A.7

If A and B are nonsingular matrices of the same dimension,
$(AB)^{-1} = B^{-1}A^{-1}$.

Property A.8 (partitioned inverse)

Suppose a matrix M is partitioned into four **sub-matrices** *as follows.*

$$M = \begin{bmatrix} A & B \\ C & D \end{bmatrix}$$

If A and $E = [D - CA^{-1}B]$ *are nonsingular, then the partitioned matrix inverse is*

$$M^{-1} = \begin{bmatrix} A & B \\ C & D \end{bmatrix}^{-1} = \begin{bmatrix} A^{-1}(I + BE^{-1}CA^{-1}) & -A^{-1}BE^{-1} \\ -E^{-1}CA^{-1} & E^{-1} \end{bmatrix}$$

Rank and Trace

A set of $n \times 1$ vectors $a_1, a_2, \ldots, a_m$ is **linearly dependent** if and only if there exist scalars $\lambda_1, \lambda_2, \ldots, \lambda_m$ not all 0 such that $\lambda_1 a_1 + \lambda_2 a_2 + \cdots + \lambda_m a_m = 0$. This means that one of the a's can be expressed as a linear combination of the other vectors. If such a set of λ's does not exist, the vectors are said to be **linearly independent**. In this case, the only solution to $\Sigma \lambda_i a_i = 0$ is $\lambda_1 = \lambda_2 = \cdots = \lambda_m = 0$. The maximum number of linearly independent columns or rows of a matrix A is known as the **rank** of A. If A is $m \times n$, then $\text{rank}(A) \leq \min(m, n)$. If $\text{rank}(A) = \min(m, n)$, then we say that the matrix has **full rank**.

The **trace** of an $n \times n$ (that is, square) matrix A is the sum of the elements in the main diagonal, that is, $\text{trace}(A) = \Sigma_1^n a_{ii}$. Several properties of rank and trace are worth noting.

Property A.9 : $rank(A) = rank(A')$.

Property A.10 : $rank(AB) \leq min[rank(A), rank(B)]$.

Property A.11

An $n \times n$ matrix is nonsingular if and only if its rank is n.

Property A.12 : *trace $(A + B) = $ trace $(A) +$ trace (B).*

Property A.13 : *trace $(A') = $ trace (A).*

Property A.14

trace $(AB) = $ trace $(BA) = \sum\limits_{i=1}^{i=m} \sum\limits_{j=1}^{j=n} a_{ij} b_{ji}$ when A is $m \times n$ and B is $n \times m$.

Property A.15

If A is symmetric and idempotent, then rank $(A) = $ trace (A).

Property A.16

If a matrix Σ is nonsingular, then there exists a nonsingular triangular matrix T such that $TT' = \Sigma$. The matrix T is often referred to as the **square root** *matrix of Σ.*

Property A.17

If a square matrix Σ of order n has less than full rank k ($0 < k < n$), that is, it is singular, then there exists a nonsingular matrix T such that TT' is a nonsingular sub-matrix of Σ.

Determinants

Consider the 2×2 matrix

$$A = \begin{bmatrix} a_{11} & a_{12} \\ a_{21} & a_{22} \end{bmatrix}$$

The quantity $(a_{11} a_{22} - a_{21} a_{12})$ is called the **determinant** of A and is denoted as det(A) or as $|A|$. If the columns of A are linearly dependent, then $a_{12} = \lambda a_{11}$ and $a_{22} = \lambda a_{21}$ for some nonzero λ. Then det$(A) = a_{11} \lambda a_{21} - a_{21} \lambda a_{11} = 0$. Thus, *if the columns of A are linearly dependent, then its determinant is zero.* If A is nonsingular, then the columns of A are not linearly dependent and det$(A) \neq 0$ *if and only if A is nonsingular.*

It is easy to verify that the inverse matrix is given by

$$A^{-1} = \frac{1}{|A|} \begin{bmatrix} a_{22} & -a_{12} \\ -a_{21} & a_{11} \end{bmatrix}$$

The concept of a determinant can be generalized to a square matrix of any order n. First consider the 3×3 matrix

$$A = \begin{bmatrix} a_{11} & a_{12} & a_{13} \\ a_{21} & a_{22} & a_{23} \\ a_{31} & a_{32} & a_{33} \end{bmatrix}$$

If we delete a row and a column, we get a 2×2 sub-matrix for each of which we can calculate the determinant. Denote by A_{ij} the sub-matrix obtained from A by deleting row i and column j and compute its determinant. We get

$$|A_{11}| = \begin{vmatrix} a_{22} & a_{23} \\ a_{32} & a_{33} \end{vmatrix} = (a_{22}\, a_{33} - a_{32}\, a_{23})$$

$$|A_{12}| = \begin{vmatrix} a_{21} & a_{23} \\ a_{31} & a_{33} \end{vmatrix} = (a_{21}\, a_{33} - a_{31}\, a_{23})$$

$$|A_{13}| = \begin{vmatrix} a_{21} & a_{22} \\ a_{31} & a_{32} \end{vmatrix} = (a_{21}\, a_{32} - a_{31}\, a_{22})$$

Next construct

$$\det(A) = |A| = \sum_{j=1}^{n} a_{ij}(-1)^{i+j}\,|A_{ij}|$$

For $n = 3$, we have $|A| = a_{11}|A_{11}| - a_{12}|A_{12}| + a_{13}|A_{13}|$.

The value $(-1)^{i+j}\,|A_{ij}|$ is known as a **cofactor** of A and the above expression is known as **expansion by cofactors**. It is readily seen that the concept of a determinant can be generalized to a square matrix of any order. A number of properties of determinants are listed below (matrices are assumed to be square).

Property A.18

If a square matrix is diagonal or triangular, then its determinant is the product of the diagonal elements.

Property A.19: *If A is singular, $|A| = 0$.*

Property A.20: $|A'| = |A|$.

Property A.21 : $|AB| = |A| \, |B|$.

Property A.22 : *If A is nonsingular, $|A^{-1}| = 1/|A|$.*

Property A.23 : *If two rows are identical, $|A| = 0$.*

Eigenvalues and Eigenvectors

Consider the square matrix A of order n and the equation $Ax = \lambda x$ or $(A - \lambda I)x = 0$, where x and λ are $n \times 1$ vectors. If the matrix $A - \lambda I$ is nonsingular, then the only solution to the equation is $x = 0$. A nonzero solution is possible if and only if $A - \lambda I$ is singular, that is, if and only if the determinant $|A - \lambda I| = 0$. This yields a polynomial of order n in λ, known as the **characteristic polynomial of A** and the equation is called the **characteristic equation**. The roots of the characteristic equation are called **characteristic roots** or **eigenvalues**. The vector x_i corresponding to the eigenvalue λ_i is known as the **characteristic vector** or **eigenvector** corresponding to λ_i. There are thus n characteristic vectors. It should be noted that the eigenvalues need not all be distinct. Some of the solutions may even be complex. We now state a number of useful results on eigenvalues and vectors.

Property A.24

Characteristic vectors corresponding to two distinct eigenvalues are orthogonal.

Property A.25

A and A' have the same characteristic polynomial and the same eigenvalues.

Property A.26

$|A| = |A'| = \lambda_1 \lambda_2 \cdots \lambda_n$ *and trace* $(A) = \lambda_1 + \lambda_2 + \cdots + \lambda_n$

Property A.27

A is nonsingular if and only if all its eigenvalues are nonzero.

Property A.28

If A is nonsingular, the eigenvectors of A^{-1} are the same as those of A, and the eigenvalues of A^{-1} are the reciprocals of those of A.

Property A.29

If A is real and symmetric, all its eigenvalues are real and its rank is equal to the number of nonzero eigenvalues.

Property A.30 (diagonalization)

Given any $n \times n$ real symmetric matrix A and its eigenvalues $\lambda_1, \lambda_2, \ldots, \lambda_n$, there exists a real matrix C such that, (a) $C'C = I$, that is, C is orthogonal; (b) $C'AC$ is diagonal with the λ's as the diagonal elements; and (c) the columns of C are the eigenvectors of A corresponding to $\lambda_1, \lambda_2, \ldots, \lambda_n$.

Property A.31

Let A be an $n \times n$ symmetric idempotent matrix. Then, (a) A is nonsingular if and only if $A = I$; (b) trace (A) = rank (A); (c) every characteristic root is either 0 or 1 and hence $|A|$ is 0 or 1; (d) rank (A) is the number of nonzero eigenvalues of A, and (e) if A is real, singular, and nonzero, then there exists an orthogonal matrix C such that $C'C = I$ and

$$C'AC = \begin{bmatrix} I & 0 \\ 0 & 0 \end{bmatrix}$$

Quadratic Forms and Definite Matrices

The double sum $\Sigma_{i=1}^{n} \Sigma_{j=1}^{n} a_{ij} x_i x_j$ can be written as $x'Ax$, where x is $n \times 1$ and $A = [a_{ij}]$ is $n \times n$, and is known as a **quadratic form**. For some matrices, the quadratic form $x'Ax$ may be always positive or always negative for any choice of x for which at least one element is not zero. These are known as **definite matrices**. In particular,

A is said to be **positive definite** if and only if $x'Ax > 0$ for all nonzero vectors x.

A is said to be **positive semi-definite** if $x'Ax \geq 0$ for all x.

A is said to be **negative definite** if and only if $x'Ax < 0$ for all nonzero vectors x.

A is said to be **negative semi-definite** if $x'Ax \leq 0$ for all x.

Property A.32

If A is an $n \times n$ positive definite matrix, then $|A| > 0$, rank $(A) = n$, and A is nonsingular. The converse is also true.

Transformations and Jacobians

Consider the two continuously differentiable transformations $u = g(x, y)$, $v = h(x, y)$, and the four partial derivatives $\partial u / \partial x$, $\partial u / \partial y$, $\partial v / \partial x$, and $\partial v / \partial y$. The determinant

$$\begin{vmatrix} \partial u / \partial x & \partial u / \partial y \\ \partial v / \partial x & \partial v / \partial y \end{vmatrix}$$

is written in the compact notation

$$\frac{\partial(u, v)}{\partial(x, y)}$$

The above determinant is known as the **Jacobian of the transformation** and is denoted by $| J |$.

Property A.33

Let the transformations $u = g(x, y)$ and $v = h(x, y)$ be continuously differentiable at all points (u, v) in the neighborhood of the point (u_0, v_0) and let the Jacobian of the transformation be nonzero at (u_0, v_0). Then there exist unique inverse functions $x = G(u, v)$ and $y = H(u, v)$ so that there is a one-to-one correspondence between (x, y) and (u, v) at (u_0, v_0). Also, the functions G and H are continuously differentiable.

Property A.34

The Jacobian of the inverse transformations $x = G(u, v)$ and $y = H(u, v)$ is the reciprocal of $|J|$.

The concepts of a Jacobian and inverse transformations are easily extended to the case of n transformations. Consider the transformations

$$u_1 = g_1(x_1, x_2, \ldots, x_n)$$
$$u_2 = g_2(x_1, x_2, \ldots, x_n)$$
$$\cdots\cdots\cdots$$
$$u_n = g_n(x_1, x_2, \ldots, x_n)$$

The inverse transformations $x_i = G_i(u_1, u_2, \ldots, u_n)$ will exist if the following Jacobian is nonzero.

$$| J | = \begin{vmatrix} \dfrac{\partial(u_1, u_2, \ldots, u_n)}{\partial(x_1, x_2, \ldots, x_n)} \end{vmatrix}$$

As in the bivariate case, the Jacobian of the inverse transformations is the reciprocal of the Jacobian of the original transformations.

Vector and Matrix Differentiation

Let A be a matrix, x and y be vectors, and z be a scalar. Then the following rules are used when differentiating vectors and matrices. It should be noted that the resulting derivatives are all matrices.

$$(1) \quad \frac{\partial A}{\partial z} = \left[\frac{\partial a_{ij}}{\partial z} \right]$$

$$(2) \quad \frac{\partial z}{\partial A} = \left[\frac{\partial z}{\partial a_{ij}} \right]$$

$$(3) \quad \frac{\partial y}{\partial x} = \left[\frac{\partial y_i}{\partial x_j} \right]$$

Property A.35: $\dfrac{\partial (Ax)}{\partial x} = A$.

Property A.36: $\dfrac{\partial (AB)}{\partial z} = \left[\dfrac{\partial A}{\partial z} \right] B + A \left[\dfrac{\partial B}{\partial z} \right]$.

Property A.37: If A is symmetric, $\dfrac{\partial (x'Ax)}{\partial x} = 2Ax$.

Property A.38

If A is nonsingular, $\dfrac{\partial A^{-1}}{\partial z} = -A^{-1} \left[\dfrac{\partial A}{\partial z} \right] A^{-1}$.

APPENDIX B: STATISTICAL TABLES

Table B.1   Binomial Distribution, Right-Hand Tail Probabilities

n	x´	.05	.10	.15	.20	.25	.30	.35	.40	.45	.50
						θ					
2	1	.0975	.1900	.2775	.3600	.4375	.5100	.5775	.6400	.6975	.7500
	2	.0025	.0100	.0225	.0400	.0625	.0900	.1225	.1600	.2025	.2500
3	1	.1426	.2710	.3859	.4880	.5781	.6570	.7254	.7840	.8336	.8750
	2	.0072	.0280	.0608	.1040	.1562	.2160	.2818	.3520	.4252	.5000
	3	.0001	.0010	.0034	.0080	.0156	.0270	.0429	.0640	.0911	.1250
4	1	.1855	.3439	.4780	.5904	.6836	.7599	.8215	.8704	.9085	.9375
	2	.0140	.0523	.1095	.1808	.2617	.3483	.4370	.5248	.6090	.6875
	3	.0005	.0037	.0120	.0272	.0508	.0837	.1265	.1792	.2415	.3125
	4	.0000	.0001	.0005	.0016	.0039	.0081	.0150	.0256	.0410	.0625
5	1	.2262	.4095	.5563	.6723	.7627	.8319	.8840	.9222	.9497	.9688
	2	.0226	.0815	.1648	.2627	.3672	.4718	.5716	.6630	.7438	.8125
	3	.0012	.0086	.0266	.0579	.1035	.1631	.2352	.3174	.4069	.5000
	4	.0000	.0005	.0022	.0067	.0156	.0308	.0540	.0870	.1312	.1875
	5	.0000	.0000	.0001	.0003	.0010	.0024	.0053	.0102	.0185	.0312
6	1	.2649	.4686	.6229	.7379	.8220	.8824	.9246	.9533	.9723	.9844
	2	.0328	.1143	.2235	.3447	.4661	.5798	.6809	.7667	.8364	.8906
	3	.0022	.0158	.0473	.0989	.1694	.2557	.3529	.4557	.5585	.6562
	4	.0001	.0013	.0059	.0170	.0376	.0705	.1174	.1792	.2553	.3438
	5	.0000	.0001	.0004	.0016	.0046	.0109	.0223	.0410	.0692	.1094
	6	.0000	.0000	.0000	.0001	.0002	.0007	.0018	.0041	.0083	.0156
7	1	.3017	.5217	.6794	.7903	.8665	.9176	.9510	.9720	.9848	.9922
	2	.0444	.1497	.2834	.4233	.5551	.6706	.7662	.8414	.8976	.9375
	3	.0038	.0257	.0738	.1480	.2436	.3529	.4677	.5801	.6836	.7734
	4	.0002	.0027	.0121	.0333	.0706	.1260	.1998	.2898	.3917	.5000
	5	.0000	.0002	.0012	.0047	.0129	.0288	.0556	.0963	.1529	.2266
	6	.0000	.0000	.0001	.0004	.0013	.0038	.0090	.0188	.0357	.0625
	7	.0000	.0000	.0000	.0000	.0001	.0002	.0006	.0016	.0037	.0078
8	1	.3366	.5695	.7275	.8322	.8999	.9424	.9681	.9832	.9916	.9961
	2	.0572	.1869	.3428	.4967	.6329	.7447	.8309	.8936	.9368	.9648
	3	.0058	.0381	.1052	.2031	.3215	.4482	.5722	.6846	.7799	.8555
	4	.0004	.0050	.0214	.0563	.1138	.1941	.2936	.4059	.5230	.6367
	5	.0000	.0004	.0029	.0104	.0273	.0580	.1061	.1737	.2604	.3633
	6	.0000	.0000	.0002	.0012	.0042	.0113	.0253	.0498	.0885	.1445
	7	.0000	.0000	.0000	.0001	.0004	.0013	.0036	.0085	.0181	.0352
	8	.0000	.0000	.0000	.0000	.0000	.0001	.0002	.0007	.0017	.0039
9	1	.3698	.6126	.7684	.8658	.9249	.9596	.9793	.9899	.9954	.9980
	2	.0712	.2252	.4005	.5638	.6997	.8040	.8789	.9295	.9615	.9805
	3	.0084	.0530	.1409	.2618	.3993	.5372	.6627	.7682	.8505	.9102
	4	.0006	.0083	.0339	.0856	.1657	.2703	.3911	.5174	.6386	.7461
	5	.0000	.0009	.0056	.0196	.0489	.0988	.1717	.2666	.3786	.5000
	6	.0000	.0001	.0006	.0031	.0100	.0253	.0536	.0994	.1658	.2539
	7	.0000	.0000	.0000	.0003	.0013	.0043	.0112	.0250	.0498	.0898

(Continued)

Table B.1   Binomial Distribution Probabilities (Continued)

n	x′	.05	.10	.15	.20	.25	.30	.35	.40	.45	.50
9	8	.0000	.0000	.0000	.0000	.0001	.0004	.0014	.0038	.0091	.0195
	9	.0000	.0000	.0000	.0000	.0000	.0000	.0001	.0003	.0008	.0020
10	1	.4013	.6513	.8031	.8926	.9437	.9718	.9865	.9940	.9975	.9990
	2	.0861	.2639	.4557	.6242	.7560	.8507	.9140	.9536	.9767	.9893
	3	.0115	.0702	.1798	.3222	.4744	.6172	.7384	.8327	.9004	.9453
	4	.0010	.0128	.0500	.1209	.2241	.3504	.4862	.6177	.7340	.8281
	5	.0001	.0016	.0099	.0328	.0781	.1503	.2485	.3669	.4956	.6230
	6	.0000	.0001	.0014	.0064	.0197	.0473	.0949	.1662	.2616	.3770
	7	.0000	.0000	.0001	.0009	.0035	.0106	.0260	.0548	.1020	.1719
	8	.0000	.0000	.0000	.0001	.0004	.0016	.0048	.0123	.0274	.0547
	9	.0000	.0000	.0000	.0000	.0000	.0001	.0005	.0017	.0045	.0107
	10	.0000	.0000	.0000	.0000	.0000	.0000	.0000	.0001	.0003	.0010
11	1	.4312	.6862	.8327	.9141	.9578	.9802	.9912	.9964	.9986	.9995
	2	.1019	.3026	.5078	.6779	.8029	.8870	.9394	.9698	.9861	.9941
	3	.0152	.0896	.2212	.3826	.5448	.6873	.7999	.8811	.9348	.9673
	4	.0016	.0185	.0694	.1611	.2867	.4304	.5744	.7037	.8089	.8867
	5	.0001	.0028	.0159	.0504	.1146	.2103	.3317	.4672	.6029	.7256
	6	.0000	.0003	.0027	.0117	.0343	.0782	.1487	.2465	.3669	.5000
	7	.0000	.0000	.0003	.0020	.0076	.0216	.0501	.0994	.1738	.2744
	8	.0000	.0000	.0000	.0002	.0012	.0043	.0122	.0293	.0610	.1133
	9	.0000	.0000	.0000	.0000	.0001	.0006	.0020	.0059	.0148	.0327
	10	.0000	.0000	.0000	.0000	.0000	.0000	.0002	.0007	.0022	.0059
	11	.0000	.0000	.0000	.0000	.0000	.0000	.0000	.0000	.0002	.0005
12	1	.4596	.7176	.8578	.9313	.9683	.9862	.9943	.9978	.9992	.9998
	2	.1184	.3410	.5565	.7251	.8416	.9150	.9576	.9804	.9917	.9968
	3	.0196	.1109	.2642	.4417	.6093	.7472	.8487	.9166	.9579	.9807
	4	.0022	.0256	.0922	.2054	.3512	.5075	.6533	.7747	.8655	.9270
	5	.0002	.0043	.0239	.0726	.1576	.2763	.4167	.5618	.6956	.8062
	6	.0000	.0005	.0046	.0194	.0544	.1178	.2127	.3348	.4731	.6128
	7	.0000	.0001	.0007	.0039	.0143	.0386	.0846	.1582	.2607	.3872
	8	.0000	.0000	.0001	.0006	.0028	.0095	.0255	.0573	.1117	.1938
	9	.0000	.0000	.0000	.0001	.0004	.0017	.0056	.0153	.0356	.0730
	10	.0000	.0000	.0000	.0000	.0000	.0002	.0008	.0028	.0079	.0193
	11	.0000	.0000	.0000	.0000	.0000	.0000	.0001	.0003	.0011	.0032
	12	.0000	.0000	.0000	.0000	.0000	.0000	.0000	.0000	.0001	.0002
13	1	.4867	.7458	.8791	.9450	.9762	.9903	.9963	.9987	.9996	.9999
	2	.1354	.3787	.6017	.7664	.8733	.9363	.9704	.9874	.9951	.9983
	3	.0245	.1339	.2704	.4983	.6674	.7975	.8868	.9421	.9731	.9888
	4	.0031	.0342	.0967	.2527	.4157	.5794	.7217	.8314	.9071	.9539
	5	.0003	.0065	.0260	.0991	.2060	.3457	.4995	.6470	.7721	.8666
	6	.0000	.0009	.0053	.0300	.0802	.1654	.2841	.4256	.5732	.7095
	7	.0000	.0001	.0013	.0070	.0243	.0624	.1295	.2288	.3563	.5000
	8	.0000	.0000	.0002	.0012	.0056	.0182	.0462	.0977	.1788	.2905

(Continued)

Table B.1 Binomial Distribution Probabilities (Continued)

n	x´	.05	.10	.15	.20	θ .25	.30	.35	.40	.45	.50
13	9	.0000	.0000	.0000	.0002	.0010	.0040	.0126	.0321	.0698	.1334
	10	.0000	.0000	.0000	.0000	.0001	.0007	.0025	.0078	.0203	.0461
	11	.0000	.0000	.0000	.0000	.0000	.0001	.0003	.0013	.0041	.0112
	12	.0000	.0000	.0000	.0000	.0000	.0000	.0000	.0001	.0005	.0017
	13	.0000	.0000	.0000	.0000	.0000	.0000	.0000	.0000	.0000	.0001
14	1	.5123	.7712	.8972	.9560	.9822	.9932	.9976	.9992	.9998	.9999
	2	.1530	.4154	.6433	.8021	.8990	.9525	.9795	.9919	.9971	.9991
	3	.0301	.1584	.3521	.5519	.7189	.8392	.9161	.9602	.9830	.9935
	4	.0042	.0441	.1465	.3018	.4787	.6448	.7795	.8757	.9368	.9713
	5	.0004	.0092	.0467	.1298	.2585	.4158	.5773	.7207	.8328	.9102
	6	.0000	.0015	.0115	.0439	.1117	.2195	.3595	.5141	.6627	.7880
	7	.0000	.0002	.0022	.0116	.0383	.0933	.1836	.3075	.4539	.6047
	8	.0000	.0000	.0003	.0024	.0103	.0315	.0753	.1501	.2586	.3953
	9	.0000	.0000	.0000	.0004	.0022	.0083	.0243	.0583	.1189	.2120
	10	.0000	.0000	.0000	.0000	.0003	.0017	.0060	.0175	.0426	.0898
	11	.0000	.0000	.0000	.0000	.0000	.0002	.0011	.0039	.0114	.0287
	12	.0000	.0000	.0000	.0000	.0000	.0000	.0001	.0006	.0022	.0065
	13	.0000	.0000	.0000	.0000	.0000	.0000	.0000	.0001	.0003	.0009
	14	.0000	.0000	.0000	.0000	.0000	.0000	.0000	.0000	.0000	.0001
15	1	.5367	.7941	.9126	.9648	.9866	.9953	.9984	.9995	.9999	1.0000
	2	.1710	.4510	.6814	.8329	.9198	.9647	.9858	.9948	.9983	.9995
	3	.0362	.1841	.3958	.6020	.7639	.8732	.9383	.9729	.9893	.9963
	4	.0055	.0556	.1773	.3518	.5387	.7031	.8273	.9095	.9576	.9824
	5	.0006	.0127	.0617	.1642	.3135	.4845	.6481	.7827	.8796	.9408
	6	.0001	.0022	.0168	.0611	.1484	.2784	.4357	.5968	.7392	.8491
	7	.0000	.0003	.0036	.0181	.0566	.1311	.2452	.3902	.5478	.6964
	8	.0000	.0000	.0006	.0042	.0173	.0500	.1132	.2131	.3465	.5000
	9	.0000	.0000	.0001	.0008	.0042	.0152	.0422	.0950	.1818	.3036
	10	.0000	.0000	.0000	.0001	.0008	.0037	.0124	.0338	.0769	.1509
	11	.0000	.0000	.0000	.0000	.0001	.0007	.0028	.0093	.0255	.0592
	12	.0000	.0000	.0000	.0000	.0000	.0001	.0005	.0019	.0063	.0176
	13	.0000	.0000	.0000	.0000	.0000	.0000	.0001	.0003	.0011	.0037
	14	.0000	.0000	.0000	.0000	.0000	.0000	.0000	.0000	.0001	.0005
	15	.0000	.0000	.0000	.0000	.0000	.0000	.0000	.0000	.0000	.0000
16	1	.5599	.8147	.9257	.9719	.9900	.9967	.9990	.9997	.9999	1.0000
	2	.1892	.4853	.7161	.8593	.9365	.9739	.9902	.9967	.9990	.9997
	3	.0429	.2108	.4386	.6482	.8029	.9006	.9549	.9817	.9934	.9979
	4	.0070	.0684	.2101	.4019	.5950	.7541	.8661	.9349	.9719	.9894
	5	.0009	.0170	.0791	.2018	.3698	.5501	.7108	.8334	.9147	.9616
	6	.0001	.0033	.0235	.0817	.1897	.3402	.5100	.6712	.8024	.8949
	7	.0000	.0005	.0056	.0267	.0796	.1753	.3119	.4728	.6340	.7228
	8	.0000	.0001	.0011	.0070	.0271	.0744	.1594	.2839	.4371	.5982
	9	.0000	.0000	.0002	.0015	.0075	.0257	.0671	.1423	.2559	.4018
	10	.0000	.0000	.0000	.0002	.0016	.0071	.0229	.0583	.1241	.2272

(Continued)

Table B.1   Binomial Distribution Probabilities (Continued)

n	x′	.05	.10	.15	.20	.25	.30	.35	.40	.45	.50
16	11	.0000	.0000	.0000	.0000	.0003	.0016	.0062	.0191	.0486	.1051
	12	.0000	.0000	.0000	.0000	.0000	.0003	.0013	.0049	.0149	.0384
	13	.0000	.0000	.0000	.0000	.0000	.0000	.0002	.0009	.0035	.0106
	14	.0000	.0000	.0000	.0000	.0000	.0000	.0000	.0001	.0006	.0021
	15	.0000	.0000	.0000	.0000	.0000	.0000	.0000	.0000	.0001	.0003
	16	.0000	.0000	.0000	.0000	.0000	.0000	.0000	.0000	.0000	.0000
17	1	.5819	.8332	.9369	.9775	.9925	.9977	.9993	.9998	1.0000	1.0000
	2	.2078	.5182	.7475	.8818	.9499	.9807	.9933	.9979	.9994	.9999
	3	.0503	.2382	.4802	.6904	.8363	.9226	.9673	.9877	.9959	.9988
	4	.0088	.0826	.2444	.4511	.6470	.7981	.8972	.9536	.9816	.9936
	5	.0012	.0221	.0987	.2418	.4261	.6113	.7652	.8740	.9404	.9755
	6	.0001	.0047	.0319	.1057	.2347	.4032	.5803	.7361	.8529	.9283
	7	.0000	.0008	.0083	.0377	.1071	.2248	.3812	.5522	.7098	.8338
	8	.0000	.0001	.0017	.0109	.0402	.1046	.2128	.3595	.5257	.6855
	9	.0000	.0000	.0003	.0026	.0124	.0403	.0994	.1989	.3374	.5000
	10	.0000	.0000	.0000	.0005	.0031	.0127	.0383	.0919	.1834	.3145
	11	.0000	.0000	.0000	.0001	.0006	.0032	.0120	.0348	.0826	.1662
	12	.0000	.0000	.0000	.0000	.0001	.0007	.0030	.0106	.0301	.0717
	13	.0000	.0000	.0000	.0000	.0000	.0001	.0006	.0025	.0086	.0245
	14	.0000	.0000	.0000	.0000	.0000	.0000	.0000	.0005	.0019	.0064
	15	.0000	.0000	.0000	.0000	.0000	.0000	.0000	.0001	.0003	.0012
	16	.0000	.0000	.0000	.0000	.0000	.0000	.0000	.0000	.0000	.0001
	17	.0000	.0000	.0000	.0000	.0000	.0000	.0000	.0000	.0000	.0000
18	1	.6028	.8499	.9464	.9820	.9944	.9984	.9996	.9999	1.0000	1.0000
	2	.2265	.5497	.7759	.9009	.9605	.9858	.9954	.9987	.9997	.9999
	3	.0581	.2662	.5203	.7287	.8647	.9400	.9764	.9918	.9975	.9993
	4	.0109	.0982	.2798	.4990	.6943	.8354	.9217	.9672	.9880	.9962
	5	.0015	.0282	.1206	.2836	.4813	.6673	.8114	.9058	.9589	.9846
	6	.0002	.0064	.0419	.1329	.2825	.4656	.6450	.7912	.8923	.9519
	7	.0000	.0012	.0118	.0513	.1390	.2783	.4509	.6257	.7742	.8811
	8	.0000	.0002	.0027	.0163	.0569	.1407	.2717	.4366	.6085	.7597
	9	.0000	.0000	.0005	.0043	.0193	.0596	.1391	.2632	.4222	.5927
	10	.0000	.0000	.0001	.0009	.0054	.0210	.0597	.1347	.2527	.4073
	11	.0000	.0000	.0000	.0002	.0012	.0061	.0212	.0576	.1280	.2403
	12	.0000	.0000	.0000	.0000	.0002	.0014	.0062	.0203	.0537	.1189
	13	.0000	.0000	.0000	.0000	.0000	.0003	.0014	.0058	.0183	.0481
	14	.0000	.0000	.0000	.0000	.0000	.0000	.0003	.0013	.0049	.015
	15	.0000	.0000	.0000	.0000	.0000	.0000	.0000	.0002	.0010	.00
	16	.0000	.0000	.0000	.0000	.0000	.0000	.0000	.0000	.0001	
	17	.0000	.0000	.0000	.0000	.0000	.0000	.0000	.0000	.0000	
	18	.0000	.0000	.0000	.0000	.0000	.0000	.0000	.0000	.0000	
19	1	.6226	.8649	.9544	.9856	.9958	.9989	.9997	.9999	1.000	.99
	2	.2453	.5797	.8015	.9171	.9690	.9896	.9969	.9992	.99	

(Con

Table B.1   Binomial Distribution Probabilities (Continued)

n	x′	.05	.10	.15	.20	.25	.30	.35	.40	.45	.50
19	3	.0665	.2946	.5587	.7631	.8887	.9538	.9830	.9945	.9985	.9996
	4	.0132	.1150	.3159	.5449	.7369	.8668	.9409	.9770	.9923	.9978
	5	.0020	.0352	.1444	.3267	.5346	.7178	.8500	.9304	.9720	.9904
	6	.0002	.0086	.0537	.1631	.3322	.5261	.7032	.8371	.9223	.9682
	7	.0000	.0017	.0163	.0676	.1749	.3345	.5188	.6919	.8273	.9165
	8	.0000	.0003	.0041	.0233	.0775	.1820	.3344	.5122	.6831	.8204
	9	.0000	.0000	.0008	.0067	.0287	.0839	.1855	.3325	.5060	.6762
	10	.0000	.0000	.0001	.0016	.0089	.0326	.0875	.1861	.3290	.5000
	11	.0000	.0000	.0000	.0003	.0023	.0105	.0347	.0885	.1841	.3238
	12	.0000	.0000	.0000	.0000	.0005	.0028	.0114	.0352	.0871	.1796
	13	.0000	.0000	.0000	.0000	.0001	.0006	.0031	.0116	.0342	.0835
	14	.0000	.0000	.0000	.0000	.0000	.0001	.0007	.0031	.0109	.0318
	15	.0000	.0000	.0000	.0000	.0000	.0000	.0001	.0006	.0028	.0096
	16	.0000	.0000	.0000	.0000	.0000	.0000	.0000	.0001	.0005	.0022
	17	.0000	.0000	.0000	.0000	.0000	.0000	.0000	.0000	.0001	.0004
	18	.0000	.0000	.0000	.0000	.0000	.0000	.0000	.0000	.0000	.0000
	19	.0000	.0000	.0000	.0000	.0000	.0000	.0000	.0000	.0000	.0000
20	1	.6415	.8784	.9612	.9885	.9968	.9992	.9998	1.0000	1.0000	1.0000
	2	.2642	.6083	.8244	.9308	.9757	.9924	.9979	.9995	.9999	1.0000
	3	.0755	.3231	.5951	.7939	.9087	.9645	.9879	.9964	.9991	.9998
	4	.0159	.1330	.3523	.5886	.7748	.8929	.9556	.9840	.9951	.9987
	5	.0026	.0432	.1702	.3704	.5852	.7625	.8818	.9490	.9811	.9941
	6	.0003	.0113	.0673	.1958	.3828	.5836	.7546	.8744	.9447	.9793
	7	.0000	.0024	.0219	.0867	.2142	.3920	.5834	.7500	.8701	.9423
	8	.0000	.0004	.0059	.0321	.1018	.2277	.3990	.5841	.7480	.8684
	9	.0000	.0001	.0013	.0100	.0409	.1133	.2376	.4044	.5857	.7483
	10	.0000	.0000	.0002	.0026	.0139	.0480	.1218	.2447	.4086	.5881
	11	.0000	.0000	.0000	.0006	.0039	.0171	.0532	.1275	.2493	.4119
	12	.0000	.0000	.0000	.0001	.0009	.0051	.0196	.0565	.1308	.2517
	13	.0000	.0000	.0000	.0000	.0002	.0013	.0060	.0210	.0580	.1316
	14	.0000	.0000	.0000	.0000	.0000	.0003	.0015	.0065	.0214	.0577
	15	.0000	.0000	.0000	.0000	.0000	.0000	.0003	.0016	.0064	.0207
	16	.0000	.0000	.0000	.0000	.0000	.0000	.0000	.0003	.0015	.0059
	17	.0000	.0000	.0000	.0000	.0000	.0000	.0000	.0000	.0003	.0013
	18	.0000	.0000	.0000	.0000	.0000	.0000	.0000	.0000	.0000	.0002
	19	.0000	.0000	.0000	.0000	.0000	.0000	.0000	.0000	.0000	.0000
	20	.0000	.0000	.0000	.0000	.0000	.0000	.0000	.0000	.0000	.0000
	1	.6594	.8906	.9671	.9908	.9976	.9994	.9999	1.0000	1.0000	1.0000
	2	.2830	.6353	.8450	.9424	.9810	.9944	.9996	.9997	.9999	1.0000
		.0849	.3516	.6295	.8213	.9255	.9729	.9914	.9976	.9994	.9999
		.0189	.1520	.3887	.6296	.8083	.9144	.9669	.9890	.9969	.9993
		.0032	.0522	.1975	.4140	.6326	.8016	.9076	.9630	.9874	.9967
		.04	.0144	.0827	.2307	.4334	.6373	.7991	.9043	.9611	.9867
		.00	.0033	.0287	.1085	.2564	.4495	.6433	.7998	.9036	.9608

(Continued)

Table B.1 Binomial Distribution Probabilities (Continued)

n	x´	.05	.10	.15	.20	.25	.30	.35	.40	.45	.50
19	3	.0665	.2946	.5587	.7631	.8887	.9538	.9830	.9945	.9985	.9996
	4	.0132	.1150	.3159	.5449	.7369	.8668	.9409	.9770	.9923	.9978
	5	.0020	.0352	.1444	.3267	.5346	.7178	.8500	.9304	.9720	.9904
	6	.0002	.0086	.0537	.1631	.3322	.5261	.7032	.8371	.9223	.9682
	7	.0000	.0017	.0163	.0676	.1749	.3345	.5188	.6919	.8273	.9165
	8	.0000	.0003	.0041	.0233	.0775	.1820	.3344	.5122	.6831	.8204
	9	.0000	.0000	.0008	.0067	.0287	.0839	.1855	.3325	.5060	.6762
	10	.0000	.0000	.0001	.0016	.0089	.0326	.0875	.1861	.3290	.5000
	11	.0000	.0000	.0000	.0003	.0023	.0105	.0347	.0885	.1841	.3238
	12	.0000	.0000	.0000	.0000	.0005	.0028	.0114	.0352	.0871	.1796
	13	.0000	.0000	.0000	.0000	.0001	.0006	.0031	.0116	.0342	.0835
	14	.0000	.0000	.0000	.0000	.0000	.0001	.0007	.0031	.0109	.0318
	15	.0000	.0000	.0000	.0000	.0000	.0000	.0001	.0006	.0028	.0096
	16	.0000	.0000	.0000	.0000	.0000	.0000	.0000	.0001	.0005	.0022
	17	.0000	.0000	.0000	.0000	.0000	.0000	.0000	.0000	.0001	.0004
	18	.0000	.0000	.0000	.0000	.0000	.0000	.0000	.0000	.0000	.0000
	19	.0000	.0000	.0000	.0000	.0000	.0000	.0000	.0000	.0000	.0000
20	1	.6415	.8784	.9612	.9885	.9968	.9992	.9998	1.0000	1.0000	1.0000
	2	.2642	.6083	.8244	.9308	.9757	.9924	.9979	.9995	.9999	1.0000
	3	.0755	.3231	.5951	.7939	.9087	.9645	.9879	.9964	.9991	.9998
	4	.0159	.1330	.3523	.5886	.7748	.8929	.9556	.9840	.9951	.9987
	5	.0026	.0432	.1702	.3704	.5852	.7625	.8818	.9490	.9811	.9941
	6	.0003	.0113	.0673	.1958	.3828	.5836	.7546	.8744	.9447	.9793
	7	.0000	.0024	.0219	.0867	.2142	.3920	.5834	.7500	.8701	.9423
	8	.0000	.0004	.0059	.0321	.1018	.2277	.3990	.5841	.7480	.8684
	9	.0000	.0001	.0013	.0100	.0409	.1133	.2376	.4044	.5857	.7483
	10	.0000	.0000	.0002	.0026	.0139	.0480	.1218	.2447	.4086	.5881
	11	.0000	.0000	.0000	.0006	.0039	.0171	.0532	.1275	.2493	.4119
	12	.0000	.0000	.0000	.0001	.0009	.0051	.0196	.0565	.1308	.2517
	13	.0000	.0000	.0000	.0000	.0002	.0013	.0060	.0210	.0580	.1316
	14	.0000	.0000	.0000	.0000	.0000	.0003	.0015	.0065	.0214	.0577
	15	.0000	.0000	.0000	.0000	.0000	.0000	.0003	.0016	.0064	.0207
	16	.0000	.0000	.0000	.0000	.0000	.0000	.0000	.0003	.0015	.0059
	17	.0000	.0000	.0000	.0000	.0000	.0000	.0000	.0000	.0003	.0013
	18	.0000	.0000	.0000	.0000	.0000	.0000	.0000	.0000	.0000	.0002
	19	.0000	.0000	.0000	.0000	.0000	.0000	.0000	.0000	.0000	.0000
	20	.0000	.0000	.0000	.0000	.0000	.0000	.0000	.0000	.0000	.0000
21	1	.6594	.8906	.9671	.9908	.9976	.9994	.9999	1.0000	1.0000	1.0000
	2	.2830	.6353	.8450	.9424	.9810	.9944	.9996	.9997	.9999	1.0000
	3	.0849	.3516	.6295	.8213	.9255	.9729	.9914	.9976	.9994	.9999
	4	.0189	.1520	.3887	.6296	.8083	.9144	.9669	.9890	.9969	.9993
	5	.0032	.0522	.1975	.4140	.6326	.8016	.9076	.9630	.9874	.9967
	6	.0004	.0144	.0827	.2307	.4334	.6373	.7991	.9043	.9611	.9867
	7	.0000	.0033	.0287	.1085	.2564	.4495	.6433	.7998	.9036	.9608

(Continued)

Table B.1 Binomial Distribution Probabilities (Continued)

n	x´	.05	.10	.15	.20	.25	.30	.35	.40	.45	.50
16	11	.0000	.0000	.0000	.0000	.0003	.0016	.0062	.0191	.0486	.1051
	12	.0000	.0000	.0000	.0000	.0000	.0003	.0013	.0049	.0149	.0384
	13	.0000	.0000	.0000	.0000	.0000	.0000	.0002	.0009	.0035	.0106
	14	.0000	.0000	.0000	.0000	.0000	.0000	.0000	.0001	.0006	.0021
	15	.0000	.0000	.0000	.0000	.0000	.0000	.0000	.0000	.0001	.0003
	16	.0000	.0000	.0000	.0000	.0000	.0000	.0000	.0000	.0000	.0000
17	1	.5819	.8332	.9369	.9775	.9925	.9977	.9993	.9998	1.0000	1.0000
	2	.2078	.5182	.7475	.8818	.9499	.9807	.9933	.9979	.9994	.9999
	3	.0503	.2382	.4802	.6904	.8363	.9226	.9673	.9877	.9959	.9988
	4	.0088	.0826	.2444	.4511	.6470	.7981	.8972	.9536	.9816	.9936
	5	.0012	.0221	.0987	.2418	.4261	.6113	.7652	.8740	.9404	.9755
	6	.0001	.0047	.0319	.1057	.2347	.4032	.5803	.7361	.8529	.9283
	7	.0000	.0008	.0083	.0377	.1071	.2248	.3812	.5522	.7098	.8338
	8	.0000	.0001	.0017	.0109	.0402	.1046	.2128	.3595	.5257	.6855
	9	.0000	.0000	.0003	.0026	.0124	.0403	.0994	.1989	.3374	.5000
	10	.0000	.0000	.0000	.0005	.0031	.0127	.0383	.0919	.1834	.3145
	11	.0000	.0000	.0000	.0001	.0006	.0032	.0120	.0348	.0826	.1662
	12	.0000	.0000	.0000	.0000	.0001	.0007	.0030	.0106	.0301	.0717
	13	.0000	.0000	.0000	.0000	.0000	.0001	.0006	.0025	.0086	.0245
	14	.0000	.0000	.0000	.0000	.0000	.0000	.0000	.0005	.0019	.0064
	15	.0000	.0000	.0000	.0000	.0000	.0000	.0000	.0001	.0003	.0012
	16	.0000	.0000	.0000	.0000	.0000	.0000	.0000	.0000	.0000	.0001
	17	.0000	.0000	.0000	.0000	.0000	.0000	.0000	.0000	.0000	.0000
18	1	.6028	.8499	.9464	.9820	.9944	.9984	.9996	.9999	1.0000	1.0000
	2	.2265	.5497	.7759	.9009	.9605	.9858	.9954	.9987	.9997	.9999
	3	.0581	.2662	.5203	.7287	.8647	.9400	.9764	.9918	.9975	.9993
	4	.0109	.0982	.2798	.4990	.6943	.8354	.9217	.9672	.9880	.9962
	5	.0015	.0282	.1206	.2836	.4813	.6673	.8114	.9058	.9589	.9846
	6	.0002	.0064	.0419	.1329	.2825	.4656	.6450	.7912	.8923	.9519
	7	.0000	.0012	.0118	.0513	.1390	.2783	.4509	.6257	.7742	.8811
	8	.0000	.0002	.0027	.0163	.0569	.1407	.2717	.4366	.6085	.7597
	9	.0000	.0000	.0005	.0043	.0193	.0596	.1391	.2632	.4222	.5927
	10	.0000	.0000	.0001	.0009	.0054	.0210	.0597	.1347	.2527	.4073
	11	.0000	.0000	.0000	.0002	.0012	.0061	.0212	.0576	.1280	.2403
	12	.0000	.0000	.0000	.0000	.0002	.0014	.0062	.0203	.0537	.1189
	13	.0000	.0000	.0000	.0000	.0000	.0003	.0014	.0058	.0183	.0481
	14	.0000	.0000	.0000	.0000	.0000	.0000	.0003	.0013	.0049	.0154
	15	.0000	.0000	.0000	.0000	.0000	.0000	.0000	.0002	.0010	.0038
	16	.0000	.0000	.0000	.0000	.0000	.0000	.0000	.0000	.0001	.0007
	17	.0000	.0000	.0000	.0000	.0000	.0000	.0000	.0000	.0000	.0001
	18	.0000	.0000	.0000	.0000	.0000	.0000	.0000	.0000	.0000	.0000
19	1	.6226	.8649	.9544	.9856	.9958	.9989	.9997	.9999	1.0000	1.0000
	2	.2453	.5797	.8015	.9171	.9690	.9896	.9969	.9992	.9998	1.0000

(Continued)

Table B.1   Binomial Distribution Probabilities (Continued)

n	x´	.05	.10	.15	.20	.25	.30	.35	.40	.45	.50
21	8	.0000	.0006	.0083	.0431	.1299	.2770	.4635	.6505	.8029	.9054
	9	.0000	.0001	.0020	.0144	.0561	.1477	.2941	.4763	.6587	.8083
	10	.0000	.0000	.0004	.0041	.0206	.0676	.1632	.3086	.4883	.6682
	11	.0000	.0000	.0001	.0010	.0064	.0264	.0772	.1744	.3210	.5000
	12	.0000	.0000	.0000	.0002	.0017	.0087	.0313	.0849	.1841	.3318
	13	.0000	.0000	.0000	.0000	.0004	.0024	.0108	.0352	.0908	.1917
	14	.0000	.0000	.0000	.0000	.0001	.0006	.0031	.0123	.0379	.0946
	15	.0000	.0000	.0000	.0000	.0000	.0001	.0007	.0036	.0132	.0392
	16	.0000	.0000	.0000	.0000	.0000	.0000	.0001	.0008	.0037	.0133
	17	.0000	.0000	.0000	.0000	.0000	.0000	.0000	.0002	.0008	.0036
	18	.0000	.0000	.0000	.0000	.0000	.0000	.0000	.0000	.0001	.0007
	19	.0000	.0000	.0000	.0000	.0000	.0000	.0000	.0000	.0000	.0001
	20	.0000	.0000	.0000	.0000	.0000	.0000	.0000	.0000	.0000	.0000
	21	.0000	.0000	.0000	.0000	.0000	.0000	.0000	.0000	.0000	.0000
22	1	.6765	.9015	.9720	.9926	.9982	.9966	.9999	1.0000	1.0000	1.0000
	2	.3018	.6608	.8633	.9520	.9851	.9959	.9990	.9998	1.0000	1.0000
	3	.0948	.3800	.6618	.8455	.9394	.9793	.9399	.9984	.9997	.9999
	4	.0222	.1719	.4248	.6680	.8376	.9319	.9755	.9924	.9980	.9996
	5	.0040	.0621	.2262	.4571	.6765	.8355	.9284	.9734	.9917	.9978
	6	.0006	.0182	.0999	.2674	.4832	.6866	.8371	.9278	.9729	.9915
	7	.0001	.0044	.0368	.1330	.3006	.5058	.6978	.8416	.9295	.9738
	8	.0000	.0009	.0114	.0561	.1615	.3287	.5264	.7102	.8482	.9331
	9	.0000	.0001	.0030	.0201	.0746	.1865	.3534	.5460	.7236	.8569
	10	.0000	.0000	.0007	.0061	.0295	.0916	.2084	.3756	.5650	.7383
	11	.0000	.0000	.0001	.0016	.0100	.0387	.1070	.2281	.3963	.5841
	12	.0000	.0000	.0000	.0003	.0029	.0140	.0474	.1207	.2457	.4159
	13	.0000	.0000	.0000	.0001	.0007	.0043	.0180	.0551	.1328	.2617
	14	.0000	.0000	.0000	.0000	.0001	.0011	.0058	.0215	.0617	.1431
	15	.0000	.0000	.0000	.0000	.0000	.0002	.0015	.0070	.0243	.0669
	16	.0000	.0000	.0000	.0000	.0000	.0000	.0003	.0019	.0080	.0262
	17	.0000	.0000	.0000	.0000	.0000	.0000	.0001	.0004	.0021	.0085
	18	.0000	.0000	.0000	.0000	.0000	.0000	.0000	.0001	.0005	.0022
	19	.0000	.0000	.0000	.0000	.0000	.0000	.0000	.0000	.0001	.0004
	20	.0000	.0000	.0000	.0000	.0000	.0000	.0000	.0000	.0000	.0001
	21	.0000	.0000	.0000	.0000	.0000	.0000	.0000	.0000	.0000	.0000
	22	.0000	.0000	.0000	.0000	.0000	.0000	.0000	.0000	.0000	.0000
23	1	.6926	.9114	.9762	.9941	.9987	.9997	1.0000	1.0000	1.0000	1.0000
	2	.3206	.6849	.8796	.9602	.9884	.9970	.9993	.9999	1.0000	1.0000
	3	.1052	.4080	.6920	.8668	.9508	.9843	.9957	.9990	1.0000	1.0000
	4	.0258	.1927	.4604	.7035	.8630	.9462	.9819	.9948	.9988	.9998
	5	.0049	.0731	.2560	.4993	.7168	.8644	.9449	.9810	.9945	.9987
	6	.0008	.0226	.1189	.3053	.5315	.7312	.8691	.9460	.9814	.9947
	7	.0001	.0058	.0463	.1598	.3463	.5601	.7466	.8760	.9490	.9827

(Continued)

Table B.1   Binomial Distribution Probabilities (Continued)

n	x'	.05	.10	.15	.20	θ .25	.30	.35	.40	.45	.50
23	8	.0000	.0012	.0152	.0715	.1963	.3819	.5864	.7627	.8848	.9534
	9	.0000	.0002	.0042	.0273	.0963	.2291	.4140	.6116	.7797	.8950
	10	.0000	.0000	.0010	.0089	.0408	.1201	.2592	.4438	.6364	.7976
	11	.0000	.0000	.0002	.0025	.0149	.0546	.1425	.2871	.4722	.6612
	12	.0000	.0000	.0000	.0006	.0046	.0214	.0682	.1636	.3135	.5000
	13	.0000	.0000	.0000	.0001	.0012	.0072	.0283	.0813	.1836	.3388
	14	.0000	.0000	.0000	.0000	.0003	.0021	.0100	.0349	.0937	.2024
	15	.0000	.0000	.0000	.0000	.0001	.0005	.0030	.0128	.0411	.1050
	16	.0000	.0000	.0000	.0000	.0000	.0001	.0008	.0040	.0153	.0466
	17	.0000	.0000	.0000	.0000	.0000	.0000	.0002	.0010	.0048	.0173
	18	.0000	.0000	.0000	.0000	.0000	.0000	.0000	.0002	.0012	.0053
	19	.0000	.0000	.0000	.0000	.0000	.0000	.0000	.0000	.0002	.0013
	20	.0000	.0000	.0000	.0000	.0000	.0000	.0000	.0000	.0000	.0002
	21	.0000	.0000	.0000	.0000	.0000	.0000	.0000	.0000	.0000	.0000
	22	.0000	.0000	.0000	.0000	.0000	.0000	.0000	.0000	.0000	.0000
	23	.0000	.0000	.0000	.0000	.0000	.0000	.0000	.0000	.0000	.0000
24	1	.7080	.9202	.9798	.9953	.9990	.9998	1.0000	1.0000	1.0000	1.0000
	2	.3391	.7075	.8941	.9669	.9910	.9978	.9995	.9999	1.0000	1.0000
	3	.1159	.4357	.7202	.8855	.9602	.9881	.9970	.9993	.9999	1.0000
	4	.0298	.2143	.4951	.7361	.8850	.9576	.9867	.9965	.9992	.9999
	5	.0060	.0851	.2866	.5401	.7534	.8889	.9578	.9866	.9964	.9992
	6	.0010	.0277	.1394	.3441	.5778	.7712	.8956	.9600	.9873	.9967
	7	.0001	.0075	.0572	.1889	.3926	.6114	.7894	.9040	.9636	.9887
	8	.0000	.0017	.0199	.0892	.2338	.4353	.6425	.8081	.9137	.9680
	9	.0000	.0003	.0059	.0362	.1213	.2750	.4743	.6721	.8270	.9242
	10	.0000	.0001	.0015	.0126	.0547	.1528	.3134	.5109	.7009	.8463
	11	.0000	.0000	.0003	.0038	.0213	.0742	.1833	.3498	.5461	.7294
	12	.0000	.0000	.0001	.0010	.0072	.0314	.0942	.2130	.3849	.5806
	13	.0000	.0000	.0000	.0002	.0021	.0115	.0423	.1143	.2420	.4194
	14	.0000	.0000	.0000	.0000	.0005	.0036	.0164	.0535	.1341	.2706
	15	.0000	.0000	.0000	.0000	.0001	.0010	.0055	.0217	.0648	.1537
	16	.0000	.0000	.0000	.0000	.0000	.0002	.0016	.0075	.0269	.0758
	17	.0000	.0000	.0000	.0000	.0000	.0000	.0004	.0022	.0095	.0320
	18	.0000	.0000	.0000	.0000	.0000	.0000	.0001	.0005	.0028	.0113
	19	.0000	.0000	.0000	.0000	.0000	.0000	.0000	.0001	.0007	.0033
	20	.0000	.0000	.0000	.0000	.0000	.0000	.0000	.0000	.0001	.0008
	21	.0000	.0000	.0000	.0000	.0000	.0000	.0000	.0000	.0000	.0001
	22	.0000	.0000	.0000	.0000	.0000	.0000	.0000	.0000	.0000	.0000
	23	.0000	.0000	.0000	.0000	.0000	.0000	.0000	.0000	.0000	.0000
	24	.0000	.0000	.0000	.0000	.0000	.0000	.0000	.0000	.0000	.0000
25	1	.7226	.9282	.9828	.9962	.9992	.9999	1.0000	1.0000	1.0000	1.0000
	2	.3576	.7288	.9069	.9726	.9930	.9984	.9997	.9999	1.0000	1.0000
	3	.1271	.4629	.7463	.9018	.9679	.9910	.9979	.9996	.9999	1.0000

(Continued)

Table B.1   Binomial Distribution Probabilities (Continued)

n	x´	.05	.10	.15	.20	.25	.30	.35	.40	.45	.50
25	4	.0341	.2364	.5289	.7660	.9038	.9668	.9903	.9976	.9995	.9999
	5	.0072	.0980	.3179	.5793	.7863	.9095	.9680	.9905	.9977	.9995
	6	.0012	.0334	.1615	.3833	.6217	.8065	.9174	.9706	.9914	.9980
	7	.0002	.0095	.0695	.2200	.4389	.6593	.8266	.9264	.9742	.9927
	8	.0000	.0023	.0255	.1091	.2735	.4882	.6939	.8464	.9361	.9784
	9	.0000	.0005	.0080	.0468	.1494	.3231	.5332	.7265	.8660	.9461
	10	.0000	.0001	.0021	.0173	.0713	.1894	.3697	.5754	.7576	.8852
	11	.0000	.0000	.0005	.0056	.0297	.0978	.2288	.4142	.6157	.7878
	12	.0000	.0000	.0001	.0015	.0107	.0442	.1254	.2677	.4574	.6550
	13	.0000	.0000	.0000	.0004	.0034	.0175	.0604	.1538	.3063	.5000
	14	.0000	.0000	.0000	.0001	.0009	.0060	.0255	.0778	.1827	.3450
	15	.0000	.0000	.0000	.0000	.0002	.0018	.0093	.0344	.0960	.2122
	16	.0000	.0000	.0000	.0000	.0000	.0005	.0029	.0132	.0440	.1148
	17	.0000	.0000	.0000	.0000	.0000	.0001	.0008	.0043	.0174	.0539
	18	.0000	.0000	.0000	.0000	.0000	.0000	.0002	.0012	.0058	.0216
	19	.0000	.0000	.0000	.0000	.0000	.0000	.0000	.0003	.0016	.0073
	20	.0000	.0000	.0000	.0000	.0000	.0000	.0000	.0001	.0004	.0020
	21	.0000	.0000	.0000	.0000	.0000	.0000	.0000	.0000	.0001	.0005
	22	.0000	.0000	.0000	.0000	.0000	.0000	.0000	.0000	.0000	.0001
	23	.0000	.0000	.0000	.0000	.0000	.0000	.0000	.0000	.0000	.0000
	24	.0000	.0000	.0000	.0000	.0000	.0000	.0000	.0000	.0000	.0000

Source: *Handbook of Tables for Mathematics*, Fourth Edition, 1970. Reprinted with permission from the CRC Press Inc., Boca Raton, Florida.

Table B.2   Poisson Distribution, Right-Hand Tail Probabilities

x'	0.1	0.2	0.3	0.4	λ 0.5	0.6	0.7	0.8	0.9	1.0
0	1.0000	1.0000	1.0000	1.0000	1.0000	1.0000	1.0000	1.0000	1.0000	1.0000
1	.0952	.1813	.2592	.3297	.3935	.4512	.5034	.5507	.5934	.6321
2	.0047	.0175	.0369	.0616	.0902	.1219	.1558	.1912	.2275	.2642
3	.0002	.0011	.0036	.0079	.0144	.0231	.0341	.0474	.0629	.0803
4	.0000	.0001	.0003	.0008	.0018	.0034	.0058	.0091	.0135	.0190
5	.0000	.0000	.0000	.0001	.0002	.0004	.0008	.0014	.0023	.0037
6	.0000	.0000	.0000	.0000	.0000	.0000	.0001	.0002	.0003	.0006
7	.0000	.0000	.0000	.0000	.0000	.0000	.0000	.0000	.0000	.0001

x'	1.1	1.2	1.3	1.4	λ 1.5	1.6	1.7	1.8	1.9	2.0
0	1.0000	1.0000	1.0000	1.0000	1.0000	1.0000	1.0000	1.0000	1.0000	1.0000
1	.6671	.6988	.7275	.7534	.7769	.7981	.8173	.8347	.8504	.8647
2	.3010	.3374	.3732	.4082	.4422	.4751	.5068	.5372	.5663	.5940
3	.0996	.1205	.1429	.1665	.1912	.2166	.2428	.2694	.2963	.3233
4	.0257	.0338	.0431	.0537	.0656	.0788	.0932	.1087	.1253	.1429
5	.0054	.0077	.0107	.0143	.0186	.0237	.0296	.0364	.0441	.0527
6	.0010	.0015	.0022	.0032	.0045	.0060	.0080	.0104	.0132	.0166
7	.0001	.0003	.0004	.0006	.0009	.0013	.0019	.0026	.0034	.0045
8	.0000	.0000	.0001	.0001	.0002	.0003	.0004	.0006	.0008	.0011
9	.0000	.0000	.0000	.0000	.0000	.0000	.0001	.0001	.0002	.0002

x'	2.1	2.2	2.3	2.4	λ 2.5	2.6	2.7	2.8	2.9	3.0
0	1.0000	1.0000	1.0000	1.0000	1.0000	1.0000	1.0000	1.0000	1.0000	1.0000
1	.8775	.8892	.8997	.9093	.9179	.9257	.9328	.9392	.9450	.9502
2	.6204	.6454	.6691	.6916	.7127	.7326	.7513	.7689	.7854	.8009
3	.3504	.3773	.4040	.4303	.4562	.4816	.5064	.5305	.5540	.5768
4	.1614	.1806	.2007	.2213	.2424	.2640	.2859	.3081	.3304	.3528
5	.0621	.0725	.0838	.0959	.1088	.1226	.1371	.1523	.1682	.1847
6	.0204	.0249	.0300	.0357	.0420	.0490	.0567	.0651	.0742	.0839
7	.0059	.0075	.0094	.0116	.0142	.0172	.0206	.0244	.0287	.0335
8	.0015	.0020	.0026	.0033	.0042	.0053	.0066	.0081	.0099	.0119
9	.0003	.0005	.0006	.0009	.0011	.0015	.0019	.0024	.0031	.0038
10	.0001	.0001	.0001	.0002	.0003	.0004	.0005	.0007	.0009	.0011
11	.0000	.0000	.0000	.0000	.0001	.0001	.0001	.0002	.0002	.0003
12	.0000	.0000	.0000	.0000	.0000	.0000	.0000	.0000	.0001	.0001

x'	3.1	3.2	3.3	3.4	λ 3.5	3.6	3.7	3.8	3.9	4.0
0	1.0000	1.0000	1.0000	1.0000	1.0000	1.0000	1.0000	1.0000	1.0000	1.0000
1	.9550	.9592	.9631	.9666	.9698	.9727	.9753	.9776	.9798	.9817
2	.8153	.8288	.8414	.8532	.8641	.8743	.8838	.8926	.9008	.9084
3	.5988	.6201	.6406	.6603	.6792	.6973	.7146	.7311	.7469	.7619

(Continued)

Table B.2 Poisson Distribution Probabilities (Continued)

x'	3.1	3.2	3.3	3.4	3.5	3.6	3.7	3.8	3.9	4.0
4	.3752	.3975	.4197	.4416	.4634	.4848	.5058	.5265	.5468	.5665
5	.2018	.2194	.2374	.2558	.2746	.2936	.3128	.3322	.3516	.3712
6	.0943	.1054	.1171	.1295	.1424	.1559	.1699	.1844	.1994	.2149
7	.0388	.0446	.0510	.0579	.0653	.0733	.0818	.0909	.1005	.1107
8	.0142	.0168	.0198	.0231	.0267	.0308	.0352	.0401	.0454	.0511
9	.0047	.0057	.0069	.0083	.0099	.0117	.0137	.0160	.0185	.0214
10	.0014	.0018	.0022	.0027	.0033	.0040	.0048	.0058	.0069	.0081
11	.0004	.0005	.0006	.0008	.0010	.0013	.0016	.0019	.0023	.0028
12	.0001	.0001	.0002	.0002	.0003	.0004	.0005	.0006	.0007	.0009
13	.0000	.0000	.0000	.0001	.0001	.0001	.0001	.0002	.0002	.0003
14	.0000	.0000	.0000	.0000	.0000	.0000	.0000	.0000	.0001	.0001

x'	4.1	4.2	4.3	4.4	4.5	4.6	4.7	4.8	4.9	5.0
0	1.0000	1.0000	1.0000	1.0000	1.0000	1.0000	1.0000	1.0000	1.0000	1.0000
1	.9834	.9850	.9864	.9877	.9889	.9899	.9909	.9918	.9926	.9933
2	.9155	.9220	.9281	.9337	.9389	.9437	.9482	.9523	.9561	.9596
3	.7762	.7898	.8026	.8149	.8264	.8374	.8477	.8575	.8667	.8753
4	.5858	.6046	.6228	.6406	.6577	.6743	.6903	.7058	.7207	.7350
5	.3907	.4102	.4296	.4488	.4679	.4868	.5054	.5237	.5418	.5595
6	.2307	.2469	.2633	.2801	.2971	.3142	.3316	.3490	.3665	.3840
7	.1214	.1325	.1442	.1564	.1689	.1820	.1954	.2092	.2233	.2378
8	.0573	.0639	.0710	.0786	.0866	.0951	.1040	.1133	.1231	.1334
9	.0245	.0279	.0317	.0358	.0403	.0451	.0503	.0558	.0618	.0681
10	.0095	.0111	.0129	.0149	.0171	.0195	.0222	.0251	.0283	.0318
11	.0034	.0041	.0048	.0057	.0067	.0078	.0090	.0104	.0120	.0137
12	.0011	.0014	.0017	.0020	.0024	.0029	.0034	.0040	.0047	.0055
13	.0003	.0004	.0005	.0007	.0008	.0010	.0012	.0014	.0017	.0020
14	.0001	.0001	.0002	.0002	.0003	.0003	.0004	.0005	.0006	.0007
15	.0000	.0000	.0000	.0001	.0001	.0001	.0001	.0001	.0002	.0002
16	.0000	.0000	.0000	.0000	.0000	.0000	.0000	.0000	.0001	.0001

x'	5.1	5.2	5.3	5.4	5.5	5.6	5.7	5.8	5.9	6.0
0	1.0000	1.0000	1.0000	1.0000	1.0000	1.0000	1.0000	1.0000	1.0000	1.0000
1	.9939	.9945	.9950	.9955	.9959	.9963	.9967	.9970	.9973	.9975
2	.9628	.9658	.9686	.9711	.9734	.9756	.9776	.9794	.9811	.9826
3	.8835	.8912	.8984	.9052	.9116	.9176	.9232	.9285	.9334	.9380
4	.7487	.7619	.7746	.7867	.7983	.8094	.8200	.8300	.8396	.8488
5	.5769	.5939	.6105	.6267	.6425	.6579	.6728	.6873	.7013	.7149
6	.4016	.4191	.4365	.4539	.4711	.4881	.5050	.5217	.5381	.5543
7	.2526	.2676	.2829	.2983	.3140	.3297	.3456	.3616	.3776	.3937
8	.1440	.1551	.1665	.1783	.1905	.2030	.2159	.2290	.2424	.2560
9	.0748	.0819	.0894	.0974	.1056	.1143	.1234	.1328	.1426	.1528

(Continued)

Table B.2 Poisson Distribution Probabilities (Continued)

x'	5.1	5.2	5.3	5.4	5.5	5.6	5.7	5.8	5.9	6.0
10	.0356	.0397	.0441	.0488	.0538	.0591	.0648	.0708	.0772	.0839
11	.0156	.0177	.0200	.0225	.0253	.0282	.0314	.0349	.0386	.0426
12	.0063	.0073	.0084	.0096	.0110	.0125	.0141	.0160	.0179	.0201
13	.0024	.0028	.0033	.0038	.0045	.0051	.0059	.0068	.0078	.0088
14	.0008	.0010	.0012	.0014	.0017	.0020	.0023	.0027	.0031	.0036
15	.0003	.0003	.0004	.0005	.0006	.0007	.0009	.0010	.0012	.0014
16	.0001	.0001	.0001	.0002	.0002	.0002	.0003	.0004	.0004	.0005
17	.0000	.0000	.0000	.0001	.0001	.0001	.0001	.0001	.0001	.0002
18	.0000	.0000	.0000	.0000	.0000	.0000	.0000	.0000	.0000	.0001

x'	6.1	6.2	6.3	6.4	6.5	6.6	6.7	6.8	6.9	7.0
0	1.0000	1.0000	1.0000	1.0000	1.0000	1.0000	1.0000	1.0000	1.0000	1.0000
1	.9978	.9980	.9982	.9983	.9985	.9986	.9988	.9989	.9990	.9991
2	.9841	.9854	.9866	.9877	.9887	.9897	.9905	.9913	.9920	.9927
3	.9423	.9464	.9502	.9537	.9570	.9600	.9629	.9656	.9680	.9704
4	.8575	.8658	.8736	.8811	.8882	.8948	.9012	.9072	.9129	.9182
5	.7281	.7408	.7531	.7649	.7763	.7873	.7978	.8080	.8177	.8270
6	.5702	.5859	.6012	.6163	.6310	.6453	.6594	.6730	.6863	.6993
7	.4098	.4258	.4418	.4577	.4735	.4892	.5047	.5201	.5353	.5503
8	.2699	.2840	.2983	.3127	.3272	.3419	.3567	.3715	.3864	.4013
9	.1633	.1741	.1852	.1967	.2084	.2204	.2327	.2452	.2580	.2709
10	.0910	.0984	.1061	.1142	.1226	.1314	.1404	.1498	.1505	.1695
11	.0469	.0514	.0563	.0614	.0668	.0726	.0786	.0849	.0916	.0985
12	.0224	.0250	.0277	.0307	.0339	.0373	.0409	.0448	.0490	.0534
13	.0100	.0113	.0127	.0143	.0160	.0179	.0199	.0221	.0245	.0270
14	.0042	.0048	.0055	.0063	.0071	.0080	.0091	.0102	.0115	.0128
15	.0016	.0019	.0022	.0026	.0030	.0034	.0039	.0044	.0050	.0057
16	.0006	.0007	.0008	.0010	.0012	.0014	.0016	.0018	.0021	.0024
17	.0002	.0003	.0003	.0004	.0004	.0005	.0006	.0007	.0008	.0010
18	.0001	.0001	.0001	.0001	.0002	.0002	.0002	.0003	.0003	.0004
19	.0000	.0000	.0000	.0000	.0001	.0001	.0001	.0001	.0001	.0001

x'	7.1	7.2	7.3	7.4	7.5	7.6	7.7	7.8	7.9	8.0
0	1.0000	1.0000	1.0000	1.0000	1.0000	1.0000	1.0000	1.0000	1.0000	1.0000
1	.9992	.9993	.9993	.9994	.9994	.9995	.9995	.9996	.9996	.9997
2	.9933	.9939	.9944	.9949	.9953	.9957	.9961	.9964	.9967	.9970
3	.9725	.9745	.9764	.9781	.9797	.9812	.9826	.9839	.9851	.9862
4	.9233	.9281	.9326	.9368	.9409	.9446	.9482	.9515	.9547	.9576
5	.8359	.8445	.8527	.8605	.8679	.8751	.8819	.8883	.8945	.9004
6	.7119	.7241	.7360	.7474	.7586	.7693	.7797	.7897	.7994	.8088
7	.5651	.5796	.5940	.6080	.6218	.6354	.6486	.6616	.6743	.6866
8	.4162	.4311	.4459	.4607	.4754	.4900	.5044	.5188	.5330	.5470

(Continued)

Table B.2 Poisson Distribution Probabilities (Continued)

x'	λ 7.1	7.2	7.3	7.4	7.5	7.6	7.7	7.8	7.9	8.0
9	.2840	.2973	.3108	.3243	.3380	.3518	.3657	.3796	.3935	.4075
10	.1798	.1904	.2012	.2123	.2236	.2351	.2469	.2589	.2710	.2834
11	.1058	.1133	.1212	.1293	.1378	.1465	.1555	.1648	.1743	.1841
12	.0580	.0629	.0681	.0735	.0792	.0852	.0915	.0980	.1048	.1119
13	.0297	.0327	.0358	.0391	.0427	.0464	.0504	.0546	.0591	.0638
14	.0143	.0159	.0176	.0195	.0216	.0238	.0261	.0286	.0313	.0342
15	.0065	.0073	.0082	.0092	.0103	.0114	.0127	.0141	.0156	.0173
16	.0028	.0031	.0036	.0041	.0046	.0052	.0059	.0066	.0074	.0082
17	.0011	.0013	.0015	.0017	.0020	.0022	.0026	.0029	.0033	.0037
18	.0004	.0005	.0006	.0007	.0008	.0009	.0011	.0012	.0014	.0016
19	.0002	.0002	.0002	.0003	.0003	.0004	.0004	.0005	.0006	.0006
20	.0001	.0001	.0001	.0001	.0001	.0001	.0002	.0002	.0002	.0003
21	.0000	.0000	.0000	.0000	.0000	.0000	.0001	.0001	.0001	.0001

x'	λ 8.1	8.2	8.3	8.4	8.5	8.6	8.7	8.8	8.9	9.0
0	1.0000	1.0000	1.0000	1.0000	1.0000	1.0000	1.0000	1.0000	1.0000	1.0000
1	.9997	.9997	.9998	.9998	.9998	.9998	.9998	.9998	.9999	.9999
2	.9972	.9975	.9977	.9979	.9981	.9982	.9984	.9985	.9987	.9988
3	.9873	.9882	.9891	.9900	.9907	.9914	.9921	.9927	.9932	.9938
4	.9604	.9630	.9654	.9677	.9699	.9719	.9738	.9756	.9772	.9788
5	.9060	.9113	.9163	.9211	.9256	.9299	.9340	.9379	.9416	.9450
6	.8178	.8264	.8347	.8427	.8504	.8578	.8648	.8716	.8781	.8843
7	.6987	.7104	.7219	.7330	.7438	.7543	.7645	.7744	.7840	.7932
8	.5609	.5746	.5881	.6013	.6144	.6272	.6398	.6522	.6643	.6761
9	.4214	.4353	.4493	.4631	.4769	.4906	.5042	.5177	.5311	.5443
10	.2959	.3085	.3212	.3341	.3470	.3600	.3731	.3863	.3994	.4126
11	.1942	.2045	.2150	.2257	.2366	.2478	.2591	.2706	.2822	.2940
12	.1193	.1269	.1348	.1429	.1513	.1600	.1689	.1780	.1874	.1970
13	.0687	.0739	.0793	.0850	.0909	.0971	.1035	.1102	.1171	.1242
14	.0372	.0405	.0439	.0476	.0514	.0555	.0597	.0642	.0689	.0739
15	.0190	.0209	.0229	.0251	.0274	.0299	.0325	.0353	.0383	.0415
16	.0092	.0102	.0113	.0125	.0138	.0152	.0168	.0184	.0202	.0220
17	.0042	.0047	.0053	.0059	.0066	.0074	.0082	.0091	.0101	.0111
18	.0018	.0021	.0023	.0027	.0030	.0034	.0038	.0043	.0048	.0053
19	.0008	.0009	.0010	.0011	.0013	.0015	.0017	.0019	.0022	.0024
20	.0003	.0003	.0004	.0005	.0005	.0006	.0007	.0008	.0009	.0011
21	.0001	.0001	.0002	.0002	.0002	.0002	.0003	.0003	.0004	.0004
22	.0000	.0000	.0001	.0001	.0001	.0001	.0001	.0001	.0002	.0002
23	.0000	.0000	.0000	.0000	.0000	.0000	.0000	.0000	.0001	.0001

x'	λ 9.1	9.2	9.3	9.4	9.5	9.6	9.7	9.8	9.9	10.0
0	1.0000	1.0000	1.0000	1.0000	1.0000	1.0000	1.0000	1.0000	1.0000	1.0000

(Continued)

Table B.2   Poisson Distribution Probabilities (Continued)

x'	9.1	9.2	9.3	9.4	λ 9.5	9.6	9.7	9.8	9.9	10.0
1	.9999	.9999	.9999	.9999	.9999	.9999	.9999	.9999	1.0000	1.0000
2	.9989	.9990	.9991	.9991	.9992	.9993	.9993	.9994	.9995	.9995
3	.9942	.9947	.9951	.9955	.9958	.9962	.9965	.9967	.9970	.9972
4	.9802	.9816	.9828	.9840	.9851	.9862	.9871	.9880	.9889	.9897
5	.9483	.9514	.9544	.9571	.9597	.9622	.9645	.9667	.9688	.9707
6	.8902	.8959	.9014	.9065	.9115	.9162	.9207	.9250	.9290	.9329
7	.8022	.8108	.8192	.8273	.8351	.8426	.8498	.8567	.8634	.8699
8	.6877	.6990	.7101	.7208	.7313	.7416	.7515	.7612	.7706	.7798
9	.5574	.5704	.5832	.5958	.6082	.6204	.6324	.6442	.6558	.6672
10	.4258	.4389	.4521	.4651	.4782	.4911	.5040	.5168	.5295	.5421
11	.3059	.3180	.3301	.3424	.3547	.3671	.3795	.3920	.4045	.4170
12	.2068	.2168	.2270	.2374	.2480	.2588	.2697	.2807	.2919	.3032
13	.1316	.1393	.1471	.1552	.1636	.1721	.1809	.1899	.1991	.2084
14	.0790	.0844	.0900	.0958	.1019	.1081	.1147	.1214	.1284	.1355
15	.0448	.0483	.0520	.0559	.0600	.0643	.0688	.0735	.0784	.0835
16	.0240	.0262	.0285	.0309	.0335	.0362	.0391	.0421	.0454	.0487
17	.0122	.0135	.0148	.0162	.0177	.0194	.0211	.0230	.0249	.0270
18	.0059	.0066	.0073	.0081	.0089	.0098	.0108	.0119	.0130	.0143
19	.0027	.0031	.0034	.0038	.0043	.0048	.0053	.0059	.0065	.0072
20	.0012	.0014	.0015	.0017	.0020	.0022	.0025	.0028	.0031	.0035
21	.0005	.0006	.0007	.0008	.0009	.0010	.0011	.0013	.0014	.0016
22	.0002	.0002	.0003	.0003	.0004	.0004	.0005	.0005	.0006	.0007
23	.0001	.0001	.0001	.0001	.0001	.0002	.0002	.0002	.0003	.0003
24	.0000	.0000	.0000	.0000	.0001	.0001	.0001	.0001	.0001	.0001

x'	11	12	13	14	λ 15	16	17	18	19	20
0	1.0000	1.0000	1.0000	1.0000	1.0000	1.0000	1.0000	1.0000	1.0000	1.0000
1	1.0000	1.0000	1.0000	1.0000	1.0000	1.0000	1.0000	1.0000	1.0000	1.0000
2	.9998	.9999	1.0000	1.0000	1.0000	1.0000	1.0000	1.0000	1.0000	1.0000
3	.9988	.9995	.9998	.9999	1.0000	1.0000	1.0000	1.0000	1.0000	1.0000
4	.9951	.9977	.9990	.9995	.9998	.9999	1.0000	1.0000	1.0000	1.0000
5	.9849	.9924	.9963	.9982	.9991	.9996	.9998	.9999	1.0000	1.0000
6	.9625	.9797	.9893	.9945	.9972	.9986	.9993	.9997	.9998	.9999
7	.9214	.9542	.9741	.9858	.9924	.9960	.9979	.9990	.9995	.9997
8	.8568	.9105	.9460	.9684	.9820	.9900	.9946	.9971	.9985	.9992
9	.7680	.8450	.9002	.9379	.9626	.9780	.9874	.9929	.9961	.9979
10	.6595	.7576	.8342	.8906	.9301	.9567	.9739	.9846	.9911	.9950
11	.5401	.6528	.7483	.8243	.8815	.9226	.9509	.9696	.9817	.9892
12	.4207	.5384	.6468	.7400	.8152	.8730	.9153	.9451	.9653	.9786
13	.3113	.4240	.5369	.6415	.7324	.8069	.8650	.9083	.9394	.9610
14	.2187	.3185	.4270	.5356	.6368	.7255	.7991	.8574	.9016	.9339
15	.1460	.2280	.3249	.4296	.5343	.6325	.7192	.7919	.8503	.8951
16	.0926	.1556	.2364	.3306	.4319	.5333	.6285	.7133	.7852	.8435

(Continued)

Table B.2   Poisson Distribution Probabilities (Continued)

x'	11	12	13	14	λ 15	16	17	18	19	20
17	.0559	.1013	.1645	.2441	.3359	.4340	.5323	.6250	.7080	.7789
18	.0322	.0630	.1095	.1728	.2511	.3407	.4360	.5314	.6216	.7030
19	.0177	.0374	.0698	.1174	.1805	.2577	.3450	.4378	.5305	.6186
20	.0093	.0213	.0427	.0765	.1248	.1878	.2637	.3491	.4394	.5297
21	.0047	.0116	.0250	.0479	.0830	.1318	.1945	.2693	.3528	.4409
22	.0023	.0061	.0141	.0288	.0531	.0892	.1385	.2009	.2745	.3563
23	.0010	.0030	.0076	.0167	.0327	.0582	.0953	.1449	.2069	.2794
24	.0005	.0015	.0040	.0093	.0195	.0367	.0633	.1011	.1510	.2125
25	.0002	.0007	.0020	.0050	.0112	.0223	.0406	.0683	.1067	.1568
26	.0001	.0003	.0010	.0026	.0062	.0131	.0252	.0446	.0731	.1122
27	.0000	.0001	.0005	.0013	.0033	.0075	.0152	.0282	.0486	.0779
28	.0000	.0001	.0002	.0006	.0017	.0041	.0088	.0173	.0313	.0525
29	.0000	.0000	.0001	.0003	.0009	.0022	.0050	.0103	.0195	.0343
28	.0000	.0001	.0002	.0006	.0017	.0041	.0088	.0173	.0313	.0525
29	.0000	.0000	.0001	.0003	.0009	.0022	.0050	.0103	.0195	.0343
30	.0000	.0000	.0000	.0001	.0004	.0011	.0027	.0059	.0118	.0218
31	.0000	.0000	.0000	.0001	.0002	.0006	.0014	.0033	.0070	.0135
32	.0000	.0000	.0000	.0000	.0001	.0003	.0007	.0018	.0040	.0081
33	.0000	.0000	.0000	.0000	.0000	.0001	.0004	.0010	.0022	.0047
34	.0000	.0000	.0000	.0000	.0000	.0001	.0002	.0005	.0012	.0027
35	.0000	.0000	.0000	.0000	.0000	.0000	.0001	.0002	.0006	.0015
36	.0000	.0000	.0000	.0000	.0000	.0000	.0000	.0001	.0003	.0008
37	.0000	.0000	.0000	.0000	.0000	.0000	.0000	.0001	.0002	.0004
38	.0000	.0000	.0000	.0000	.0000	.0000	.0000	.0000	.0001	.0002
39	.0000	.0000	.0000	.0000	.0000	.0000	.0000	.0000	.0000	.0001
40	.0000	.0000	.0000	.0000	.0000	.0000	.0000	.0000	.0000	.0001

Source: *Handbook of Tables for Mathematics*, Fourth Edition, 1970. Reprinted with permission from the CRC Press Inc., Boca Raton, Florida.

Table B.3 Normal Distribution, Right-Hand Tail Probabilities

z	.00	.01	.02	.03	.04	.05	.06	.07	.08	.09
0.0	.5000	.4960	.4920	.4880	.4840	.4801	.4761	.4721	.4681	.4641
0.1	.4602	.4562	.4522	.4483	.4443	.4404	.4364	.4325	.4286	.4247
0.2	.4207	.4168	.4129	.4090	.4052	.4013	.3974	.3936	.3897	.3859
0.3	.3821	.3783	.3745	.3707	.3669	.3632	.3594	.3557	.3520	.3483
0.4	.3446	.3409	.3372	.3336	.3300	.3264	.3228	.3192	.3156	.3121
0.5	.3085	.3050	.3015	.2981	.2946	.2912	.2877	.2843	.2810	.2776
0.6	.2743	.2709	.2676	.2643	.2611	.2578	.2546	.2514	.2483	.2451
0.7	.2420	.2389	.2358	.2327	.2296	.2266	.2236	.2206	.2177	.2148
0.8	.2119	.2090	.2061	.2033	.2005	.1977	.1949	.1922	.1894	.1867
0.9	.1841	.1814	.1788	.1762	.1736	.1711	.1685	.1660	.1635	.1611
1.0	.1587	.1562	.1539	.1515	.1492	.1469	.1446	.1423	.1401	.1379
1.1	.1357	.1335	.1314	.1292	.1271	.1251	.1230	.1210	.1190	.1170
1.2	.1151	.1131	.1112	.1093	.1075	.1056	.1038	.1020	.1003	.0985
1.3	.0968	.0951	.0934	.0918	.0901	.0885	.0869	.0853	.0838	.0823
1.4	.0808	.0793	.0778	.0764	.0749	.0735	.0721	.0708	.0694	.0681
1.5	.0668	.0655	.0643	.0630	.0618	.0606	.0594	.0582	.0571	.0559
1.6	.0548	.0537	.0526	.0516	.0505	.0495	.0485	.0475	.0465	.0455
1.7	.0446	.0436	.0427	.0418	.0409	.0401	.0392	.0384	.0375	.0367
1.8	.0359	.0351	.0344	.0336	.0329	.0322	.0314	.0307	.0301	.0294
1.9	.0287	.0281	.0274	.0268	.0262	.0256	.0250	.0244	.0239	.0233
2.0	.0228	.0222	.0217	.0212	.0207	.0202	.0197	.0192	.0188	.0183
2.1	.0179	.0174	.0170	.0166	.0162	.0158	.0154	.0150	.0146	.0143
2.2	.0139	.0136	.0132	.0129	.0125	.0122	.0119	.0116	.0113	.0110
2.3	.0107	.0104	.0102	.0099	.0096	.0094	.0091	.0089	.0087	.0084
2.4	.0082	.0080	.0078	.0075	.0073	.0071	.0069	.0068	.0066	.0064
2.5	.0062	.0060	.0059	.0057	.0055	.0054	.0052	.0051	.0049	.0048
2.6	.0047	.0045	.0044	.0043	.0041	.0040	.0039	.0038	.0037	.0036
2.7	.0035	.0034	.0033	.0032	.0031	.0030	.0029	.0028	.0027	.0026
2.8	.0026	.0025	.0024	.0023	.0023	.0022	.0021	.0021	.0020	.0019
2.9	.0019	.0018	.0018	.0017	.0016	.0016	.0015	.0015	.0014	.0014
3.0	.0013	.0013	.0013	.0012	.0012	.0011	.0011	.0011	.0010	.0010
3.1	.0010	.0009	.0009	.0009	.0008	.0008	.0008	.0008	.0007	.0007
3.2	.0007	.0007	.0006	.0006	.0006	.0006	.0006	.0005	.0005	.0005
3.3	.0005	.0005	.0005	.0004	.0004	.0004	.0004	.0004	.0004	.0003
3.4	.0003	.0003	.0003	.0003	.0003	.0003	.0003	.0003	.0003	.0002

Source: *Statistical Inference*, by G. Casella and R.L. Berger. Copyright © 1990 by Wadsworth, Inc. Permission to reprint by Brooks/Cole Publishing Company, Pacific Grove, California.

Table B.4 Percentage Points of the *t*-Distribution

d.f	1T=0.4† 2T=0.8	0.25 0.5	0.1 0.2	0.05 0.1	0.025 0.05	0.01 0.02	0.005 0.01	0.0025 0.005	0.002 0.002	0.0005 0.001
1	0.325	1.000	3.078	6.314	12.706	31.821	63.657	127.320	318.310	636.620
2	.289	0.816	1.886	2.920	4.303	6.965	9.925	14.089	22.327	31.598
3	.277	.765	1.638	2.353	3.182	4.541	5.841	7.453	10.214	12.924
4	.271	.741	1.533	2.132	2.776	3.747	4.604	5.598	7.173	8.610
5	0.267	0.727	1.476	2.015	2.571	3.365	4.032	4.773	5.893	6.869
6	.265	.718	1.440	1.943	2.447	3.143	3.707	4.317	5.208	5.959
7	.263	.711	1.415	1.895	2.365	2.998	3.499	4.029	4.785	5.408
8	.262	.706	1.397	1.860	2.306	2.896	3.355	3.833	4.501	5.041
9	.261	.703	1.383	1.833	2.262	2.821	3.250	3.690	4.297	4.781
10	0.260	0.700	1.372	1.812	2.228	2.764	3.169	3.581	4.144	4.587
11	.260	.697	1.363	1.796	2.201	2.718	3.106	3.497	4.025	4.437
12	.259	.695	1.356	1.782	2.179	2.681	3.055	3.428	3.930	4.318
13	.259	.694	1.350	1.771	2.160	2.650	3.012	3.372	3.852	4.221
14	.258	.692	1.345	1.761	2.145	2.624	2.977	3.326	3.787	4.140
15	0.258	0.691	1.341	1.753	2.131	2.602	2.947	3.286	3.733	4.073
16	.258	.690	1.337	1.746	2.120	2.583	2.921	3.252	3.686	4.015
17	.257	.689	1.333	1.740	2.110	2.567	2.898	3.222	3.646	3.965
18	.257	.688	1.330	1.734	2.101	2.552	2.878	3.197	3.610	3.922
19	.257	.688	1.328	1.729	2.093	2.539	2.861	3.174	3.579	3.883

(Continued)

Table B.4   Percentage Points of the *t*-Distribution (Continued)

d.f	1T=0.4 / 2T=0.8	0.25 / 0.5	0.1 / 0.2	0.05 / 0.1	0.025 / 0.05	0.01 / 0.02	0.005 / 0.01	0.0025 / 0.005	0.001 / 0.002	0.0005 / 0.001
20	0.257	0.687	1.325	1.725	2.086	2.528	2.845	3.153	3.552	3.850
21	.257	.686	1.323	1.721	2.080	2.518	2.831	3.135	3.527	3.819
22	.256	.686	1.321	1.717	2.074	2.508	2.819	3.119	3.505	3.792
23	.256	.685	1.319	1.714	2.069	2.500	2.807	3.104	3.485	3.767
24	.256	.685	1.318	1.711	2.064	2.492	2.797	3.091	3.467	3.745
25	0.256	0.684	1.316	1.708	2.060	2.485	2.787	3.078	3.450	3.725
26	.256	.684	1.315	1.706	2.056	2.479	2.779	3.067	3.435	3.707
27	.256	.684	1.314	1.703	2.052	2.473	2.771	3.057	3.421	3.690
28	.256	.683	1.313	1.701	2.048	2.467	2.763	3.047	3.408	3.674
29	.256	.683	1.311	1.699	2.045	2.462	2.756	3.038	3.396	3.659
30	0.256	0.683	1.310	1.697	2.042	2.457	2.750	3.030	3.385	3.646
40	.255	.681	1.303	1.684	2.021	2.423	2.704	2.971	3.307	3.551
60	.254	.679	1.296	1.671	2.000	2.390	2.660	2.915	3.232	3.460
120	.254	.677	1.289	1.658	1.980	2.358	2.617	2.860	3.160	3.373
∞	.253	.674	1.282	1.645	1.960	2.326	2.576	2.807	3.090	3.291

† 1 T = area under one tail; 2 T = area under both tails.

Example: For 25 degrees of freedom (d.f.), $P(t > 2.060) = 0.025$ and $P(t < -2.060) = 0.025$ and $P(t < -2.060 \text{ or } t > 2.060) = 0.05$.

Source: *Biometrika Table for Statisticians*, Vol. I, Edited by E.S. Pearson and H.O. Hartley, Third edition, 1966. Reprinted with the permission of the Biometrika Trustees.

Table B.5   Upper Percentage Points of the Chi-Square Distribution[†]

Q ν	0.975	0.950	0.250	0.100	0.050	0.025	0.010
1	0.00	0.00	1.32	2.71	3.84	5.02	6.63
2	0.05	0.10	2.77	4.61	5.99	7.38	9.21
3	0.22	0.35	4.11	6.25	7.81	9.35	11.34
4	0.48	0.71	5.39	7.78	9.49	11.14	13.28
5	0.83	1.15	6.63	9.24	11.07	12.83	15.09
6	1.24	1.64	7.84	10.64	12.59	14.45	16.81
7	1.69	2.17	9.04	12.02	14.07	16.01	18.48
8	2.18	2.73	10.22	13.36	15.51	17.53	20.09
9	2.70	3.33	11.39	14.68	16.92	19.02	21.67
10	3.25	3.94	12.55	15.99	18.31	20.48	23.21
11	3.82	4.57	13.70	17.28	19.68	21.92	24.73
12	4.40	5.23	14.85	18.55	21.03	23.34	26.22
13	5.01	5.89	15.98	19.81	22.36	24.74	27.69
14	5.63	6.57	17.12	21.06	23.68	26.12	29.14
15	6.26	7.26	18.25	22.31	25.00	27.49	30.58
16	6.91	7.96	19.37	23.54	26.30	28.85	32.00
17	7.56	8.67	20.49	24.77	27.59	30.19	33.41
18	8.23	9.39	21.60	25.99	28.87	31.53	34.81
19	8.91	10.32	22.72	27.20	30.15	32.85	36.19
20	9.59	10.85	23.83	28.41	31.41	34.17	37.57
21	10.28	11.59	24.93	29.62	32.67	35.48	38.93
22	10.98	12.34	26.04	30.81	33.92	36.78	40.29
23	11.69	13.09	27.14	32.01	35.17	38.08	41.64
24	12.40	13.85	28.24	33.20	36.42	39.36	42.98
25	13.12	14.61	29.34	34.38	37.65	40.65	44.31
26	13.84	15.38	30.43	35.56	38.89	41.92	45.64
27	14.57	16.15	31.53	36.74	40.11	43.19	46.96
28	15.31	16.93	32.62	37.92	41.34	44.46	48.28
29	16.05	17.71	33.71	39.09	42.56	45.72	49.59
30	16.79	18.49	34.80	40.26	43.77	46.98	50.89
40	24.43	26.51	45.62	51.81	55.76	59.34	63.69
50	32.36	34.76	56.33	63.17	67.50	71.42	76.15
60	40.48	43.19	66.98	74.40	79.08	83.30	88.38
70	48.75	51.70	77.58	85.53	90.53	95.02	100.43
80	57.15	60.39	88.13	96.58	101.88	106.63	112.33
90	65.64	69.12	98.65	107.57	113.15	118.14	124.12
100	74.22	77.93	109.14	118.50	124.34	129.56	135.81

[†] ν is the degrees of freedom and Q is the area in the right tail. As an example, for 25 d.f., $P(\chi^2 > 37.6525) = 0.05$.

Source: Compiled from the ECSLIB program.

Table B.6a   Upper 1% Points of the F-Distribution[†]

m / n	1	2	3	4	5	6	7	8	9	10	12	15	20	24	30	40	60	120	∞
1	4052	4999.5	5403	5625	5764	5859	5928	5981	6022	6056	6106	6157	6209	6235	6261	6287	6313	6339	6366
2	98.50	99.00	99.17	99.25	99.30	99.33	99.36	99.37	99.39	99.40	99.42	99.43	99.45	99.46	99.47	99.47	99.48	99.49	99.50
3	34.12	30.82	29.46	28.71	28.24	27.91	27.67	27.49	27.35	27.23	27.05	26.87	26.69	26.60	26.50	26.41	26.32	26.22	26.13
4	21.20	18.00	16.69	15.98	15.52	15.21	14.98	14.80	14.66	14.55	14.37	14.20	14.02	13.93	13.84	13.75	13.65	13.56	13.46
5	16.26	13.27	12.06	11.39	10.97	10.67	10.46	10.29	10.16	10.05	9.89	9.72	9.55	9.47	9.38	9.29	9.20	9.11	9.02
6	13.75	10.92	9.78	9.15	8.75	8.47	8.26	8.10	7.98	7.87	7.72	7.56	7.40	7.31	7.23	7.14	7.06	6.97	6.88
7	12.25	9.55	8.45	7.85	7.46	7.19	6.99	6.84	6.72	6.62	6.47	6.31	6.16	6.07	5.99	5.91	5.82	5.74	5.65
8	11.26	8.65	7.59	7.01	6.63	6.37	6.18	6.03	5.91	5.81	5.67	5.52	5.36	5.28	5.20	5.12	5.03	4.95	4.86
9	10.56	8.02	6.99	6.42	6.06	5.80	5.61	5.47	5.35	5.26	5.11	4.96	4.81	4.73	4.65	4.57	4.48	4.40	4.31
10	10.04	7.56	6.55	5.99	5.64	5.39	5.20	5.06	4.94	4.85	4.71	4.56	4.41	4.33	4.25	4.17	4.08	4.00	3.91
11	9.65	7.21	6.22	5.67	5.32	5.07	4.89	4.74	4.63	4.54	4.40	4.25	4.10	4.02	3.94	3.86	3.78	3.69	3.60
12	9.33	6.93	5.95	5.41	5.06	4.82	4.64	4.50	4.39	4.30	4.16	4.01	3.86	3.78	3.70	3.62	3.54	3.45	3.36
13	9.07	6.70	5.74	5.21	4.86	4.62	4.44	4.30	4.19	4.10	3.96	3.82	3.66	3.59	3.51	3.43	3.34	3.25	3.17
14	8.86	6.51	5.56	5.04	4.69	4.46	4.28	4.14	4.03	3.94	3.80	3.66	3.51	3.43	3.35	3.27	3.18	3.09	3.00
15	8.68	6.36	5.42	4.89	4.56	4.32	4.14	4.00	3.89	3.80	3.67	3.52	3.37	3.29	3.21	3.13	3.05	2.96	2.87
16	8.53	6.23	5.29	4.77	4.44	4.20	4.03	3.89	3.78	3.69	3.55	3.41	3.26	3.18	3.10	3.02	2.93	2.84	2.75
17	8.40	6.11	5.18	4.67	4.34	4.10	3.93	3.79	3.68	3.59	3.46	3.31	3.16	3.08	3.00	2.92	2.83	2.75	2.65
18	8.29	6.01	5.09	4.58	4.25	4.01	3.84	3.71	3.60	3.51	3.37	3.23	3.08	3.00	2.92	2.84	2.75	2.66	2.57
19	8.18	5.93	5.01	4.50	4.17	3.94	3.77	3.63	3.52	3.43	3.30	3.15	3.00	2.92	2.84	2.76	2.67	2.58	2.49
20	8.10	5.85	4.94	4.43	4.10	3.87	3.70	3.56	3.46	3.37	3.23	3.09	2.94	2.86	2.78	2.69	2.61	2.52	2.42
21	8.02	5.78	4.87	4.37	4.04	3.81	3.64	3.51	3.40	3.31	3.17	3.03	2.88	2.80	2.72	2.64	2.55	2.46	2.36

(Continued)

Table B.6a Upper 1% Points of the *F*-Distribution (Continued)

m \ n	1	2	3	4	5	6	7	8	9	10	12	15	20	24	30	40	60	120	∞
22	7.95	5.72	4.82	4.31	3.99	3.76	3.59	3.45	3.35	3.26	3.12	2.98	2.83	2.75	2.67	2.58	2.50	2.40	2.31
23	7.88	5.66	4.76	4.26	3.94	3.71	3.54	3.41	3.30	3.21	3.07	2.93	2.78	2.70	2.62	2.54	2.45	2.35	2.26
24	7.82	5.61	4.72	4.22	3.90	3.67	3.50	3.36	3.26	3.17	3.03	2.89	2.74	2.66	2.58	2.49	2.40	2.31	2.21
25	7.77	5.57	4.68	4.18	3.85	3.63	3.46	3.32	3.22	3.13	2.99	2.85	2.70	2.62	2.54	2.45	2.36	2.27	2.17
26	7.72	5.53	4.64	4.14	3.82	3.59	3.42	3.29	3.18	3.09	2.96	2.81	2.66	2.58	2.50	2.42	2.33	2.23	2.13
27	7.68	5.49	4.60	4.11	3.78	3.56	3.39	3.26	3.15	3.06	2.93	2.78	2.63	2.55	2.47	2.38	2.29	2.20	2.10
28	7.64	5.45	4.57	4.07	3.75	3.53	3.36	3.23	3.12	3.03	2.90	2.75	2.60	2.52	2.44	2.35	2.26	2.17	2.06
29	7.60	5.42	4.54	4.04	3.73	3.50	3.33	3.20	3.09	3.00	2.87	2.73	2.57	2.49	2.41	2.33	2.23	2.14	2.03
30	7.56	5.39	4.51	4.02	3.70	3.47	3.30	3.17	3.07	2.98	2.84	2.70	2.55	2.47	2.39	2.30	2.21	2.11	2.01
40	7.31	5.18	4.31	3.83	3.51	3.29	3.12	2.99	2.89	2.80	2.66	2.52	2.37	2.29	2.20	2.11	2.02	1.92	1.80
60	7.08	4.98	4.13	3.65	3.34	3.12	2.95	2.82	2.72	2.63	2.50	2.35	2.20	2.12	2.03	1.94	1.84	1.73	1.60
120	6.85	4.79	3.95	3.48	3.17	2.96	2.79	2.66	2.56	2.47	2.34	2.19	2.03	1.95	1.86	1.76	1.66	1.53	1.38
∞	6.63	4.61	3.78	3.32	3.02	2.80	2.64	2.51	2.41	2.32	2.18	2.04	1.88	1.79	1.70	1.59	1.47	1.32	1.00

† m = degrees of freedom for the numerator and n = degrees of freedom for the denominator.

Source: *Handbook of Tables for Mathematics*, Fourth Edition, 1970. Reprinted with permission from the CRC Press Inc., Boca Raton, Florida.

Table B.6b   Upper 5% Points of the F-Distribution[†]

m \ n	1	2	3	4	5	6	7	8	9	10	12	15	20	24	30	40	60	120	∞
1	161.4	199.5	215.7	224.6	230.2	234.0	236.8	238.9	240.5	241.9	243.9	245.9	248.0	249.1	250.1	251.1	252.2	253.3	254.3
2	18.51	19.00	19.16	19.25	19.30	19.33	19.35	19.37	19.38	19.40	19.41	19.43	19.45	19.45	19.46	19.47	19.48	19.49	19.50
3	10.13	9.55	9.28	9.12	9.01	8.94	8.89	8.85	8.81	8.79	8.74	8.70	8.66	8.64	8.62	8.59	8.57	8.55	8.53
4	7.71	6.94	6.59	6.39	6.26	6.16	6.09	6.04	6.00	5.96	5.91	5.86	5.80	5.77	5.75	5.72	5.69	5.66	5.63
5	6.61	5.79	5.41	5.19	5.05	4.95	4.88	4.82	4.77	4.74	4.68	4.62	4.56	4.53	4.50	4.46	4.43	4.40	4.36
6	5.99	5.14	4.76	4.53	4.39	4.28	4.21	4.15	4.10	4.06	4.00	3.94	3.87	3.84	3.81	3.77	3.74	3.70	3.67
7	5.59	4.74	4.35	4.12	3.97	3.87	3.79	3.73	3.68	3.64	3.57	3.51	3.44	3.41	3.38	3.34	3.30	3.27	3.23
8	5.32	4.46	4.07	3.84	3.69	3.58	3.50	3.44	3.39	3.35	3.28	3.22	3.15	3.12	3.08	3.04	3.01	2.97	2.93
9	5.12	4.26	3.86	3.63	3.48	3.37	3.29	3.23	3.18	3.14	3.07	3.01	2.94	2.90	2.86	2.83	2.79	2.75	2.71
10	4.96	4.10	3.71	3.48	3.33	3.22	3.14	3.07	3.02	2.98	2.91	2.85	2.77	2.74	2.70	2.66	2.62	2.58	2.54
11	4.84	3.98	3.59	3.36	3.20	3.09	3.01	2.95	2.90	2.85	2.79	2.72	2.65	2.61	2.57	2.53	2.49	2.45	2.40
12	4.75	3.89	3.49	3.26	3.11	3.00	2.91	2.85	2.80	2.75	2.69	2.62	2.54	2.51	2.47	2.43	2.38	2.34	2.30
13	4.67	3.81	3.41	3.18	3.03	2.92	2.83	2.77	2.71	2.67	2.60	2.53	2.46	2.42	2.38	2.34	2.30	2.25	2.21
14	4.60	3.74	3.34	3.11	2.96	2.85	2.76	2.70	2.65	2.60	2.53	2.46	2.39	2.35	2.31	2.27	2.22	2.18	2.13
15	4.54	3.68	3.29	3.06	2.90	2.79	2.71	2.64	2.59	2.54	2.48	2.40	2.33	2.29	2.25	2.20	2.16	2.11	2.07
16	4.49	3.63	3.24	3.01	2.85	2.74	2.66	2.59	2.54	2.49	2.42	2.35	2.28	2.24	2.19	2.15	2.11	2.06	2.01
17	4.45	3.59	3.20	2.96	2.81	2.70	2.61	2.55	2.49	2.45	2.38	2.31	2.23	2.19	2.15	2.10	2.06	2.01	1.96
18	4.41	3.55	3.16	2.93	2.77	2.66	2.58	2.51	2.46	2.41	2.34	2.27	2.19	2.15	2.11	2.06	2.02	1.97	1.92
19	4.38	3.52	3.13	2.90	2.74	2.63	2.54	2.48	2.42	2.38	2.31	2.23	2.16	2.11	2.07	2.03	1.98	1.93	1.88
20	4.35	3.49	3.10	2.87	2.71	2.60	2.51	2.45	2.39	2.35	2.28	2.20	2.12	2.08	2.04	1.99	1.95	1.90	1.84
21	4.32	3.47	3.07	2.84	2.68	2.57	2.49	2.42	2.37	2.32	2.25	2.18	2.10	2.05	2.01	1.96	1.92	1.87	1.81

(Continued)

Table B.6b Upper 5% Points of the F-Distribution (Continued)

m \ n	1	2	3	4	5	6	7	8	9	10	12	15	20	24	30	40	60	120	∞
22	4.30	3.44	3.05	2.82	2.66	2.55	2.46	2.40	2.34	2.30	2.23	2.15	2.07	2.03	1.98	1.94	1.89	1.84	1.78
23	4.28	3.42	3.03	2.80	2.64	2.53	2.44	2.37	2.32	2.27	2.20	2.13	2.05	2.01	1.96	1.91	1.86	1.81	1.76
24	4.26	3.40	3.01	2.78	2.62	2.51	2.42	2.36	2.30	2.25	2.18	2.11	2.03	1.98	1.94	1.89	1.84	1.79	1.73
25	4.24	3.39	2.99	2.76	2.60	2.49	2.40	2.34	2.28	2.24	2.16	2.09	2.01	1.96	1.92	1.87	1.82	1.77	1.71
26	4.23	3.37	2.98	2.74	2.59	2.47	2.39	2.32	2.27	2.22	2.15	2.07	1.99	1.95	1.90	1.85	1.80	1.75	1.69
27	4.21	3.35	2.96	2.73	2.57	2.46	2.37	2.31	2.25	2.20	2.13	2.06	1.97	1.93	1.88	1.84	1.79	1.73	1.67
28	4.20	3.34	2.95	2.71	2.56	2.45	2.36	2.29	2.24	2.19	2.12	2.04	1.96	1.91	1.87	1.82	1.77	1.71	1.65
29	4.18	3.33	2.93	2.70	2.55	2.43	2.35	2.28	2.22	2.18	2.10	2.03	1.94	1.90	1.85	1.81	1.75	1.70	1.64
30	4.17	3.32	2.92	2.69	2.53	2.42	2.33	2.27	2.21	2.16	2.09	2.01	1.93	1.89	1.84	1.79	1.74	1.68	1.62
40	4.08	3.23	2.84	2.61	2.45	2.34	2.25	2.18	2.12	2.08	2.00	1.92	1.84	1.79	1.74	1.69	1.64	1.58	1.51
60	4.00	3.15	2.76	2.53	2.37	2.25	2.17	2.10	2.04	1.99	1.92	1.84	1.75	1.70	1.65	1.59	1.53	1.47	1.39
120	3.92	3.07	2.68	2.45	2.29	2.17	2.09	2.02	1.96	1.91	1.83	1.75	1.66	1.61	1.55	1.50	1.43	1.35	1.25
∞	3.84	3.00	2.60	2.37	2.21	2.10	2.01	1.94	1.88	1.83	1.75	1.67	1.57	1.52	1.46	1.39	1.32	1.22	1.00

† m = degrees of freedom for the numerator and n = degrees of freedom for the denominator.

Source: *Handbook of Tables for Mathematics*, Fourth Edition, 1970. Reprinted with permission from the CRC Press Inc., Boca Raton, Florida.

Table B.6c   Upper 10% Points of the F-Distribution†

m \ n	1	2	3	4	5	6	7	8	9	10	12	15	20	24	30	40	60	120	∞
1	39.86	49.50	53.59	55.83	57.24	58.20	58.91	59.44	59.86	60.19	60.71	61.22	61.74	62.00	62.26	62.53	62.79	63.06	63.33
2	8.53	9.00	9.16	9.24	9.29	9.33	9.35	9.37	9.38	9.39	9.41	9.42	9.44	9.45	9.46	9.47	9.47	9.48	9.49
3	5.54	5.46	5.39	5.34	5.31	5.28	5.27	5.25	5.24	5.23	5.22	5.20	5.18	5.18	5.17	5.16	5.15	5.14	5.13
4	4.54	4.32	4.19	4.11	4.05	4.01	3.98	3.95	3.94	3.92	3.90	3.87	3.84	3.83	3.82	3.80	3.79	3.78	3.76
5	4.06	3.78	3.62	3.52	3.45	3.40	3.37	3.34	3.32	3.30	3.27	3.24	3.21	3.19	3.17	3.16	3.14	3.12	3.10
6	3.78	3.46	3.29	3.18	3.11	3.05	3.01	2.98	2.96	2.94	2.90	2.87	2.84	2.82	2.80	2.78	2.76	2.74	2.72
7	3.59	3.26	3.07	2.96	2.88	2.83	2.78	2.75	2.72	2.70	2.67	2.63	2.59	2.58	2.56	2.54	2.51	2.49	2.47
8	3.46	3.11	2.92	2.81	2.73	2.67	2.62	2.59	2.56	2.54	2.50	2.46	2.42	2.40	2.38	2.36	2.34	2.32	2.29
9	3.36	3.01	2.81	2.69	2.61	2.55	2.51	2.47	2.44	2.42	2.38	2.34	2.30	2.28	2.25	2.23	2.21	2.18	2.16
10	3.29	2.92	2.73	2.61	2.52	2.46	2.41	2.38	2.35	2.32	2.28	2.24	2.20	2.18	2.16	2.13	2.11	2.08	2.06
11	3.23	2.86	2.66	2.54	2.45	2.39	2.34	2.30	2.27	2.25	2.21	2.17	2.12	2.10	2.08	2.05	2.03	2.00	1.97
12	3.18	2.81	2.61	2.48	2.39	2.33	2.28	2.24	2.21	2.19	2.15	2.10	2.06	2.04	2.01	1.99	1.96	1.93	1.90
13	3.14	2.76	2.56	2.43	2.35	2.28	2.23	2.20	2.16	2.14	2.10	2.05	2.01	1.98	1.96	1.93	1.90	1.88	1.85
14	3.10	2.73	2.52	2.39	2.31	2.24	2.19	2.15	2.12	2.10	2.05	2.01	1.96	1.94	1.91	1.89	1.86	1.83	1.80
15	3.07	2.70	2.49	2.36	2.27	2.21	2.16	2.12	2.09	2.06	2.02	1.97	1.92	1.90	1.87	1.85	1.82	1.79	1.76
16	3.05	2.67	2.46	2.33	2.24	2.18	2.13	2.09	2.06	2.03	1.99	1.94	1.89	1.87	1.84	1.81	1.78	1.75	1.72
17	3.03	2.64	2.44	2.31	2.22	2.15	2.10	2.06	2.03	2.00	1.96	1.91	1.86	1.84	1.81	1.78	1.75	1.72	1.69
18	3.01	2.62	2.42	2.29	2.20	2.13	2.08	2.04	2.00	1.98	1.93	1.89	1.84	1.81	1.78	1.75	1.72	1.69	1.66
19	2.99	2.61	2.40	2.27	2.18	2.11	2.06	2.02	1.98	1.96	1.91	1.86	1.81	1.79	1.76	1.73	1.70	1.67	1.63
20	2.97	2.59	2.38	2.25	2.16	2.09	2.04	2.00	1.96	1.94	1.89	1.84	1.79	1.77	1.74	1.71	1.68	1.64	1.61
21	2.96	2.57	2.36	2.23	2.14	2.08	2.02	1.98	1.95	1.92	1.87	1.83	1.78	1.75	1.72	1.69	1.66	1.62	1.59
22	2.95	2.56	2.35	2.22	2.13	2.06	2.01	1.97	1.93	1.90	1.86	1.81	1.76	1.73	1.70	1.67	1.64	1.60	1.57

(Continued)

Table B.6c Upper 10% Points of the *F*-Distribution (Continued)

m \ n	1	2	3	4	5	6	7	8	9	10	12	15	20	24	30	40	60	120	∞
23	2.94	2.55	2.34	2.21	2.11	2.05	1.99	1.95	1.92	1.89	1.84	1.80	1.74	1.72	1.69	1.66	1.62	1.59	1.55
24	2.93	2.54	2.33	2.19	2.10	2.04	1.98	1.94	1.91	1.88	1.83	1.78	1.73	1.70	1.67	1.64	1.61	1.57	1.53
25	2.92	2.53	2.32	2.18	2.09	2.02	1.97	1.93	1.89	1.87	1.82	1.77	1.72	1.69	1.66	1.63	1.59	1.56	1.52
26	2.91	2.52	2.31	2.17	2.08	2.01	1.96	1.92	1.88	1.86	1.81	1.76	1.71	1.68	1.65	1.61	1.58	1.54	1.50
27	2.90	2.51	2.30	2.17	2.07	2.00	1.95	1.91	1.87	1.85	1.80	1.75	1.70	1.67	1.64	1.60	1.57	1.53	1.49
28	2.89	2.50	2.29	2.16	2.06	2.00	1.94	1.90	1.87	1.84	1.79	1.74	1.69	1.66	1.63	1.59	1.56	1.52	1.48
29	2.89	2.50	2.28	2.15	2.06	1.99	1.93	1.89	1.86	1.83	1.78	1.73	1.68	1.65	1.62	1.58	1.55	1.51	1.47
30	2.88	2.49	2.28	2.14	2.05	1.98	1.93	1.88	1.85	1.82	1.77	1.72	1.67	1.64	1.61	1.57	1.54	1.50	1.46
40	2.84	2.44	2.23	2.09	2.00	1.93	1.87	1.83	1.79	1.76	1.71	1.66	1.61	1.57	1.54	1.51	1.47	1.42	1.38
60	2.79	2.39	2.18	2.04	1.95	1.87	1.82	1.77	1.74	1.71	1.66	1.60	1.54	1.51	1.48	1.44	1.40	1.35	1.29
120	2.75	2.35	2.13	1.99	1.90	1.82	1.77	1.72	1.68	1.65	1.60	1.55	1.48	1.45	1.41	1.37	1.32	1.26	1.19
∞	2.71	2.30	2.08	1.94	1.85	1.77	1.72	1.67	1.63	1.60	1.55	1.49	1.42	1.38	1.34	1.30	1.24	1.17	1.00

† *m* = degrees of freedom for the numerator and *n* = degrees of freedom for the denominator.

Source: *Handbook of Tables for Mathematics*, Fourth Edition, 1970. Reprinted with permission from the CRC Press Inc., Boca Raton, Florida.

Table B.7a   Durbin-Watson Statistic (d)

5 % Significance Points in Two-Tailed Tests[†]

n	$k'=1$		$k'=2$		$k'=3$		$k'=4$		$k'=5$	
	d_L	d_U	d_L	d_U	d_L	d_U	d_L	d_U	d_L	d_U
15	0.95	1.23	0.83	1.40	0.71	1.61	0.59	1.84	0.48	2.09
16	0.98	1.24	0.86	1.40	0.75	1.59	0.64	1.80	0.53	2.03
17	1.01	1.25	0.90	1.40	0.79	1.58	0.68	1.77	0.57	1.98
18	1.03	1.26	0.93	1.40	0.82	1.56	0.72	1.74	0.62	1.93
19	1.06	1.28	0.96	1.41	0.86	1.55	0.76	1.72	0.66	1.90
20	1.08	1.28	0.99	1.41	0.89	1.55	0.79	1.70	0.70	1.87
21	1.10	1.30	1.01	1.41	0.92	1.54	0.83	1.69	0.73	1.84
22	1.12	1.31	1.04	1.42	0.95	1.54	0.86	1.68	0.77	1.82
23	1.14	1.32	1.06	1.42	0.97	1.54	0.89	1.67	0.80	1.80
24	1.16	1.33	1.08	1.43	1.00	1.54	0.91	1.66	0.83	1.79
25	1.18	1.34	1.10	1.43	1.02	1.54	0.94	1.65	0.86	1.77
26	1.19	1.35	1.12	1.44	1.04	1.54	0.96	1.65	.0.88	1.76
27	1.21	1.36	1.13	1.44	1.06	1.54	0.99	1.64	0.91	1.75
28	1.22	1.37	1.15	1.45	1.08	1.54	1.01	1.64	0.93	1.74
29	1.24	1.38	1.17	1.45	1.10	1.54	1.03	1.63	0.96	1.73
30	1.25	1.38	1.18	1.46	1.12	1.54	1.05	1.63	0.98	1.73
31	1.26	1.39	1.20	1.47	1.13	1.55	1.07	1.63	1.00	1.72
32	1.27	1.40	1.21	1.47	1.15	1.55	1.08	1.63	1.02	1.71
33	1.28	1.41	1.22	1.48	1.16	1.55	1.10	1.63	1.04	1.71
34	1.29	1.41	1.24	1.48	1.17	1.55	1.12	1.63	1.06	1.70
35	1.30	1.42	1.25	1.48	1.19	1.55	1.13	1.63	1.07	1.70
36	1.31	1.43	1.26	1.49	1.20	1.56	1.15	1.63	1.09	1.70
37	1.32	1.43	1.27	1.49	1.21	1.56	1.16	1.62	1.10	1.70
38	1.33	1.44	1.28	1.50	1.23	1.56	1.17	1.62	1.12	1.70
39	1.34	1.44	1.29	1.50	1.24	1.56	1.19	1.63	1.13	1.69
40	1.35	1.45	1.30	1.51	1.25	1.57	1.20	1.63	1.15	1.69
45	1.39	1.48	1.34	1.53	1.30	1.58	1.25	1.63	1.21	1.69
50	1.42	1.50	1.38	1.54	1.34	1.59	1.30	1.64	1.26	1.69
55	1.45	1.52	1.41	1.56	1.37	1.60	1.33	1.64	1.30	1.69
60	1.47	1.54	1.44	1.57	1.40	1.61	1.37	1.65	1.33	1.69
65	1.49	1.55	1.46	1.59	1.43	1.62	1.40	1.66	1.36	1.69
70	1.51	1.57	1.48	1.60	1.45	1.63	1.42	1.66	1.39	1.70
75	1.53	1.58	1.50	1.61	1.47	1.64	1.45	1.67	1.42	1.70
80	1.54	1.59	1.52	1.62	1.49	1.65	1.47	1.67	1.44	1.70
85	1.56	1.60	1.53	1.63	1.51	1.65	1.49	1.68	1.46	1.71
90	1.57	1.61	1.55	1.64	1.53	1.66	1.50	1.69	1.48	1.71
95	1.58	1.62	1.56	1.65	1.54	1.67	1.52	1.69	1.50	1.71
100	1.59	1.63	1.57	1.65	1.55	1.67	1.53	1.70	1.51	1.72

[†] n = number of observations and k' = number of explanatory variables excluding constant.

urce: "Testing for Serial Correlation in Least Squares Regression," by Durbin and G.S. Watson, *Biometrika*, Vol. 38 (1951), pp. 159–177. rinted with the permission of the Biometrika Trustees.

Table B.7b   Durbin-Watson Statistic (d)
5 % Significance Points in One-Tailed Tests[†]

n	$k'=1$ d_L	$k'=1$ d_U	$k'=2$ d_L	$k'=2$ d_U	$k'=3$ d_L	$k'=3$ d_U	$k'=4$ d_L	$k'=4$ d_U	$k'=5$ d_L	$k'=5$ d_U
15	1.08	1.36	0.95	1.54	0.82	1.75	0.69	1.97	0.56	2.21
16	1.10	1.37	0.98	1.54	0.86	1.73	0.74	1.93	0.62	2.15
17	1.13	1.38	1.02	1.54	0.90	1.71	0.78	1.90	0.67	2.10
18	1.16	1.39	1.05	1.53	0.93	1.69	0.82	1.87	0.71	2.06
19	1.18	1.40	1.08	1.53	0.97	1.68	0.86	1.85	0.75	2.02
20	1.20	1.41	1.10	1.54	1.00	1.68	0.90	1.83	0.79	1.99
21	1.22	1.42	1.13	1.54	1.03	1.67	0.93	1.81	0.83	1.96
22	1.24	1.43	1.15	1.54	1.05	1.66	0.96	1.80	0.86	1.94
23	1.26	1.44	1.17	1.54	1.08	1.66	0.99	1.79	0.90	1.92
24	1.27	1.45	1.19	1.55	1.10	1.66	1.01	1.78	0.93	1.90
25	1.29	1.45	1.21	1.55	1.12	1.66	1.04	1.77	0.95	1.89
26	1.30	1.46	1.22	1.55	1.14	1.65	1.06	1.76	0.98	1.88
27	1.32	1.47	1.24	1.56	1.16	1.65	1.08	1.76	1.01	1.86
28	1.33	1.48	1.26	1.56	1.18	1.65	1.10	1.75	1.03	1.85
29	1.34	1.48	1.27	1.56	1.20	1.65	1.12	1.74	1.05	1.84
30	1.35	1.49	1.28	1.57	1.21	1.65	1.14	1.74	1.07	1.83
31	1.36	1.50	1.30	1.57	1.23	1.65	1.16	1.74	1.09	1.83
32	1.37	1.50	1.31	1.57	1.24	1.65	1.18	1.73	1.11	1.82
33	1.38	1.51	1.32	1.58	1.26	1.65	1.19	1.73	1.13	1.81
34	1.39	1.51	1.33	1.58	1.27	1.65	1.21	1.73	1.15	1.81
35	1.40	1.52	1.34	1.58	1.28	1.65	1.22	1.73	1.16	1.80
36	1.41	1.52	1.35	1.59	1.29	1.65	1.24	1.73	1.18	1.80
37	1.42	1.53	1.36	1.59	1.31	1.66	1.25	1.72	1.19	1.80
38	1.43	1.54	1.37	1.59	1.32	1.66	1.26	1.72	1.21	1.79
39	1.43	1.54	1.38	1.60	1.33	1.66	1.27	1.72	1.22	1.79
40	1.44	1.54	1.39	1.60	1.34	1.66	1.29	1.72	1.23	1.79
45	1.48	1.57	1.43	1.62	1.38	1.67	1.34	1.72	1.29	1.78
50	1.50	1.59	1.46	1.63	1.42	1.67	1.38	1.72	1.34	1.77
55	1.53	1.60	1.49	1.64	1.45	1.68	1.41	1.72	1.38	1.77
60	1.55	1.62	1.51	1.65	1.48	1.69	1.44	1.73	1.41	1.77
65	1.57	1.63	1.54	1.66	1.50	1.70	1.47	1.73	1.44	1.77
70	1.58	1.64	1.55	1.67	1.52	1.70	1.49	1.74	1.46	1.77
75	1.60	1.65	1.57	1.68	1.54	1.71	1.51	1.74	1.49	1.77
80	1.61	1.66	1.59	1.69	1.56	1.72	1.53	1.74	1.51	1.77
85	1.62	1.67	1.60	1.70	1.57	1.72	1.55	1.75	1.52	1.77
90	1.63	1.68	1.61	1.70	1.59	1.73	1.57	1.75	1.54	1.78
95	1.64	1.69	1.62	1.71	1.60	1.73	1.58	1.75	1.56	1.78
100	1.65	1.69	1.63	1.72	1.61	1.74	1.59	1.76	1.57	1.78

[†] n = number of observations and k' = number of explanatory variables excluding constant.

Source: "Testing for Serial Correlation in Least Squares Regression," by J. Durbin and G.S. Watson, *Biometrika*, Vol. 38 (1951), pp. 159–177. Reprinted with the permission of the Biometrika Trustees.

REFERENCES

Aitchison, J. and Brown, J. A. C. [1966], *The Lognormal Distribution with Special Reference to its Uses in Economics*, Cambridge University Press, Cambridge, England.

Aitchison, J., and Silvey, D. S. [1958], "Maximum Likelihood Estimation of Parameters Subject to Restraints," *Annals of Math. Stat.*, Vol. 29.

Akaike, H. [1970], "Statistical Predictor Identification," *Annals of Institute. Stat. Math.*, Vol. 22, pp. 203–217.

Akaike, H. [1974], "A New Look at Statistical Model Identification," *IEEE Trans. Auto. Control*, Vol. 19, pp. 716–723.

Amemiya, T. [1981], "Qualitative Response Models: A Survey," *Journal of Economic Literature*, Vol. 19, pp. 1488–1536.

Amemiya, T. [1984], "Tobit Models: A Survey," *Journal of Econometrics*, Vol. 24, pp. 3–61.

Amemiya, T. [1985], *Advanced Econometrics*, Harvard Univ. Press, Cambridge, Massachusetts.

Barnett, V. [1973], *Comparative Statistical Inference*, Wiley, London.

Beach, C., and MacKinnon, J. [1978a], "A Maximum Likelihood Procedure for Regression with Autocorrelated Errors," *Econometrica*, pp. 51–58.

Beach, C., and MacKinnon, J. [1978b], "Full Maximum Likelihood Estimation of Second Order Autoregressive Error Models," *J. Econometrics*, pp. 187–198.

Bellman, R.E. [1970], *Introduction to Matrix Analysis*, McGraw-Hill, New York.

Belsley, D. A., Kuh, E., and Welsch, R. E. [1980], *Regression Diagnostics, Identifying Influential Data and Sources of Collinearity*, Wiley, New York.

Breusch, T., and Pagan, A. [1978], "A Maximum Likelihood Procedure for Regression with Autocorrelated Errors,", *Econometrica*, pp. 51–58.

Breusch, T., and Pagan, A. [1979], "A Simple Test for Heteroscedasticity and Random Coefficient Variation,", *Econometrica*, pp. 1287–1294.

Broadcasting Year Book, Broadcasting Publications, Inc., Washington, D. C.

Buse, A. [1982], "The Likelihood ratio, Wald, and Lagrange Multiplier Test: An Expository Note," *The American Statistician*, pp. 153–157.

Casella, G., and Berger, R. L. [1990], *Statistical Inference*, Wadsworth and Brooks/Cole, Pacific Grove, California.

Chow, G.C. [1960], "Tests of Equality Between Sets of Coefficients in Two Linear Regressions," *Econometrica*, Vol. 28, pp. 591–605.

Chung, K. L. [1968], *A Course in Probability Theory*, Harcourt Brace and World, New York.

Chung, K. L. [1974], *A Course in Probability Theory*, Second Edition, Academic Press, New York.

Cochrane, D., and Orcutt, G. [1949], "Application of Least Squares Regression to Relationships Containing Autocorrelated Error Terms," *J. Amer. Stat. Assoc.*, pp. 32–61.

County and City Data Book, U. S. Government Printing Office, Washington, D. C.

Craven, P., and Wahba, G. [1979], "Smoothing Noisy Data with Spline Functions," *Num. Math.*, Vol. 13, pp. 377–403.

Crow, E. L., and Shimizu, K. (Eds.) [1988], *Lognormal Distributions: Theory and Applications*, Marcel Dekker, New York.

Cyert, R.M., and DeGroot, M.H. [1987], *Bayesian Analysis and Uncertainty in Economic Theory*, Rowman & Littlefield, Totowa.

Dhrymes, P. [1970], *Econometrics: Statistical Foundations and Applications*, Harper and Row, New York.

Economic Report of the President [1987], U. S. Government Printing Office, Washington, D. C.

Engle, R. F. [1982] "Autoregressive Conditional Heteroscedasticity with Estimates of the Variance of United Kingdom Inflations," *Econometrica*, pp. 987–1008.

Engle, R. F. [1982], "A General Approach to Lagrangian Multiplier Diagnostics," *Annals of Econometrics*, pp. 83–104.

Engle, R. F. [1983], "Estimates of the Variance of U.S. Inflation Based on the ARCH Model," *J. Money, Credit, and Banking*, pp. 286–301.

Engle, R. F. [1984], "Wald, Likelihood–Ratio and Lagrangian Multiplier Tests in Econometrics," *Handbook of Econometrics*, (Ed.) Z. Griliches and M.D. Intriligator, North–Holland, New York.

Engle, R. F., and Brown, S. [1985], "Model Selection for Forecasting," *J. Computation in Statistics*.

Farebrother, R. [1980], "The Durbin–Watson Test for Serial Correlation when there is no Intercept in the Regression," *Econometrica*, pp. 1553–1563.

Fisher, R.A. [1922], "On the Mathematical Foundations of Theoretical Statistics," *Phil. Trans. Royal Society*, Series A, Vol. 222, pp. 309–368.

Fisher, R.A. [1925], "Theory of Statistical Estimation," *Proc. Cambridge Phil. Soc.*, Vol. 22, pp. 700–725.

Fisz, M. [1963], *Probability Theory and Mathematical Statistics*, Wiley, New York.

Foutz, R.V. [1977], "On the Unique Consistent Solution to the Likelihood Equations," *J. Amer. Stat. Assoc.*, pp. 147–148.

Fraser, D.A.S. [1976], *Probability and Statistics: Theory and Applications*, Duxbury Press, North Scituate, Massachusetts.

Freund, J. E. [1962], *Mathematical Statistics*, First Edition, Prentice Hall, Englewood Cliffs, New Jersey.

Freund, J. E. [1992], *Mathematical Statistics*, Second Edition, Wadsworth & Brooks/Cole, Belmont, California.

Glesjer, H. [1969], "A New Test for Heteroscedasticity," *J. Amer. Stat. Assoc.*, pp. 316–323.

Gnedenko, B. W., and Kolmogorov, A. N. [1954], *Limit Distributions for Sums of Independent Random Variables*, Addison–Wesley, Cambridge, Massachusetts.

Godfrey, L. [1978], "Testing for Multiplicative Heteroscedasticity," *J. Econometrics*, pp. 227–236.

Goldfeld, S.M., and Quandt, R.E. [1965], "Some Tests for Homoscedasticity," *J. Amer. Stat. Assoc.*, pp. 539–547.

Goldfeld, S.M., and Quandt, R.E. [1972], *Nonlinear Methods in Econometrics*, North–Holland, Amsterdam.

Gourieroux, C., Monfort, A., and Trognon, A. [1984], "Pseudo-maximum Likelihood Methods: Theory," *Econometrica*, pp. 424–438.

Greene, W.H. [1990], *Econometric Analysis*, Macmillan, New York.

Grogger, J. T., and Carson, R. T. [1991], "Models for Truncated Counts," *Journal of Applied Econometrics*, pp. 225–238.

Hadley, G. [1973], *Linear Algebra*, Addison-Wesley, Reading, Massachusetts.

Halmos, P. R. [1950], *Measure Theory*, Van Nostrand, New York.

Hannan, E. J., and Quinn, B. [1979], "The Determination of the Order of an Autoregression," *J. Royal Stat. Society*, Series B, Vol. 41, pp. 190–195.

Hildreth, C. and Lu, J. [1960], "Demand Relations with Autocorrelated Disturbances," Technical Bulletin Number 276, Michigan State University Agricultural Experiment Station.

Hinkley, D. V. [1977], "Jackknifing in Unbalanced Situations," *Technometrics*, Vol. 19, pp. 285–292.

Hoerl, A. and Kennard, R. [1970], "Ridge Regression: Biased Estimation for Nonorthogonal Problems," *Technometrics*, pp. 55–67.

Jennrich, R. I. [1969], "Asymptotic Properties of Nonlinear Least Squares Estimators," *Annals Math. Stat.*, pp. 633–643.

Johnston, J. [1984], *Econometric Methods*, McGraw–Hill, New York.

Judge, G. G, Griffiths, W. E., Hill, R. C., Lütkepohl, H. and Lee, T. C. [1985], *The Theory and Practice of Econometrics*, Wiley, New York.

Kmenta, J. [1986], *Elements of Econometrics*, Mammillan Publishing Company, New York.

Koenker, R. [1981], "A Note on Studentizing a Test for Heteroscedasticity," *J. Econometrics*, pp. 107–112.

Leamer, E. [1978], *Specification Searches: Ad Hoc Inferences with Nonexperimental Data*, Wiley, New York.

Lehmann, E.L. [1959], *Testing Statistical Hypothesis*, Wiley, New York.

Lehmann, E.L. [1991], *Testing Statistical Hypothesis*, Second Edition, Wadsworth & Brooks/Cole, Pacific Grove, California.

Lehmann, E.L. and Scheffe, H. [1950], "Completeness, Similar Regions, and Unbiased Estimation," Part I, *Sankhya*, Vol. 10, pp. 305–340.

Lindgren, B. [1976], *Statistical Theory*, Macmillan, New York.

Loève, M. [1955], *Probability Theory: Foundations, Random Sequences*, Van Nostrand, New York.

Loève, M. [1960], *Probability Theory*, Van Nostrand, Princeton, New Jersey.

Lukacs, E. [1968], *Stochastic Convergence*, D. C. Health, Lexington, Massachusetts.

Lukacs, E. [1970], *Characteristic Functions*, Griffin, London.

MacKinnon, J. G., and White, H. [1985], "Some Heteroscedasticity-Consistent Covariance Matrix Estimators with Improved Finite Sample Properties'", *J. Econometrics*, Vol. 29, pp. 305–325.

Maddala, G. S. [1977], *Econometrics*, McGraw–Hill, New York.

Maddala, G. S. [1983], *Limited Dependent and Qualitative Variables in Econometrics*, Cambridge University Press, Cambridge, England.

McFadden, D. [1984], "Econometric Analysis of Qualitative Response Models," in *Handbook of Econometrics*, Vol. 2, (Z. Griliches and M.D. Intriligator, Eds.), North–Holland, Amsterdam.

Meyer, J. and Kuh, E. [1957], "How Extraneous are Extraneous Estimates?" *Review of Econ. and Stat.*, pp. 380–393.

Mood, A. M., Graybill, F. A., and Boes, D .C. [1974], *Introduction to the Theory of Statistics*, McGraw–Hill, New York.

Press, W. H., Flannery, B. P., Teukolsky, S. A., and Vetterling, W.T. [1988], *Numerical Recipes in C*, Cambridge University Press, New York.

Pudney, S. [1989], *Modelling Individual Choice: The Econometrics of Corners, Kinks, and Holes*, Blackwell, Oxford, England.

Raiffa, H., and Schlaifer, R. [1961], *Applied Statistical Decision Theory*, Harvard Business School, Boston, Massachusetts.

Ramanathan, R. [1982], *Introduction to the Theory of Economic Growth*, Springer–Verlag, Berlin.

Ramanathan, R. [1992], *Introductory Econometrics with Applications*, Harcourt Brace Jovanovich, Fort Worth, Texas.

Rao, C. R. [1948], "Large Sample Tests of Statistical Hypotheses Concerning Several Parameters With Applications to Problems of Estimation," *Proc. Cambridge Phil. Soc.*, pp. 50–57.

Rao, C. R. [1965], *Linear Statistical Inference and Its Applications*, Wiley, New York.

Rao, C. R. [1973], *Linear Statistical Inference and Its Applications*, Wiley, New York.

Rice, J. [1984], "Bandwidth Choice for Nonparametric Kernel Regression," *Annals of Stat.*, Vol. 12, pp. 1215–1230.

Rudin, W. [1964], *Principles of Mathematical Analysis*, McGraw–Hill, New York.

Savin, E., and White, K. [1977], "The Durbin–Watson Test for Serial Correlation with Extreme Sample Sizes or Many Regressors," *Econometrica*, pp. 1989–1996.

Scheffé, H. [1970], "Practical Solutions of the Behrens–Fisher Problem," *J. Amer. Stat. Assoc.*, pp. 1501–1504.

Schwarz, G. [1978], "Estimating the Dimension of a Model," *Annals of Stat.*, Vol. 6.

Shibata, R. [1981], "An Optimal Selection of Regression Variables," *Biometrika*, Vol. 68.

Silvey, D. S. [1959], "The Lagrange Multiplier Test," *Annals of Math. Stat.*.

Spanos, A. [1989], *Statistical Foundations of Econometric Modelling*, Cambridge University Press, Cambridge, England.

State and Metropolitan Area Data Book, U. S. Government Printing Office, Washington, D. C.

Statistical Abstracts of the United States, U. S. Government Printing Office, Washington, D. C.

Strang, G. [1976], *Linear Algebra and its Applications*, Academic Press, New York.

Taylor, A. E. and Mann, W. R. [1972], *Advanced Calculus*, Xerox College Publishing, Lexington.

Wald, A. [1943], "Tests of Statistical Hypotheses Concerning Several Parameters When the Number of Observations is Large," *Transactions of the Amer. Math. Society*, Vol. 54.

White, H. [1980], "A Heteroscedasticity-Consistent Covariance Matrix Estimator and a Direct Test for Heteroscedasticity," *Econometrica*, pp. 817–838.

White, H. [1984], *Asymptotic Theory for Econometricians*, Academic Press, Orlando.

Wilks, S. S. [1962], *Mathematical Statistics*, Wiley, New York.

Zellner, A. [1971], *An Introduction to Bayesian Inference in Econometrics*, Wiley, New York.

COPYRIGHT ACKNOWLEDGMENTS

396

AUTHOR INDEX

SUBJECT INDEX